P9-ECY-750

About Island Press

Island Press is the only nonprofit organization in the United States whose principal purpose is the publication of books on environmental issues and natural resource management. We provide solutions-oriented information to professionals, public officials, business and community leaders, and concerned citizens who are shaping responses to environmental problems.

In 1999, Island Press celebrates its fifteenth anniversary as the leading provider of timely and practical books that take a multidisciplinary approach to critical environmental concerns. Our growing list of titles reflects our commitment to bringing the best of an expanding body of literature to the environmental community throughout North America and the world.

Support for Island Press is provided by The Jenifer Altman Foundation, The Bullitt Foundation, The Mary Flagler Cary Charitable Trust, The Nathan Cummings Foundation, The Geraldine R. Dodge Foundation, The Charles Engelhard Foundation, The Ford Foundation, The Vira I. Heinz Endowment, The W. Alton Jones Foundation, The John D. and Catherine T. MacArthur Foundation, The Andrew W. Mellon Foundation, The Charles Stewart Mott Foundation, The Curtis and Edith Munson Foundation, The National Fish and Wildlife Foundation, The National Science Foundation, The New-Land Foundation, The David and Lucile Packard Foundation, The Pew Charitable Trusts, The Surdna Foundation, The Winslow Foundation, and individual donors.

Special funding for this book was provided by The Bullitt Foundation and The National Fish and Wildlife Foundation.

Salmon Without *Rivers*

"These facts, when associated with new simplified methods of salmon egg incubation [and] predator and hydraulic control in water areas, plus the impoundment of migrating salmon at or near the rearing ponds for the artificial taking of spawn, may provide the reality—salmon without a river."

—Washington Department of Fisheries, 1960

A History of the Pacific Salmon Crisis

Jim Lichatowich

ISLAND PRESS
Washington, D.C. • Covelo, California

Parts of pages 1–6 first appeared in *Peninsula* magazine.
Parts of pages 42–44 and 228–229 first appeared in *The RiverKeeper,* a publication of Oregon Trout, Portland, Oregon.

Library of Congress Cataloging-in-Publication Data
Lichatowich, Jim.
Salmon without rivers : a history of the pacific salmon crisis / Jim Lichatowich.
p. cm.
Includes bibliographical references (p.).
ISBN 1–55963–360–3 (cloth)
1. Pacific salmon—Northwest, Pacific—History. 2. Fishery conservation—Northwest, Pacific History. I. Title.
SH348.L53 1999 99–16798
333.95'656'09795—dc21 CIP

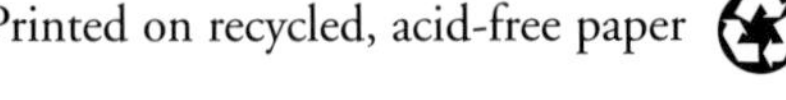
Printed on recycled, acid-free paper

Manufactured in the United States of America
10 9 8 7 6 5 4 3 2

For Paulette, Jim and Sue,
and Tim and Charlane

Contents

Preface

Why have the salmon presented the Pacific Northwest with such an intractable problem? Why have we been unable to arrest their slide toward extinction and begin rebuilding their populations to some part of their former abundance? It's not that we haven't made heroic efforts. Although we have been trying for three-quarters of a century and have spent a huge amount of money—$3 billion on the Columbia River alone, by some accounts—we have failed to reverse the salmon's decline.

I have talked to biologists, environmentalists, corporate executives, fishermen, Native Americans, journalists, lawyers, and interested citizens about these questions for the past several years. Through all these discussions, it has become clear that most people generally fail to understand that the problem has a long history. Most believe that the salmon's precarious state is due to events of the last twenty or thirty years—perhaps not surprisingly, since during those few decades we have witnessed the rapid disappearance of forests, construction of the last big dams, rapid urbanization of large areas of the landscape, and a host of other assaults on the salmon and their rivers. But the salmon's dilemma goes back much further—back to the arrival of the first Euro-Americans in the Northwest. The whites brought cattle, plows, seeds, axes, and other essentials they needed to survive; unfortunately for the salmon, the settlers also brought their worldview and their industrial economy.

Viewing the salmon and their rivers through the lens of different assumptions about nature, the Euro-Americans saw a different landscape than the one the Indians had been living in with a high degree of harmony for several thousand years. They saw a fearful wilderness that had to be tamed, simplified, and controlled. They saw the vast resources they needed to feed their voracious industrial economy. Their vision naturally directed them to reconstruct the Northwest to make it more like the places they came from than the place it was. The purpose of this book is to present the history of the impact of that vision on the salmon. It's an important history because it has been shown very clearly that the salmon restoration efforts to date, even though they have been well funded, have failed. They have failed because they are largely derived from the same worldview and assumptions that created the problem in the first place. Unless we recognize the real roots of the salmon's problem and deal with it at that fundamental level, the fish will continue their slide toward extinction. Viewing the historical relationship between people and the salmon, especially over the last 150 years, leaves little room for optimism. The view from this point into the future could be optimistic, but that is up to you.

The book is divided into two principal parts. Chapters 1 and 2 describe the evolutionary history of the salmon and the development of the Northwest Indians' gift economy, which was largely based on the salmon. Chapters 3 through 9 describe the effects of the Euro-Americans' industrial economy on the salmon: Chapter 3 covers the transition from the gift economy to the industrial economy. Chapter 4 details the destruction of the salmon's habitat through about 1930. Chapter 5 describes the development of the commercial fishery and the salmon-canning industry. The history of the salmon hatcheries, used to maintain salmon abundance in the face of overharvest and habitat destruction, is presented in Chapter 6. The beginning of our scientific understanding of the salmon's biology is presented in Chapter 7. The ideas given in Chapters 3 through 7 are tied together in Chapter 8 in a brief history of the region's two greatest salmon rivers—the Fraser and the Columbia. Chapter 9 covers the last few decades and the intervention of the federal Endangered Species Act. The epilogue suggests ways to begin building a salmon-friendly culture.

Acknowledgments

This book could not have been completed without the help, encouragement, and constructive criticism from my best friend and wife, Paulette. I owe more than I can ever describe to Charles Warren, who has been a friend and mentor for these last thirty years. My son Tim, his wife Charlane, and Carrie Hoffman spent many hours in the library xeroxing and doing other tasks that were critical to the project. Thanks for the special moments of encouragement from my son Jim. I owe a special debt of gratitude to Margy and Dave Buchanan of Tyee Winery for their encouragement, support, and loan of their beach cabin, where several chapters of this book were written. Thanks to Ron Hirschi for giving me the book's title and much more.

Dave Buchanan, Dan Bottom, and Charles Warren read the manuscript at various stages in its development and gave me constructive suggestions for improving the book.

Ann Vileisis was instrumental in helping me turn the draft of this book into its final form. I appreciate her help. I also appreciate the encouragement of Barbara Dean at Island Press, who suffered patiently through one missed deadline after another.

Many of the ideas in this book evolved from thousands of hours of discussion with some very fine biologists and salmon advocates over the past thirty years: Dan Bottom, Dave Buchanan, Reg Reisenbichler, Charles Warren, Steve Johnson, Homer Campbell, Harry Wagner, Jack McIntyre, Jay Nicholas, Bill Bakke, Tom Lichatowich, Bob Mullen, Bob Hooton, Lars Mobrand, Chris Frissell, Bill Liss, Jim Hall, Kurt Beardslee, Ron Hirschi, Tom Jay, Chip McConnaha, and many, many others.

Special thanks to the staff at the Sequim branch of the North Olympic Library, who found many obscure documents through interlibrary loan. Also very helpful were the staffs of the National Archives, Washington, D.C., and Seattle; the University of Washington library; and the Oregon State University library.

INTRODUCTION

The Salmon's Problem

It's late December, and I am about to take part in the annual count of spawning salmon. Each winter since the late 1940s, biologists all over the Northwest have counted the salmon that escape the gauntlet of nets and hooks and return to their home streams to spawn. The men and women who walk the rivers each winter are the first to know if the spawning run is strong or weak, but more important, they are among the few humans who witness the birth of a new generation of wild salmon.

I turn off the highway and follow a deeply rutted logging road into the watershed of the Pysht River, a small stream in the foothills of the Olympic Peninsula (Figure I.1). The truck bumps to a stop on the shoulder of the road, and I drink the last of the coffee from my thermos before stepping out into the icy air. My leaky chest-waders feel cold, especially the soggy insoles. But despite the protests of my feet, I slide over the stream bank and into the frigid water. A flock of kinglets and chickadees swarms through the naked alders. Their voices sound like crystalline chimes in the frozen air. The December sun is low on the horizon, but a few golden rays of weak light filter through the canopy. A thin curtain of steam rises from the river, giving the scene a primal quality—fitting for the ancient rite of the wild salmon I am about to observe and record.

In earlier times, Indians lived at the mouth of the Pysht River. Several small settlements were clustered near the Pysht, Hoko, and Clallam Rivers at the western end of the Strait of Juan de Fuca. Indians occupied some of the villages 3,000 years ago. Archaeologists estimate that the native people living near these small streams harvested about 45,000 salmon each year.[1] If

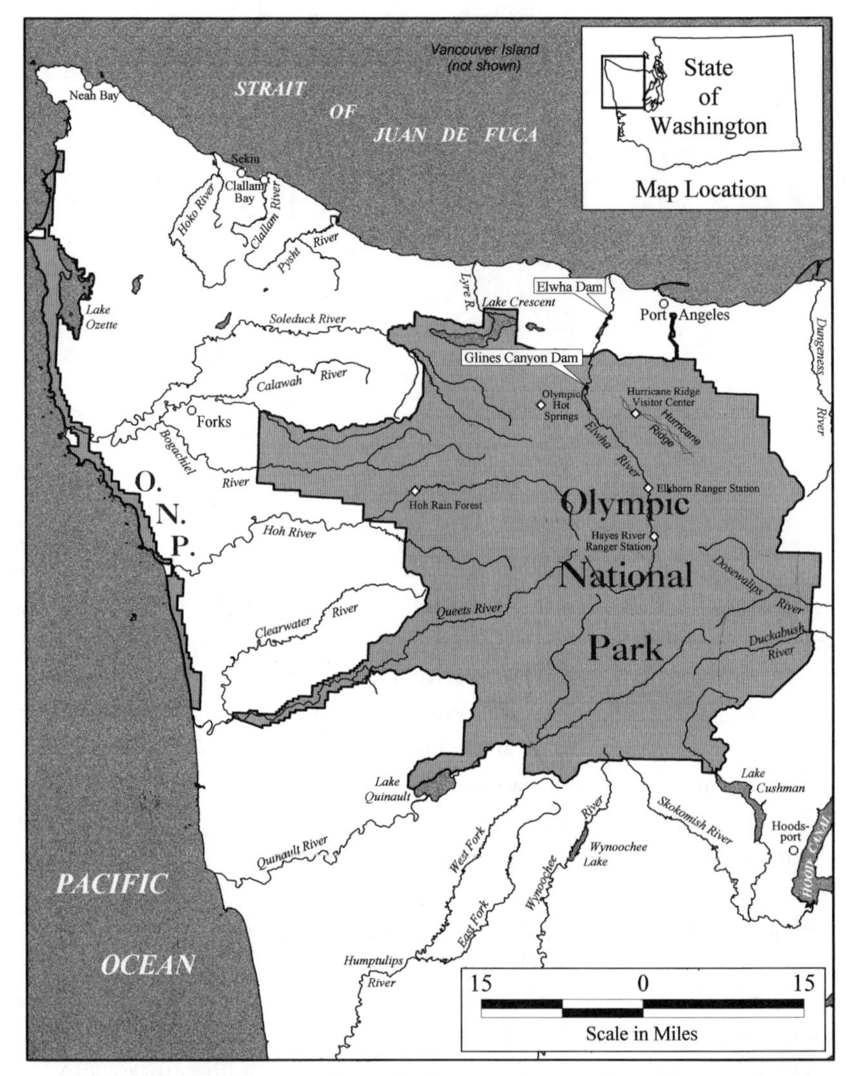

Figure I.1. Olympic Peninsula and Olympic National Park in Washington State. (Source: Randall McCoy, Geographic Information Systems [GIS] Technician, Lower Elwha Klallam Tribe)

they harvested 50 percent of the salmon, then the total run to these small streams would have been about 90,000 fish annually. Today the wild coho have been reduced to only a remnant of their former abundance, and the wild chinook in the Pysht River are nearly extinct.

There are at least three accounts of the origin of the name "Pysht." Some people think a ship by that name wrecked nearby, and others think the name originated from a S'Klallam Indian word roughly meaning "where the wind blows from all directions." But the origin I prefer says that Pysht comes from Chinook jargon, a pidgin language developed for communication between the native peoples and Euro-Americans, and it means "fish."[2]

After four hours I've made twenty-one marks in my waterproof notebook. Twenty-one coho salmon made it back from their ocean travels to bury their eggs in the gravel of this little tributary of the Pysht. My marks will be converted to numbers, sent to a large office building, and made into "data." The biologists working there will analyze all the numbers from all the streams; then they will attend endless meetings to interpret and reinterpret the data.

In the controlled environment of the office building, soggy waders, icy air, chickadees, rivers, and even the salmon are excluded. In these offices, the primary contact with the natural, outside world is through reams and reams of computer printouts. The biologists talk a lot about salmon, but the salmon they refer to are a dry, lifeless representation of these magnificent fish. The salmon they see in the data are not the same fish I've come to know. In the "reality" constructed in conference rooms, the numbers are abstracted from the living salmon, the salmon are abstracted from their habitat, the habitat is abstracted from the river, and the river is abstracted from the ecosystem. The world of computers, numbers, and meetings is an essential part of salmon management, but having lived and worked in that world, I worry that it has lost too many connections to the rivers and the salmon.

Once I asked a biologist who was very good at mathematical analysis what he had learned about the salmon in the river he had been studying for about a decade. I wanted to know what he had learned from being there every day, watching the river and working with its silver fish. Had he learned things that he couldn't put into numbers? Were there questions that his models couldn't answer? What special insight had he acquired? He handed me several sheets of paper filled with mathematical equations and told me all he knew was there in those numbers. Technically he had done his job, but his myopic approach blinded him to the possibility of other insights, understandings, or hypotheses about the salmon that could have enriched the story his equations told.

Fisheries biologists have generally focused on populations that can be described by numbers. Biologists tend to treat the statistics as though they were independent of the ecological systems that produce them, and regard the fish as though they were under the complete control of the humans who manage them.[3] Biologists relying solely on models treat salmon populations like chessmen being moved around on a board rather than as living animals continuously responding and connected to their environment.

When I finish the entire survey reach, I stop a few minutes, sit on a log next to a large, dark pool, and peel an orange. In the rapidly fading light, the pool is a bowl of black ink. I'm sure my presence has disrupted the nest-building of a pair of coho salmon hiding in the darkness of the pool. If I am lucky, in the dying light of this December day I will be able to watch the mating pair move from their hiding place and onto the riffle to dig their redd. The coho are here for a single purpose: here and throughout the Northwest, adult salmon are giving their last days and hours of life to extend the fragile thread of life to the next generation. It is a thread that trails back in time at least 10,000 years, back to an era when the raw landscape of the Olympic Peninsula and its rivers emerged from the melting glaciers of the Wisconsin ice age.

It is in the gravel, pools, and riffles of this stream, and all the small streams and rivers, that the few remaining wild salmon make the difficult leap from one generation to the next. Already thousands of salmon eggs are incubating in the gravel of this stream, billions in all the rivers. They will remain in the redd for several weeks during this particularly dangerous early life stage. Some will be eaten by predators such as the sculpins that lurk on the stream bottom or the mergansers that poke their bills into the gravel looking for eggs.

Extensive logging in the Pysht watershed further heightens the danger to salmon eggs in their gravel incubators. Recently, for instance, a large section of logging road collapsed and slid one-third of a mile into a tributary, bringing along large chunks of the hill slope and smothering several salmon redds.[4] In the Pacific Northwest, road failures and landslides are common by-products of logging practices. Floods, aggravated by the loss of forests on the hill slopes, scour the eggs from the streambed and grind them up in the moving gravel. The death of thousands or perhaps millions of incubating salmon eggs is a cost of doing business that does not show up in the accounting ledgers of corporations and natural resource agencies. There are no records to tell us how many salmon eggs are killed under layers of suffocating mud or in the churning rocks. It's a cost that is quietly passed on to the salmon, and eventually to the families who fish for salmon for sport or profit.

Once the surviving eggs hatch, the young coho leave the gravel and become free-swimming fry. They will remain in the Pysht River until they go to sea in the spring of their second year. During their stream residence, the small salmon must survive additional threats. The same landslides that smother incubating eggs also destroy the salmon's nursery habitat by filling pools and disrupting aquatic food webs. Removal of trees and shrubs from stream banks allows the sun to warm the water to temperatures beyond the tolerance of the young fish. Even under the best of habitat conditions, most salmon that start life as little pink spheres in gravel nests will not survive to enter the sea. Under the pressure from human inroads into their habitat, fewer and fewer juvenile salmon survive their river residence and migrate to sea.

The short December afternoon gives way to near darkness before the coho salmon move into the riffle above the pool. I can't see them, but I can hear their splashing. I imagine their bruised and battered bodies thrashing against the gravel in their race to create life before death overtakes them. After several minutes of listening, I climb out of the river and fight my way through a tangle of branches and stumps on the cutover land until I reach the logging road and start the long walk back to my truck.

It has been a cold, dreary afternoon. There were too few salmon; the forests are rapidly disappearing; and the habitat, the salmon's home, is being devastated. But in the darkness, as I stumble over frozen ruts, I think about the wild salmon splashing in the darkness, and it gives me hope. The thread of life, however fragile it may be in the ravaged streams of the Northwest, is not yet broken. Wild salmon are survivors—living through volcanic eruptions, ice ages, mountain building, fires, floods, and droughts. I feel certain they will persist if we can control our behavior and give back what we've taken from them—rivers that retain some of their healthy ecological function. As I drive home, my thoughts keep returning to the river flowing in the darkness and the hopeful splashing of coho salmon preparing the way for the next generation.

Halfway around the world, a man who owns a large part of the Pysht River Basin sits in a London office and decides the fate of this little river and its salmon.[5] Like the biologists and their computer printouts, the man in London is connected to the Pysht River only through dry, lifeless columns of numbers. As he makes decisions to liquidate timber in the Pysht Basin, he thinks about the price of lumber in various world markets, about the profits from different deals, about hundreds of factors. Among the piles of facts and figures on his desk, there is nothing that tells him about the wild salmon splashing in the fading light of a December day and their dangerous leap to the next generation.[6]

As a salmon biologist, I have watched wild salmon spawning many times, but I have also studied columns of numbers. For twenty-eight years, I have worked in both the real world of the salmon's rivers and the abstract management world, and I have observed a growing disconnect between the two. This separation is an artifact of an attitude that denies the existence of a relationship between humans and salmon, a relationship between humans and any part of nature. This dominant worldview defines ecosystems as warehouses for the storage and production of commodities, insists that humans stand apart from those ecosystems, and demands that they control, manipulate, and "improve" them. It teaches even resource managers that the divide between humans and their ecosystems must be maintained and strengthened. It's a worldview that has created conflict and controversy and depletion of the salmon. But all this escaped me until I searched beyond the controversy, beyond the conflict, and beyond the attempts by everyone, including the salmon managers, to shift blame for the salmon's condition to someone else. When I studied the history of the relationship between salmon and people, I came to realize the power of our dominant worldview, the false assumptions it has caused us to make about nature and its role in the destruction of the salmon. What I discovered is the subject of this book.

The salmon are among the oldest natives of the Pacific Northwest, and over millions of years they learned to inhabit and use nearly all the region's freshwater, estuarine, and marine habitats. Chinook salmon, for example, thrive in streams flowing through rain forests as well as through deserts; they spawn in tributaries just a few miles from the sea and in Rocky Mountain streams 900 miles from salt water. From a mountaintop where an eagle carries a salmon carcass to feed its young, out to the distant oceanic waters of the California current and the Alaskan Gyre, the salmon have penetrated the Northwest to an extent unmatched by any other animal. They are like silver threads woven deep into the fabric of the Northwest ecosystem. The decline of salmon to the brink of extinction is a clear sign of serious problems. The beautiful ecological tapestry that northwesterners call home is unraveling; its silver threads are frayed and broken.

For millions of years the salmon's great strength was their ability to reach and reproduce in the wide variety of habitats found throughout the region. But since the arrival of Euro-Americans, that same trait has made the salmon vulnerable to destruction. Their ubiquitous distribution brings them into contact with a wide range of human economic activities: mining and timber cutting in the headwaters; grazing, irrigation, and other agricultural operations farther downstream; industrial and residential development in the lower river reaches and the estuary; and large-scale commercial fishing in the ocean.

Solving the salmon's problem has proven difficult because their extensive migrations create an ideal situation for obfuscation. Each industry, institution, or individual that contributes to the salmon's depletion at some place in their extended ecosystem can readily point to some other industry, institution, or individual that affects the salmon at some other place in their ecosystem as the cause of the problem. For the last half-century, the salmon's rapid decline has generated endless attempts to shift blame and prompted disingenuous evasion. Because the depletion is the cumulative effect of many human activities over a wide geographic area, proof that absolves or implicates a particular factor is impossible to obtain. Though it has sparked debate and study, the misguided search for a singular cause of salmon decline has wasted a great deal of time, effort, and money. More important, it accomplishes little in solving the salmon's problem, which has truly become everyone's problem.

The people of the Pacific Northwest must answer some important questions: Do they value the salmon enough to restore habitat and pull the fish back from the brink of extinction? Are they willing to save the salmon even if it means changing the way the industrial economy uses the region's land and water? And if they are not willing to make the necessary changes, will a Pacific Northwest without salmon retain its appeal as a high-quality environment for people?

For many of the region's residents, restoring the Pacific salmon is an important goal. Anglers enjoy fishing and the fine taste of fresh salmon. Native Americans harvest salmon to satisfy commercial, cultural, and religious needs. Commercial fishermen want to continue fishing as a way of life. Many northwesterners view the salmon as an important symbol of a clean and healthy environment. Still others want to see the salmon restored to avoid the regulations that will accompany enforcement of the Endangered Species Act. Those who express concerns about the salmon and a desire to restore their productivity have a wide array of values, but all share the same goal of returning healthy salmon runs to the streams of the Northwest.

Some people believe that the century-long decline of the Pacific salmon following the arrival of Euro-Americans resulted from inept management, greed, and political power grabs on the part of those who benefited from access to the West's natural resources. But that explanation is too simple and misleading. Clearly, the intent of the people, their political leaders, and their institutions was not to destroy the salmon. As early as 1848, Oregon prohibited the erection of barriers to salmon migration. And in the 1870s, both Oregon and Washington tried to protect salmon habitat by passing laws to prohibit the dumping of sawdust in rivers. If we want to truly understand what destroyed the salmon's productivity, we must look deeper—beyond the greed that drove the exploitation of the Northwest's watersheds, and beyond the weakness of bureaucrats too timid to enforce laws protecting the salmon.

Fundamentally, the salmon's decline has been the consequence of a vision based on flawed assumptions and unchallenged myths—a vision that has guided the relationship between salmon and humans for the past 150 years. We assumed we could control the biological productivity of salmon and "improve" upon natural processes that we didn't even try to understand. We assumed we could have salmon without rivers. As a direct outcome of these assumptions, we believed that human economic activities such as mining, logging, and fishing were unrelated to the ecological processes that produced fish. The natural limits of ecosystems seemed irrelevant because people believed they could circumvent them through technology. Placing misguided confidence in technological solutions, salmon managers accepted the myth that controlling salmon reproduction in hatcheries would ultimately lead to increased productivity. Despite the best of intentions, these hard-working people produced disaster because their efforts were based on false assumptions. Their reasoning is a classic example of what historian Donald Worster called "instrumental thinking": "Thinking carefully and systematically about means while ignoring the problem of ends."[7]

Though the ends are what we face today, we are still hung up on the means and hamstrung by mythic assumptions. Currently, hatcheries remain the primary means of restoring salmon even though such programs have clearly failed to achieve their purpose for more than a century.

Author Tim Egan described the Pacific Northwest as "any place salmon can get to," and by that definition the region has been shrinking for the last 150 years.[8] Since the turn of the twentieth century, the natural productivity of salmon in Oregon, Washington, California, and Idaho has declined by 80 percent as riverine habitat has been destroyed. To confront this loss, we need a different vision, a different story to guide the relationship between salmon and humans. To give the salmon any hope of recovery, we have to break free of the myths that have brought us to the point of crisis.

Chapter 1 Hooknose

Before the Endangered Species Act; before shopping centers covered streams with asphalt; before dams and dynamos harnessed the energy of wild rivers; before irrigation sucked rivers dry; before timber harvest robbed rivers of their protective forests; before fishermen's nets swept through the rivers and bays; before humans walked across the Bering Strait and into the Pacific Northwest; before glaciers gouged out Puget Sound; before the Oregon coast migrated away from Idaho; before all this, there were the salmon.

In the Linnean system for classification of plants and animals, the Pacific salmon fall into the family Salmonidae and the genus *Oncorhynchus,* the Pacific salmon and trout (Appendix A). The name "Oncorhynchus," from the Russian term for "hooknose," refers to the hooked upper jaw that males develop during mating. There are seven species of Pacific salmon within the genus *Oncorhynchus.* Five are found in North America: pink (*O. gorbuscha*), chum (*O. keta*), sockeye (*O. nerka*), coho (*O. kisutch*), and chinook (*O. tshawytscha*). Two are found only in Asia: masu (*O. masou*) and amago (*O. rhodurus*). Pacific trout within the genus *Oncorhynchus* include the anadromous steelhead, *O. mykiss,* and sea-run cutthroat trout, *O. clarkii.* The five species of Pacific salmon and the two anadromous trout have evolved a rich array of life histories and local and regional distributions, which are summarized in Appendix B.

Native Americans sometimes called the salmon "lightning following one another," a name that evokes an image of the large silver fish flashing through swift water.[1] Our culture's most common image of the salmon is of a fish climbing the face of a seemingly impassable falls, wiggling and fighting for purchase to launch another jump. We associate the salmon with a strong, fighting spirit and an unstoppable determination to return to the stream of their birth to create the next generation. Our image fosters the belief that the salmon possess an inherent ability to persist regardless of obstacles. But their persistence is more than just legend. Nature and, more recently, humans have repeatedly created conditions that threatened the salmon's survival, yet they have persisted.

All organisms are historical phenomena, and the salmon are no exception. For thousands or even millions of years, the salmon have accumulated the history of their species and retained it in the gene pools of individuals and populations.[2] As a result, the salmon's genetic program, coded in its DNA, is a textbook containing thousands of years of evolutionary experience—lessons on how to survive in a harsh, changing world. That the Pacific salmon have survived our largely unrestrained assault on them and their habitats over the past 150 years proves the value of the lessons each salmon carries in its genes. And it is precisely because the salmon are such tough, persistent animals that their catastrophic decline in California, Oregon, Washington, and Idaho is so tragic. Once we understand the salmon's story, we can see their collapse as a clear testimony to the failure of our management institutions and to the rapid and massive habitat destruction over the past century. To fully appreciate the salmon's resilience, we must understand their evolutionary history and the episodes of cataclysmic environmental change they have survived.

A Rough Trip through Evolutionary Time

The Pacific salmon belong to a group of fishes whose branch on the evolutionary tree sprouted relatively late, about 400 million years ago. Within a short 40 million years these newcomers, the ray-finned fishes, came to dominate many freshwater habitats and were well on their way to dominating the sea. The ray-finned fishes comprise three main groups, and each achieved prominence for a time. Sturgeons and paddlefishes are the surviving members of the oldest group, and the gars give us a glimpse at the second group. Teleosts are the last and still prevalent group, which includes familiar fishes such as salmon, trout, perches, pikes, and catfishes.[3] The fossil record shows that teleosts arrived relatively late in western North America. They did not dominate western fresh waters until about 55 million years ago.[4]

Teleosts, like the other fishes before them, rose to dominance because they evolved improved structural and physiological features that gave them a competitive edge. The teleosts acquired a more efficient respiratory system, and their body form and musculature changed to permit more rapid and complex movement—important survival traits for both prey and predator. Among these efficiently designed fishes is the family Salmonidae, which contains salmon, trout, whitefishes, graylings, and ciscoes.

No one knows just when the Salmonidae split off from other ancient teleost fishes, but the family might be 100 million years old.[5] That means ancestral salmon could have swum in the same waters used by dinosaurs during the last few million years of their reign on earth. The earliest confirmed Salmonidae fossil is *Eosalmo driftwoodensis,* whose bones were found in Eocene sediments of British Columbia (Appendix C). It resembled the modern grayling and lived in ancient lakes of western Canada about 40 to 50 million years ago. Based on the fossil record, Salmonidae's prehistoric range in western North America extended as far north as the Yukon Basin (66° north latitude) and as far south as the Lake Chapala Basin of Mexico (31° north latitude).[6]

Native fish fauna in the Pacific Northwest survived a rough trip through evolutionary time. No other region in North America has been as geologically active as the Pacific Northwest, which means no other region has experienced the same degree of habitat and environmental transformation. Mountains rose, the coastline migrated, the climate changed drastically, and volcanoes flooded large areas of the region with thick layers of molten lava. Many fish species—pikes, freshwater catfishes, bass, sunfish—did not survive the monumental geologic and climatic changes. But the tough and tenacious salmon endured, and their resilience is still a source of hope for the future.

Origins of Anadromy

Salmon are anadromous. Adult salmon spawn, their eggs incubate, and juveniles rear for a few weeks to several years in fresh water. Then the juveniles migrate to the saltwater sea, where they spend a few months to several years before returning to their home stream to spawn in fresh water and repeat the cycle. For about a century, scientists have debated whether the salmon originated in fresh or salt water; the weight of the evidence now suggests a freshwater origin.[7] Paleontologists have found fossils of ancient trout only in freshwater sediments, and the bones of primitive salmonids such as the graylings and whitefishes are found only in fresh water (although some graylings may enter estuaries). Since the extinct members of the family and the older existing Salmonidae reside exclusively in fresh water, we can conclude, tentatively at least, that the salmon's life in the rivers and lakes goes back further in time than their life in the ocean.[8]

What advantage did the salmon gain by crossing from fresh to salt water and back? Whatever the advantage, it had to outweigh heavy physiological costs, because moving from fresh to salt water puts tremendous stress on the fish and requires much preparation. Before entering the sea, the salmon change from stream-dwelling parrs to smolts, a transformation that involves physiological and behavioral adaptations to the saltwater environment. For example, when a parr becomes a smolt, it increases the purine and guanine in its scales.[9] As a result, the young salmon's natural camouflage, which is well suited to hiding in a stream, gives way to a uniform silver on its sides and undersides—a coloration better suited to life in the sea. Because oxygen concentrations are lower in sea water, the salmon smolt must also produce a different, more efficient hemoglobin to cope with the decrease in oxygen.[10] Furthermore, salt pumps in the gill membranes must reverse. In fresh water these pumps prevent dilution of plasma electrolytes, but in salt water the pumps must keep electrolytes out to prevent concentration above normal levels.[11] In addition to the physiological changes, the salmon must undergo major behavioral changes. Life under an overhanging bank in a small stream is quite different than life in the ocean, where the habitat is open and predators abound.

Migration across the fresh–saltwater boundary occurs in two distinct patterns. Anadromous fishes, such as the salmon, breed in fresh water and then migrate to sea to feed and mature. Catadromous fishes, such as the eel, breed in the sea and then migrate to fresh water to rear and mature. Anadromous fishes are found most often in northern latitudes, while catadromous fishes are found most often in the tropics. In the mid-1980s, biologist Mart Gross and his colleagues looked at these patterns and discovered an important clue to the anadromy puzzle. The clue turned out to be food. In northern latitudes the oceans are more productive than the adjacent fresh waters, but in southern latitudes the reverse is true. Apparently, the Pacific salmon abandoned fresh water to rear in the more nourishing oceanic pastures of the northern latitudes.[12]

This makes sense in the framework of evolutionary time because the oceans off the Pacific Northwest were not always as cold and productive as they are today. Prior to the Oligocene epoch, about 40 million years ago, the oceans were 10°C (about 18°F) warmer and therefore probably less productive. Then about 25 million years ago, they began to cool, reaching their current temperature regime about 8 million years ago.[13] As the seas cooled, their productivity increased. According to current theory, the Pacific salmon developed anadromy to take advantage of this richness.

The salmon grew rapidly in the oceans, and their larger size at maturity improved their fitness in the rivers of the Northwest in at least three important ways. Anadromous salmon could produce more offspring because the larger females carried more eggs to the spawning grounds. Furthermore, the larger, stronger fish were more able to muscle their way over falls or through strong currents, extending their distribution throughout the Northwest's network of rivers. Finally, with their greater bulk, the salmon could dig deeper redds, giving greater protection to their incubating eggs.

Anadromy also benefited the rivers of the Northwest. The migration of large salmon into the interior of the Northwest was, in effect, a mass transfer of nutrients from the sea upstream into the headwaters. Those nutrients fertilized the less productive river ecosystems, increasing the growth and survival of juvenile salmon.[14]

If the salmon did originate in fresh water, then the hooknose probably developed anadromy in stages beginning during the late Oligocene, when the oceans started to cool (Appendix C). The current fossil record can prove only that salmonids similar to extant anadromous species lived in coastal rivers at least 10 million years ago. Confirming an older origin for anadromy will have to wait until fossils are found in Oligocene sediments in sites with access to the sea.[15]

Upheavals and Eruptions

Biologists believe the ancestors of the modern salmon were lake dwellers because fossils of the earliest salmonids are usually found in fine-grained sediments associated with lakes or slow-moving, low-gradient rivers. According to this hypothesis, the early Salmonidae colonized interior lake systems, where they lived in isolation and evolved a rich species diversity.[16] By the late Miocene (about 10 to 15 million years ago), however, salmonid fossils appear in coarse gravels, suggesting that the ancient fish expanded their range into the high-energy habitats of rivers.[17] According to the current fossil record, *Oncorhynchus,* the genus containing the Pacific salmon, probably emerged at this time.

The most remarkable stream-dwelling salmonid from this period is the now-extinct saber-toothed salmon (*Smilodonichthys rastrosus*).[18] This formidable six-foot-long fish had a pair of enormous, curved breeding teeth, but despite its ferocious appearance, it fed primarily on plankton, much like the modern but smaller sockeye salmon. Fossil remains of the saber-toothed salmon have been found in coarse gravels, indicating that it spent part of its life in riverine habitats of central Oregon and coastal California. Its bones have been found along with other *Oncorhynchus* fossils that resemble the modern coho salmon. Since a diet of plankton suggests lake or ocean rearing habitats, *S. rastrosus* likely used the rivers only for spawning. Judging from the location of the fossils found in central Oregon, the saber-toothed salmon probably spawned in the Columbia River watershed.[19]

When the bones of the earliest salmonid, *Eosalmo,* settled into lake sediments in British Columbia 40 to 50 million years ago, the coastal mountains of California, Oregon, and Washington lay submerged beneath the Pacific, and the ocean's waves washed beaches near the Idaho border. A series of volcanic islands broke the sea's surface well offshore between the future states of Oregon and California. Those islands would later rise and become part of the Klamath Mountains. The landscape of the northern California and Oregon coastal mountains began to take on its current form about 12 million years ago.[20]

Anyone traveling U.S. Highway 101 along the West Coast would view the coastal mountains as stationary, but in geologic time the coastal ranges undertook a migration no less spectacular than the migration of salmon. Fifteen million years ago, the Oregon Coast Range, which hugged the western edge of the Cascade Mountains, rotated out and 32° north to its present location. As the coastal mountains rotated and uplifted, rivers kept pace by downcutting or eroding their channels through the rising land. By the Pliocene, about 2 to 5 million years ago, the coastal rivers had cut their catchment basins into the drainage patterns we see today. The formation of the Cascade Mountains also influenced the development of rivers in the Northwest during this same period. Runoff from these peaks enabled a few large rivers—the Fraser, Columbia, Umpqua, Siuslaw, Rogue, Klamath, Sacramento, and Chehalis—to erode rapidly enough to maintain their connections with the interior regions.[21]

If the hooknose came into existence before the mountains rose and the coast shifted to its present position, as hypothesized, then the early members of the genus had to survive millions of years of cataclysmic habitat disruption. For example, one consequence of active mountain building and river downcutting would have been landslides large enough to block streams or create impassable waterfalls. Mountain building would have resulted in a continuous round of habitat destruction and creation. Individual populations of salmon would have become extinct when they were cut off from their spawning habitat or when their habitat was rendered unlivable. At the same time, stray salmon would have colonized new habitats, and the process of adaptation would begin again as new populations took hold. Local extinction balanced by recolonization on an evolutionary time scale has long been an important survival mechanism of the Pacific salmon.

While the coastal rivers cut watersheds out of the rising western edge of the Northwest, farther east the earth opened and spilled lava over the gently rolling hills of interior Oregon, Washington, and British Columbia. Between 12 and 17 million years ago, massive flows of hot lava moving at twenty-five to thirty miles per hour repeatedly covered the land. Each successive eruption entombed salmon habitat with a layer of lava 100 feet thick and set the ecological clock back to near zero. Rivers had to erode new channels, soils had to re-form, and plants and animals from outside the region had to repopulate the barren landscape.

The ancient lava flows of the Columbia Basin are clearly visible today as basalt, especially where rivers have cut through and exposed the individual strata like layers of a gigantic cake. Two of the most spectacular such exposures can be seen in the canyon of the Grande Ronde River in eastern Oregon, and along the seventy-two-mile gorge of the Columbia River. The lava flowed into and plugged the ancient Columbia River, forcing it to move north and cut a new channel. Since then, the Columbia has remained in its present channel.[22]

By about 2 million years ago, watersheds in the Northwest had carved their current drainage patterns (Figure 1.1). Genetic analysis of the Pacific salmon suggests that the modern species evolved from three ancestral lines around this time. The first two diverged a little more than 2 million years ago; one produced rainbow trout, coho, and chinook salmon, while the second yielded sockeye salmon. The third line diverged about 1.25 million years ago, giving rise to pink and chum salmon.[23] According to this analysis, if the hooknose came into existence during the Miocene, these fish underwent 10 million years of evolutionary development before the appearance of the ancestral lines that produced the modern salmon.

Stabilized watersheds and the appearance of modern species did not signal the end of the cataclysmic changes for salmon. A period of intense glaciation soon followed and drastically altered the region's climate, rivers, and salmon habitat, especially in coastal Washington, Puget Sound, and the Columbia River Basin.

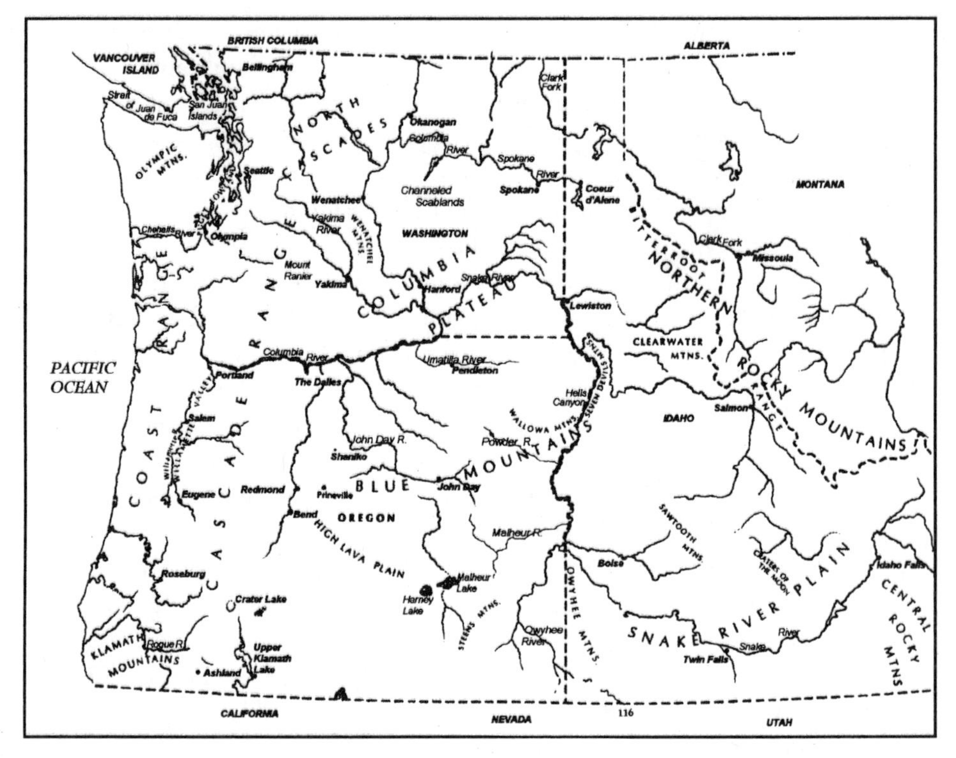

Figure 1.1. The major rivers and mountain regions of the Pacific Northwest.

An Onslaught of Ice

The most recent ice age, known as the Wisconsin, peaked about 18,000 years ago, covering North America with two great ice sheets. The Laurentide ice sheet blanketed most of north-central and eastern North America, while the Cordilleran ice sheet covered the Pacific Northwest (Figure 1.2). During the Wisconsin's peak, the Cordilleran ice sheet spread into the Puget lowlands, where it covered the present site of Seattle with a layer of ice 4,000 feet thick. At its southern extent, the ice sheet reached the area around the city of Olympia, Washington. The slow movement of the thick glaciers carved many of the features of Puget Sound.

While the rivers of Puget Sound, British Columbia, and southeast Alaska were frozen, the salmon took refuge in watersheds to the south and, surprisingly, to the north of the Cordilleran ice sheet. They lived in ice-free rivers, including the Sacramento, Columbia, and Chehalis, as well as in the streams of Beringia—a large, ice-free area in northwest Alaska that extended into

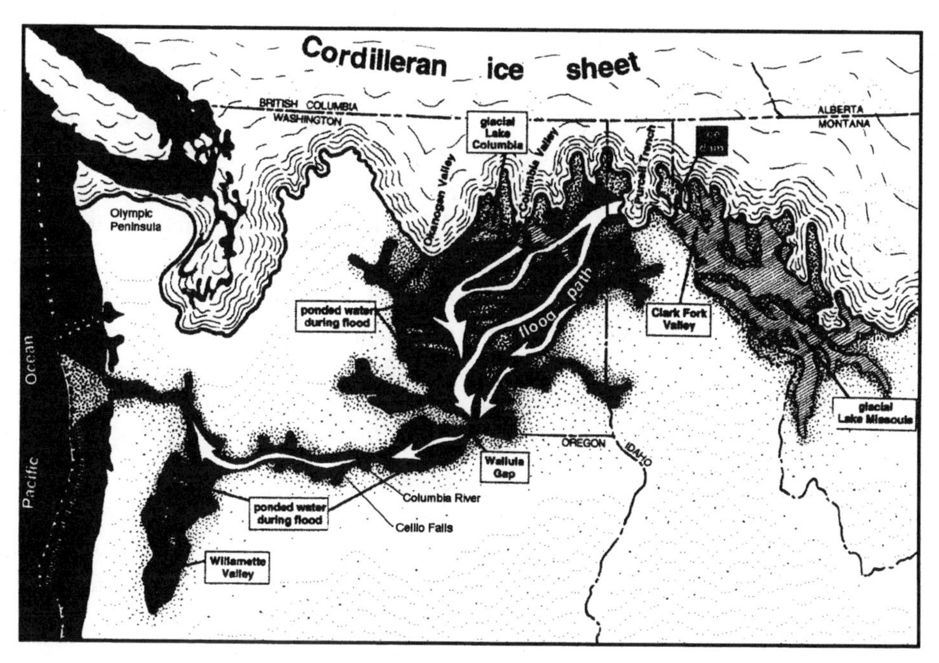

Figure 1.2. Glacial Lake Missoula and the path of its floods about 15,000 years ago, during the late ice age in the Pacific Northwest. (After Alt and Hyndman 1995)

Siberia. South of the Columbia River, glaciers formed only in the higher elevations, leaving the drainage patterns of rivers intact.[24]

The Columbia River was the largest watershed close to the southern edge of the Cordilleran ice sheet. Although the river's lower two-thirds remained ice-free, it did not escape the cataclysmic side effects of glaciation. After the glaciers began their retreat, the melting ice generated gigantic floods that shaped the landscape in northeastern Washington and left visible scars down through the lower reaches of the Columbia.

Toward the end of the Wisconsin ice age, the glacier's leading edge advanced and retreated many times. During several advances, a lobe of ice pushed down from British Columbia, past the present site of Lake Pend Oreille, Idaho, and up the Clark Fork River. Each time the ice lobe moved up the Clark Fork, it spread across the valley floor, plugging it with a wall of ice that at times reached 2,500 feet in height. The Clark Fork glacier impounded water that drained throughout western Montana, creating the ancient Lake Missoula, which measured 3,000 square miles. With a depth of 2,000 feet in places, the lake contained about half the volume of Lake Michigan.

Eventually Lake Missoula's water level rose enough to float the ice dam off the valley floor, releasing a huge flood. A giant wall of water raced across Idaho at fifty miles per hour. It scoured northeastern Washington, creating the topography known today as the channeled scablands. Then the water overflowed the Columbia River channel, reaching depths of 1,000 feet in places like Wallula Gap, near the mouth of the Snake River. In the lower Columbia River the flood covered the present site of Portland, Oregon, with over 200 feet of water and transformed the Willamette Valley into a lake. The Lake Missoula floods took place at least forty times between 15,000 and 12,800 years ago, or roughly every fifty-five years during the final stages of the Wisconsin glaciation.[25]

The largest of these floods released an energy equivalent to 106 percent of all the electricity generated in the United States in 1970. The erosive power of water mixed with ice, boulders, gravel, and silt carved the Columbia Gorge and the series of channels and falls in the mid-Columbia known as Celilo.[26] The salmon probably survived these torrents in tributary refugia such as the Deschutes, John Day, and Snake Rivers.

Even in the streams south of the farthest advance of the ice field, salmon habitat changed due to the indirect effects of glaciation and the associated climatic change. South of the main Cordilleran ice sheet, smaller glaciers covered the upper elevations of the Cascade, Blue, Steens, and Klamath Mountain Ranges.[27] During the short period of rapid melting, these glaciers also unleashed high flows that shaped salmon habitat, but the ensuing floods were far less severe than those generated by the Cordilleran ice sheet. Farther south in the Sierra Nevada, a 2,000-foot-thick glacier covered Yosemite Valley, creating the flat meadows of the valley floor. [28]

During the cooler temperatures of the Wisconsin ice age, it's likely that the salmon's range expanded southward; however, a period of intensive warming followed the retreat of the glaciers. During this hot, dry period, the salmon's southern range moved north. Indirect evidence from Indian middens suggests that the streams in northern California and southern Oregon supported only very small runs—likely because they became too warm for the salmon.[29] About 4,000 years ago the Northwest began cooling to its current climate; as it did, the salmon recolonized the southern end of their current range. Steelhead moved as far south as the Mexican border, and chinook salmon reached the Ventura River just north of Los Angeles.

The salmon that survived the ice age in the refugia of the Columbia River

Basin repopulated streams as far north as the Stikine River in British Columbia and southeast Alaska. The colonizing salmon also moved south from rivers in Beringia. Ten millennia after the retreat of the ice, the Columbia River would become the great engine of human industrial economies in the region. But in the period following deglaciation, the Columbia was a great "organic machine," the mother river of the Pacific salmon, which filled the rivers that had been released from the ice.[30]

After the Ice

The melting ice left behind a desolate landscape composed of glacial outwash, boulders, gravel, and rock flour produced by the grinding action of the moving glacier. Frequent floods washed over this mineral landscape as ice dams formed and broke during the erratic retreat of the glaciers. A period of rapid warming began about 14,000 years ago, reached a peak about 6,000 to 5,000 years ago, and lasted until about 4,000 years ago. This hot, dry climate slowed revegetation, so that the lodgepole pine and fir that colonized the barren landscape were only small, stunted versions of their modern relatives. The small trees and shrubs could hold neither the rocky soil nor the streams in stabilized channels.

Alternating between roaring floods and impoverished trickles, the recently deglaciated rivers flowed in broad, shallow channels over the glacial outwash. Over several thousand years, the slowly moving glaciers had excavated millions of tons of gravel, soil, and boulders and locked it into the ice. When the melting ice released this material, the newly forming rivers were engulfed by sediment loads sixty times that of modern Pacific Northwest rivers.[31] Not confined to stable channels, the rivers moved back and forth across wide valleys, depositing thick layers of sediment. During floods, the gray rivers frequently jumped out of their shallow channels and whipped across the landscape like loose fire hoses.

The broad, shallow rivers, flowing thick with mud, sand, and gravel, were hostile to any stray salmon that swam into them to spawn. But eventually a few eggs did manage to survive the shifting, grinding gravel and smothering blanket of silt. Once the eggs hatched, the salmon fry found no quiet water or deep pools where they could rest, hide, and feed. The sparse vegetation along the banks of the shifting rivers could neither shade the streams nor stabilize the stream banks. The lack of shade sent summer stream temperatures soaring in the hot, dry climate. Aside from shade, well-developed riparian communities also yield an important source of food for juvenile salmon; insects dropping off trees and shrubs onto the stream make up a large part of the salmon's summer diet. In the postglacial rivers, that food was not available. Furthermore, the rivers split into many small, braided channels, which flowed around piles of gravel and hummocks of land barely held together by the pioneering vegetation. In these shallow braids, juvenile salmon were highly vulnerable to predatory birds. From time to time, the smaller of the channels dried up, stranding any fish trapped in them. The raw postglacial rivers were killing grounds for salmon.

Between 5,000 and 4,000 years ago, the climate began to shift toward the cool, moist maritime conditions that persist today. This change in the environment enabled forests to grow. Early in the cooling period, Douglas fir, which favors drier conditions, dominated; then western hemlock and western red cedar took hold, especially in the wetter areas. Eventually, what we now call old-growth forests emerged, creating stable, well-shaded streams and high-quality salmon habitat. The extensive root systems of the large trees held the streams to defined channels and prevented erosion, while the thick foliage filtered out the sun's heat, helping to maintain cool water temperatures. When the large trees died and fell across stream channels, they created pools and added structure and diversity to riverine habitats. The logs accumulated and eventually backed up water, creating productive side channels and sloughs. The thousands of trees falling across streams also functioned as small check dams, holding back sediment and stabilizing streambeds, thus enabling streams to run clear. As a result, these rivers carried far less sediment than the unshaded rivers had. Old-growth forests created complex and stable stream habitats, and it's possible that salmon populations first reached levels of eighteenth- and nineteenth-century abundance around 5,000 years ago.[32]

Evolutionary Legacy

Over the last several million years, the salmon and the landscape of the Northwest underwent considerable change. About 3,000 to 5,000 years ago, the changing habitat and the evolving salmon converged into an ecological harmony that produced an abundance ranked as one of the natural wonders of the world. To achieve that harmony, the salmon had to survive a rough trip through evolutionary time. They had to continuously adapt to a landscape with high levels of spatial and temporal diversity. Not only did the landscape change through time, but at any particular time it was also composed of a patchwork of diverse geologic, climatic, and biotic conditions.

As the various species of Pacific salmon evolved, they developed traits that maximized their survival in different habitats within a watershed. The result of this evolutionary process was a rough division of a watershed's habitat among the different species. Figure 1.3 shows how the spawning habitats of a hypothetical watershed are partitioned by chum, pink, coho, chinook, and sockeye salmon. Pink and chum salmon generally spawn in the lower stream reaches, and the juveniles migrate to sea after a brief period in fresh water. Coho salmon spawn and the juveniles rear in the upper, smaller tributaries. Chinook salmon prefer the mainstem and larger tributaries. The spring chinook spawn in the headwaters. Summer and fall chinook make use of the middle and lower river, respectively. When sockeye salmon emerge from the gravel, they migrate to a lake, where they complete their freshwater rearing.

The generalized spawning distributions shown in Figure 1.3, as well as the species life histories in Appendix B, are not rigid laws that are cast in concrete. Each species has a characteristic life history with unique attributes that separate it from the other species. Among those attributes are age structure, length of freshwater residence, and their spawning and rearing distribution within a watershed. But the generalized life histories are really central themes

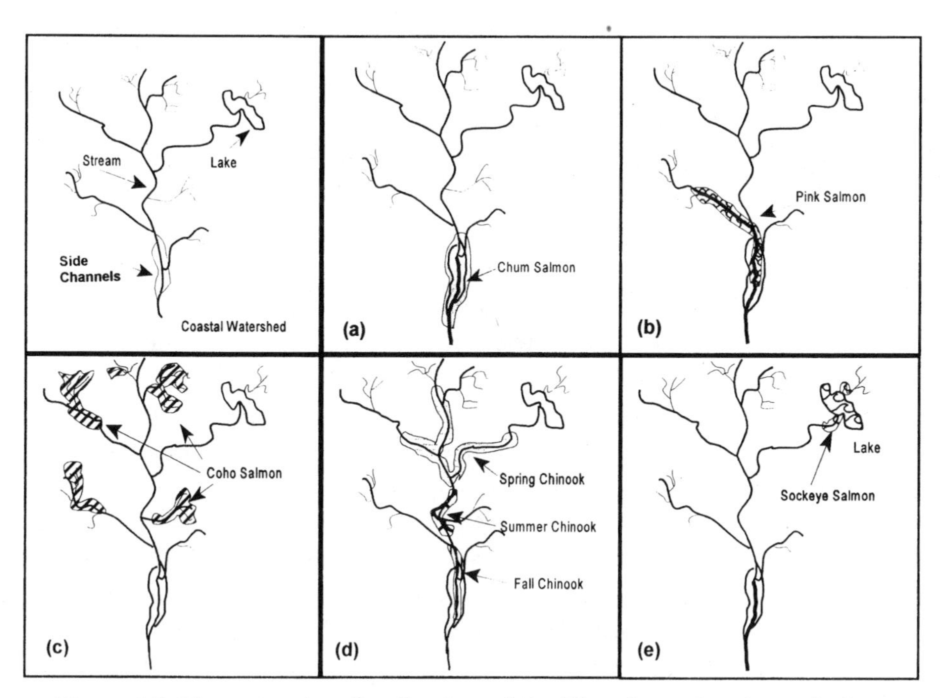

Figure 1.3. The spawning distribution of Pacific salmon in a hypothetical watershed. Typical distribution of chum (a), pink (b), coho (c), chinook (d), and sockeye (e).

around which each population has developed a rich diversity in response to its local habitat. For example, chum salmon normally spawn in the lower reaches of rivers, including the tidal zone; in the Amur River of Russia and the Yukon River of Alaska and Canada, however, chum salmon migrate up to 1,560 miles from the sea to spawn. Juvenile coho salmon usually rear in small streams and tributaries throughout their freshwater life history, but in some watersheds in southwestern Oregon, they spend most of their freshwater residence in coastal lakes. Conversely, sockeye salmon almost always spend their juvenile rearing period in freshwater lakes; but in a few populations in Alaska and Canada, they may migrate to sea shortly after emergence from the gravel, or rear in the river or side channels until they go to sea, without ever residing in a lake.[33]

The variation in the salmon's life history is an important legacy of their survival and evolution through the geologic and climatic events that marked the history of the Northwest. Life-history diversity is the salmon's response to the old adage about not putting all one's eggs in the same basket. It is the salmon's solution to the problem of survival in changing environments. All of this variation also reveals that the salmon and their habitats are inextricably linked, that they have to be considered as a single unit, especially when we attempt to manage or restore them.[34]

Biologist W. F. Thompson described the salmon's ecosystem as a "chain of favorable environments connected within a definite season in time and place"; within a watershed, he wrote, there are "various possible combinations or chains of this sort which can be connected in time and space in a variety of ways."[35] Each of those chains is a life-history variant, a different pathway the salmon follow through the chain of favorable environments that collectively make up their ecosystem. The survival of salmon following a given life history or pathway will vary over time in response to natural or man-made changes in the habitat. As the climate or habitat changes, some life histories will be favored, while others will be disadvantaged; thus the survival value of a given life history may vary over time.

I will return to the significance of the salmon's life-history diversity again, but for now it is important to recognize that life-history diversity is a critical legacy of the salmon's evolutionary history, and it is the salmon's key to survival in the complex landscapes of the Pacific Northwest. As an essential part of the salmon's evolutionary legacy, life-history diversity must be a prominent consideration in all management and restoration programs. Unfortunately, our management and restoration programs have actually contributed to the loss of life-history diversity.

My study of the evolutionary history of the salmon and the geologic history of the Pacific Northwest changed what I see when I look at a salmon. Now I see more than a silver fish sitting at the center of a regional crisis. Instead, when I look at a salmon today, I am reminded of the region's long history. The large, ocean-fed salmon recalls the cooling sea 10 million years ago; how the rise of mountains and the advance of ice shaped the salmon's ability to colonize new habitat; and how the diversity of the salmon's life history matches the highly variable landscape, climate, and vegetation patterns. I see too the physical beauty of the land: the old-growth forests that protect and nurture the salmon's habitat; the wild and free rivers that flow through those forests; the mountains that contribute gravel for the salmon to spawn in; and the cool, drippy climate that keeps the rivers flowing. And finally, the salmon remind me of people: the Indian fishermen and their 10,000-year relationship with the salmon, the commercial fishermen and their unique communities and way of life, and the sport fishermen who come from all over the Northwest to hunt the salmon. When I look at a salmon, I don't just see a silver fish, I see the Northwest.

Chapter 2 The Five Houses of Salmon

A man and his family lived under an overhanging ledge of basalt perched high above the river. The shallow cave sheltered them from rain, wind, and sun, and gave them easy access to the river below and to the grasslands on the plateau above. Although lush grass covered the uplands, game had become scarce. Other families living on the plateau had already moved east in search of the disappearing mammoths and other large mammals, but this man and his family remained and turned to the river for subsistence.

One day 12,000 years ago, the family made its way down a narrow trail to the canyon floor. The large river dominated all senses. A constant roar filled their ears as tons of water dropped over cataracts and churned through rock chutes carved in the basalt. Mist rising from the frothing river imbued the land with a cool dampness, and the air carried a musty taste that came from the 1,000-year-old water released by melting glaciers far upstream.

As the family walked along one of the smaller channels, the man saw a large silver fish leap from a pool, push through the water spilling over a three-foot fall, and swim away to the river above. Soon another salmon jumped from the pool and thrashed through the cascading water to the main river above the falls. The man was ready when the third salmon leaped out of the pool; he impaled it with his spear. The family quickly butchered the fish and placed it in a crude basket, and the man went back for more.

With the whole family busy spearing and butchering the salmon, no one noticed the freshening breeze, and they didn't notice an ominous rumble, lost in the roar of the river. When the man had caught enough fish to make a full load for each family member, he set down his spear. The rumble from upstream now drowned out the sound of the cascading river. He could feel the cold, brisk wind and smell its strong, earthy odor. He looked up just in time to see a wall of mud, large boulders, and house-sized blocks of ice turn the bend in the river. The thunderous mass filled the canyon, tearing downstream at forty miles per hour. Two minutes later the fishing site was under 300 feet of muddy water.[1]

The arrival of humans in the Pacific Northwest coincided with the last of the great floods from ancient Lake Missoula. Although some salmon were present at this time, the first people did not find streams teeming with them because the effects of glaciation and climate change had combined to limit their number. Over the next several thousand years, however, the Northwest's streams, rivers, and lakes would evolve into one of the world's richest salmon-producing ecosystems. During the same period, those first North Americans would build a rich culture and economy based on the annual return of the silver fish to their natal streams.

Humans Enter the Pacific Northwest

During the peak of the Wisconsin ice age, shrinking oceans left behind a land bridge between Siberia and Alaska that was used by the first people to reach North America. Because the land bridge and large parts of northwest Alaska remained ice free during the Wisconsin glaciation, archaeologists suspect that this area, known as Beringia, was the place where the earliest humans set foot on North America, sometime between 15,000 and 13,000 years ago.[2] The Cordilleran and the Laurentide ice sheets moved toward each other but never met. This left an ice-free corridor through present-day Alberta, which was one of the possible routes people used to enter the Pacific Northwest. The other possible route followed the coast. By at least 12,000 years ago, people were radiating out through the ice-free areas of North America.

The first North Americans' journey along the edge of mile-high glaciers to the ice-free areas of the Pacific Northwest must have been one of the greatest historical and environmental adventures of all time—an amazing story that will never be completely told. The weather alone would have made the experience a nightmare. Sharp temperature gradients along the glacier's edge would have created horrendous storms and severe winds.[3] But once they reached the ice-free area, the travelers probably stood in awe of the rich natural productivity they found in northwestern North America. Not only did the landscape have a different cover of vegetation then, but it also supported an abundance of large mammals, including mammoths, mastodons, ground sloths, camels, horses, and giant beavers.[4]

The large herbivores of North America had no history of contact with humans, so they had no defensive behaviors to help them avoid these new predators. Because the animals had no fear of the fragile-looking humans, the first people found hunting easy, even with stone-tipped spears, and the result was the widespread extinction of large mammals. In a relatively short period of time, North America lost 74 percent of its megafauna.[5]

The first people in the Pacific Northwest did not leave written or oral records of this decimation, but we can piece together a picture of their first contact with North American megafauna from indirect evidence. Some ecosystems were not invaded by humans until the eighteenth century, and in those cases explorers did leave extensive records of their first contact with defenseless animals. For example, when Arthur Bowes Smyth landed on Lord Howe Island in the South Pacific in 1788, he wrote, "I [could] not help picturing to myself the Golden Age as described by Ovid to see the Fowls . . . in thousands . . . walking totally fearless & unconcern'd in all part around us, so we had nothing more to do than to stand still a minute or two & knock down as many as we pleas'd wt. a short stick—if you throwed at them and missed them, or even hit them without killing them, they never made the least attempt to fly away & indeed they would only run a few yards from you & be as quiet & unconcerned as if nothing happened."[6]

Charles Darwin, who had similar experiences during his voyage on the *Beagle,* described the naïveté of animals with no previous human contact in this way: "We may infer from these facts, what havoc the introduction of any new beast of prey must cause in a country, before the instincts of the indigenous inhabitants have become adapted to the stranger's craft or power."[7]

Because the Bering land bridge remained open for several thousand years, multiple migrations occurred as people from different cultural traditions crossed from Asia to North America. According to anthropologists, four basal cultures made the trip across the Bering land bridge, each identified by the design of its spear points and knives: the Stemmed Point, Fluted Point, Pebble Tool, and Micro Blade cultures. In the Pacific Northwest, the Stemmed and Fluted Point cultures hunted and foraged near pluvial lakes in Oregon and California. By 10,000 years ago, these two hunting cultures had moved eastward in pursuit of the disappearing mammoths, their principal food source.[8] As a result, the Stemmed Point culture rapidly expanded, occupying most of North America and ultimately reaching the tip of South America.[9]

The other two cultures were more closely associated with riverine (Pebble) and coastal (Micro Blade) ecosystems, and it was those cultures that eventually inhabited the Pacific Northwest. Although they ate fish and shellfish, initially their economies did not depend on the salmon. Living in small groups of twenty-five to forty, the early inhabitants of the Northwest coast continuously moved in search of food. They were opportunists, harvesting whatever was available on the land or in the water, including various rodents, elk, deer, seals, porpoises, shellfish, staghorn sculpin, flounder, tomcod, herring, hake, sturgeon, eulachon, and salmon. Where salmon was a part of their diet, it was generally only a small part.[10]

Anthropologists offer different explanations of why the salmon were unimportant to the earliest northwesterners. According to one theory, the value of a resource as human food depends in part on its availability and distribution, and the salmon's availability is highly seasonal, usually lasting only a few weeks or months during the spawning migration. Value also depends on the technology needed to capture the fish during their short migration period and to process the flesh for later consumption. To catch large numbers of salmon, people had to create tools such as weirs, traps, and seines. In addition, the salmon could be used as food for only a relatively short time (a few weeks or months at most) until people discovered ways to preserve and store it for a year or more. Finally, the capture, processing, and distribution of large numbers of salmon required social organization and cooperation on a large scale and a management system that regulated fishing to avoid overharvest and starvation. Developing a salmon-based economy required a significant investment in technology and social organization.

Anthropologists who subscribe to this theory conclude that early humans in North America would have invested in the technology and social organization needed to exploit the salmon only as a last resort, probably in response to pressures from a growing population with few other available food sources. Under this scenario, salmon-based economies developed slowly no matter how abundant the fish were, and they developed only to alleviate problems of scarcity caused by population growth.[11] The assumption inherent in this argument is that human activities are determined largely by social, economic, and demographic factors.

On the other hand, some anthropologists believe that human activities are determined primarily by their biological and physical environments. Proponents of this view argue that the salmon played a minor role in early Indian economies because environmental conditions during the postglacial period created unstable and unproductive salmon habitat. Since the salmon were not abundant, they could not form the basis for a regional economy. According to this theory, the salmon-based economies coevolved with the postglacial emergence of productive habitats and the increase in abundance and stability of annual salmon runs.[12] Important events in the formation of those habitats were the stabilization of sea levels; forest succession that led to old-growth communities of western hemlock, western red cedar, and Douglas fir; and the shift from a hot, dry, postglacial climate to the cool, wet climate that prevails today. (The postglacial development of salmon habitat was described in Chapter 1.)

The actual process by which salmon-based economies evolved probably included elements of both theories. Whatever the true explanation, we do know the process was long and slow. The climate, landscape, forest communities, salmon populations, stream habitats, and human economies engaged in a long ecological dance of coevolutionary change. At first, the dance was probably clumsy and out of step. There's little doubt that Native Americans made mistakes and learned hard lessons through starvation. Over thousands of years, the dance improved as humans adapted their economies to the changing landscape and resources. By the time Euro-Americans arrived in the Northwest, the indigenous people living on the Northwest coast had created an economy tightly choreographed to the seasonal cycles of the salmon. This complex biological and cultural dance persisted for 2,000 to 5,000 years, depending on the area, ending only when Euro-Americans arrived in large numbers about 150 years ago.

The Rise of Salmon-Based Economies

According to archaeologists, the relationship between people and salmon in the Northwest began to evolve around 9,000 years ago. About that time, men and women sat on the banks of the Fraser River near Milliken, ate wild cherries, and flipped the pits into their fires. Anthropologists discovered the charred pits, but they found no direct evidence that the cherry eaters were also enjoying salmon. Wild cherries ripen in August and September, however—the same time the sockeye salmon migrate up the Fraser River—so it seems likely that it was fish, rather than ripe cherries, that attracted humans to Milliken 9,000 years ago. The fragile salmon bones simply did not survive the passage of time at that location.[13]

Because the salmon took refuge in the Columbia River during the Wisconsin ice age, it is not surprising that some of the earliest evidence of salmon harvest by humans has been found there. Archaeological findings suggest that indigenous people caught salmon at a site near The Dalles on the lower Columbia, and at the Benard Creek Rock Shelter on the Snake

River, roughly 7,000 to 8,000 years ago—about the same time as the earliest confirmed harvest on the Fraser River.[14]

It makes sense that the earliest evidence of human salmon use comes from the Columbia and the Fraser, the region's largest rivers. In both waterways, the earliest known fishing sites are located in canyons where the mountains physically constrict flows, concentrating the salmon and making them easy to catch even if overall abundance is low. Picture a large river basin as a double-ended funnel with a narrow neck in the middle (the canyon) and wide mouths at either end (the estuary and nearshore ocean at the lower end of the funnel, and dispersed tributaries at the upper end). When the salmon left their distant ocean feeding grounds and arrived at the mouth of a river, they entered the wide mouth of the funnel. They then swam upstream into the constricted canyons, where turbulent, high-velocity flows slowed their migration, concentrated them, and forced them through predictable pathways. Once past the narrow canyon, the salmon dispersed into the tributaries. When salmon abundance was low during the immediate postglacial period, the river canyons were the only places where the fish were concentrated enough to make harvesting them a profitable activity.

In these canyon areas, fishermen perched over the churning water on small platforms that hung precariously from rock walls. They lowered their dip nets into the eddies and pools where the salmon held up to rest before fighting their way upstream through whitewater rapids. A large hoop held the dip net open, and both were attached to a handle that may have reached

Early days of dipnet fishing on the Columbia River near Celilo Falls. (Photo courtesy Oregon Historical Society, OrHi 156.5)

twenty feet in length. The Indian fisherman lowered the net into the water and positioned it so that its open end faced the migrating salmon. When he felt a salmon hit the net, he quickly pulled it up and retrieved the fish. Dip nets were used widely throughout the Northwest but were most important in the Fraser, Columbia, and Klamath River canyons.[15]

Although indigenous peoples harvested the salmon in river canyons 8,000 years ago, the fish were not yet a staple food item throughout the entire Northwest. They were important only locally: within a watershed, salmon might be a critical food source near a good fishing site, but not in tributaries a short distance away. Archaeological evidence of the local nature of the salmon harvest comes from the Thompson River, a tributary of the Fraser. In the lower Thompson River near the town of Spences Bridge, the oldest confirmed salmon fishing site in British Columbia dates back at least 7,500 years. However, chemical analysis of the remains of a boy buried in a mudslide 8,000 years ago on the south Thompson River shows that he ate land animals rather than fish from the ocean.[16]

By identifying the bones of animals found in middens and hearths of ancient campsites, archaeologists have reconstructed the diets of prehistoric humans in other areas as well. Analysis of middens at three campsites—one at Glenrose in the lower Fraser River, another at Bear Cove in northern Vancouver Island, and a third at Tahkenitch Lake on the central Oregon coast—revealed that 9,000 to 8,000 years ago, Native Americans primarily hunted and killed land mammals. They were just beginning to exploit coastal food sources such as shellfish, marine mammals, and fishes, including small amounts of salmon. For example, at Glenrose people ate mostly elk, deer, and seals, but they also consumed a small amount of salmon. At Bear Cove, salmon made up 10 percent of the faunal remains recovered; but at the Tahkenitch site on the Oregon coast, archaeologists found no salmon, although other fish remains were abundant.[17]

North of Vancouver Island on the outer coast of British Columbia, the settlement of Namu sits near the mouth of the Bella Coola River (Figures 2.1 and 2.2). Namu means "whirlwind," which aptly describes the winter storms that can pound this stretch of coast. Because humans have lived at Namu continuously for 10,000 years, the place is important to the story of human settlement of the Northwest and to the evolution of the strong relationship between salmon and people. The remains at Namu reveal details about daily life on the Northwest coast almost from the time that humans first arrived there.

According to the archaeological record, it wasn't until 6,000 years ago that

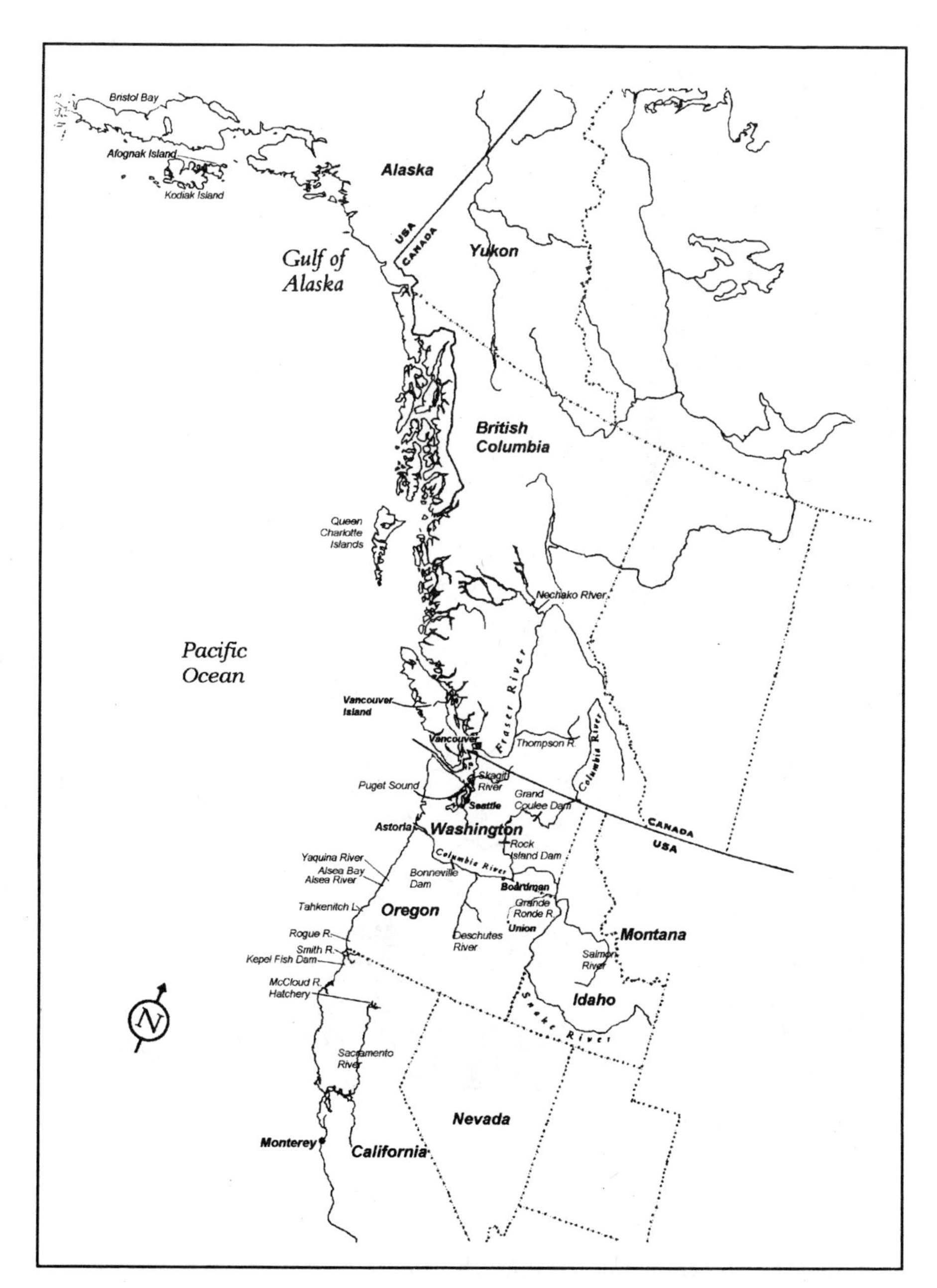

Figure 2.1. Pacific Northwest, British Columbia, and Southeast Alaska. (After Lewis 1994, courtesy Raincoast Books)

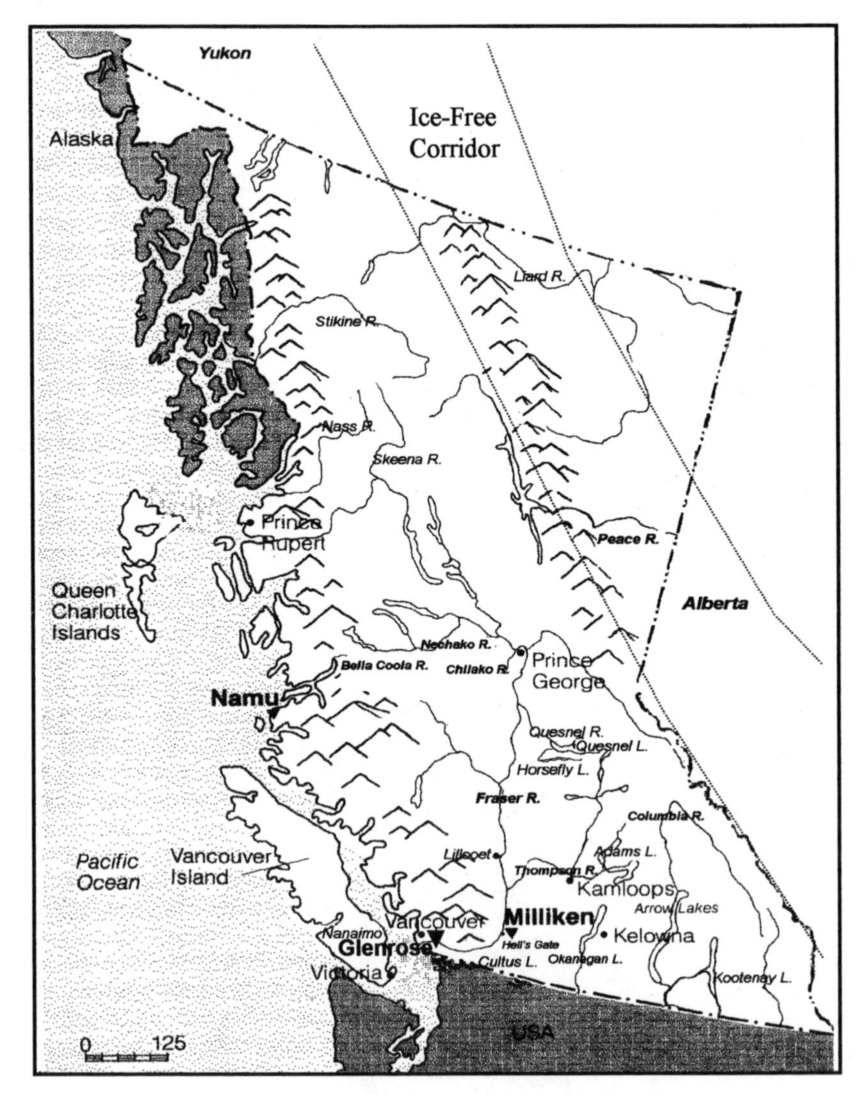

Figure 2.2. British Columbia's major rivers and archaeological sites. (After Fladmark 1986, courtesy Canadian Museum of Civilization)

salmon and other marine creatures became fundamental parts of Namu's economy and that salmon came to make up 80 percent of the settlement's animal remains. As the community's adaptation to coastal resources continued, the importance of salmon in Namu's economy increased. By 3,500 years ago, salmon comprised 94 percent of the faunal remains found at the site.[18]

While the economy of Namu depended on the salmon 6,000 years ago, farther south the situation was different. For example, at Tahkenitch Lake on Oregon's coast, the Indians were also learning to harvest food from the sea, but of a large collection containing 17,000 pieces of fish remains, less than 1 percent were salmon or trout. Excavations at several other sites in Oregon and northern California paint the same picture. Although Native Americans grew more dependent on riverine and coastal resources, salmon remained a minor part of their diet. Fully developed salmon-based economies did not appear in northern California and southern Oregon until about 1,500 years ago, probably because salmon at the southern end of their range took longer to recover from the postglacial conditions, particularly the hot, dry climate, than salmon in the northern parts of their range.[19]

About 3,000 years ago the discovery and mastery of preservation technology, including drying and smoking, increased the value of the salmon in Indian economies in the Northwest. The ability to preserve the salmon captured during a season of abundance so they could be consumed during the lean winter and spring seasons significantly boosted the region's human carrying capacity, or the number of people the ecosystem could support.[20] Preservation also created a basis for accumulating wealth and promoted the development of complex cultural customs, such as community cooperation in the harvesting and processing of salmon. Once the salmon became a reliable source of food for the entire year, aboriginal peoples took up permanent or semipermanent residence in large, multifamily houses. As the Indians became more dependent on the salmon, the economy and the rules governing the relationship between salmon and people grew more complex.

The Gift Economy

When Juan Francisco de la Bodega and George Vancouver explored the Pacific Northwest coast for Spain and England, respectively, in the 1790s, they found Native Americans in possession of a well-developed economy, specialized technology, and a rich mythology that provided rules for the wise use of their principal resource, the Pacific salmon. A fully developed Northwest coastal culture existed for at least 1,500 years (and far longer in some areas) before Europeans "discovered" the region.

The long-term sustainability of this salmon-based economy has led some to conclude that Northwest Indians lived within the constraints of the natural economy—an economy that exhibited, at least in its mature stages of development, a high degree of ecological harmony between humans and nature. But there is a danger in a direct transfer of ecological principles from our culture to another culture whose beliefs and assumptions about the world they lived in were very different. The Indians developed a sustainable economy based on the salmon through the power of their own worldview—a way of thinking that probably could not have accommodated modern ecological concepts even if they had been available. The sustainable relationship between the Pacific salmon and Native Americans derived not from ecology but from an economy based on the age-old concept of the gift and a belief system that treated all parts of the earth—plants, animals, rocks—as equal members of a community.

In ancient economies in such disparate parts of the world as southeast Alaska, Polynesia, and early Germanic Europe, the gift formed the basis for exchange and commerce. Nowadays we think of a gift as a commodity: once it is given, the receiver takes possession, and it becomes his or her property. But in ancient economies, a gift was not a possession; in fact, it could not be owned. Rather, it had to be passed on. The very act of accepting a gift meant the receiver also accepted an obligation to return it in kind to the giver. Rather than transfer goods through a market, the gift-based economies worked through a cycle of obligatory returns. The giving of gifts generated a record of obligations that defined the credit structure of society. In ancient economies, the gift and the obligation it incurred were more like a bank loan and its repayment than like our exchange of gifts at Christmastime. Anthropologist Marcel Mauss equated the gift in ancient economies to Adam Smith's "invisible hand" in modern, market-driven economies.[21]

In gift economies, people or tribes attained prestige and stature through the size of their gifts, not through the amount of wealth they accumulated (as in our market-driven economy). The tribe, village, or family maintained an oral history of the gifts given, the generosity with which they were returned, and the obligations still outstanding. These stories were equivalent to an accountant's balance sheet, and tribal elders often repeated them during potlatch celebrations. The gift stories were so important that the Coast Salish peoples of Puget Sound and the Strait of Georgia believed that those who lost track of the gifts they had given and received—essentially their history—were second-class members of the tribe.[22] Failing to return gifts with generosity was like failing to repay debts in our economy, with at least a similar if not a greater loss of status.

On the Northwest coast, one of the gift economy's most important institutions was the potlatch, a formal gathering with a well-defined set of customs and procedural rules. A tribe or village hosted invited guests—usually from other tribes or villages, although some groups held potlatches among families within a single tribe or village. Potlatches were initiated to celebrate a variety of village events, including marriages, funerals, the ascension of a new chief, the naming of a child, or the need to remove shame from some past wrongdoing. During the potlatch, the host tribe recited its history through songs, dances, masks, and displays of hereditary possessions. The history also included an accounting of past gifts received and given. Through their participation, the guests witnessed and validated the marriage, the claim of rank, or the removal of shame. The potlatch included much feasting, so food played a major role in the ritual: a tribe might spend months collecting and preparing salmon, clams, berries, and deer. Although the potlatch served many purposes, the gift was always at its core. The host tribe showered the invited guests with furs, canoes, baskets, and cedar boxes that were purchased from neighboring tribes with surplus dried salmon.[23]

Because gifts established an obligation to return even greater gifts, the system of potlatches effectively redistributed wealth within a region. From an economic perspective, the potlatch functioned much like our own tax and welfare system. But from an ecological perspective, it accomplished far more.

Salmon abundance varies naturally; fires, landslides, drought, floods, and shifting ocean currents can reduce the number of adult salmon returning to a river. Because the salmon do not swarm indiscriminately from the ocean to the closest river but return to the stream of their birth, a river and its people might be blessed with a large run of salmon and surplus wealth or confronted with starvation, depending on conditions in that specific watershed. Excess and scarcity could occur among different rivers within the same region. The potlatch tempered the effects of natural highs and lows in salmon abundance. Giving gifts through the potlatch promoted stability in an economy where the supply of the principal resource could fluctuate widely.[24]

Aside from regulating economic relations among native people, the gift economy also influenced the relationship between Indians and the natural world. According to Native American beliefs, people lived in a world of equivalent beings; the trees, rocks, deer, ravens, and humans all had equal standing as citizens of the same community. Consequently, when an Indian hunter killed a deer or bear or salmon, the death did not signify human domination over nature. Rather, the killing was understood as a gift of food or fur, given by the animal to the man. The salmon or deer allowed the human to kill it, and in accordance with the rules of the gift economy, the hunter assumed an obligation to treat the animal (the gift) with respect. Animals would retaliate if they were not treated with respect, it was thought, and many taboos regarding the disposal of animal remains reinforced that belief. In this way, the rules of the gift economy prevented wasteful practices and tended to limit the kill.[25]

Paying respect to the salmon was a particularly well articulated aspect of the coastal people's gift economies. According to the beliefs of these Native Americans, the salmon were a race of supernatural beings who lived under the sea in five great houses. Each species—pink, chum, coho, sockeye, and chinook—lived in a separate house. While under the sea, the salmon people took on human form. Each year, however, at the direction of the salmon king, the salmon people donned silver skins and presented themselves to the Indians as a gift from the king. The tribe treated the first salmon caught each year with the respect accorded a royal visitor, greeting it with an elaborate ceremony intended to demonstrate the tribe's appreciation for the gift.[26]

From central British Columbia to northern California and inland to the Lemhi River of Idaho, tribes performed this First Salmon Ceremony. Although each tribe had its own unique practices, the ceremony generally included special handling of the first fish and a ritual butchering, followed by roasting and eating the salmon's flesh in a shared feast. The first salmon might be placed on a new cedar-bark tray decorated with feathers, and butchered with a specially designated mussel-shell knife. When moving the salmon, the bearers always kept the fish's head pointed upstream to show their respect. During the ceremony people made speeches entreating the salmon to return to its king with a favorable report on the treatment it received.[27] Eventually the cooked fish was shared among all the tribal members, except those considered unclean. After the salmon was eaten, the tribe completed the gift cycle by carefully returning the fish's bones to the river. The Indians believed that the salmon were immortal. If they returned all the bones to the river, the salmon would reconstitute itself and return to its great house under the sea until the next year.[28]

It's easy to understand how the Indians derived the concept of immortality from the salmon's life cycle. Each year they watched the salmon migrate upriver; then they saw the spawned-out carcasses drifting downstream past their village, back to the sea. Yet the next year they found the salmon alive and back in the river.

The First Salmon Ceremony renewed and reinforced the belief that the salmon would remain abundant if they were treated with the respect due a gift. By ritualizing the exchange of benefits and obligations between equal parties, the ceremony strengthened the relationship between people and salmon. But the ceremony was more than just a ritual. The exchange of gifts between salmon and man—the giving of food in exchange for respectful treatment—actually assured a continuous supply of food for Native Americans. By causing men and women to treat the salmon with respect, the ceremony reinforced behaviors that minimized waste and reduced the chance of overharvest. Both the potlatch and the First Salmon Ceremony illustrate how Northwest Indians melded the concept of the gift and the biology of the Pacific salmon into a sustainable economy prior to contact with Europeans.

Because the Indians were keen observers of nature, I suspect that the potlatch, with its massive gift giving, and the gift economy in general might have actually been suggested by the salmon's generous gift to the natural economy of the rivers. The potlatch bears a remarkable correspondence to the life cycle of the Pacific salmon. Juvenile salmon migrate to the sea, where they graze the ocean's pastures and accumulate their biological wealth in rapidly growing bodies. The mature salmon return to their home stream and, in an orgy of spawning and death, give their bodies—their accumulated wealth—back to the stream of their birth. As a result, a healthy salmon population transfers nutrients from the ocean to the watershed. The salmon's gift benefits the whole ecosystem, including at least twenty-two species of mammals and birds that feed on salmon flesh, such as bears, eagles, and even little winter wrens.[29] The decaying carcasses release nutrients back into the river and the surrounding forest. When a bear pulls a salmon from the river and leaves the partially eaten body under a cedar tree, the fish fertilizes the cedar, which in turn shades the stream and keeps it cool for the juvenile salmon. The gift economy of the native northwesterners was fundamentally in harmony with nature's own economy.

The First Salmon Management

The Pacific salmon, their habitat, and the salmon-based gift economy coevolved over several millennia. As part of this coevolutionary process, native peoples developed the knowledge, skills, and technology to harvest fish efficiently. At the same time, they also developed a set of customs that functioned to limit their catch. It is useful to think of the Indians' technology and culture as a system of salmon management—a system that furnished a sustainable salmon-based economy for at least 1,500 years at the southern end of the salmon's range and for 4,000 years or more near the center of their range.

Northwest Indians developed many different methods for catching salmon. Indians caught Pacific salmon with hooks, gaffs, spears, nets, weirs, and traps of many designs. But the varied implements shared one common feature: all were constructed from native materials. While this technology enabled the Indians to catch all the fish they needed, their customs, ceremonies, myths, and taboos produced a management system that defined and limited the scale of their fishing.

For example, Indians used whale sinew and braided plant fibers, including bull kelp, inner cedar bark, and nettle stems, to make strong fishing lines and to weave different types of nets.[30] In the estuary of the lower Columbia River, the Chinook people fished with beach seines made of these materials. According to James Swan, who lived near the mouth of the Columbia in 1852, the Chinook fished their nets by starting "at the top of high-water, just as the tide begins to ebb. . . . Two persons get into the canoe, on the stern of which is coiled a net on a frame. . . . She is then paddled up the stream, close in to the beach, where the current is not so strong. A tow-line, with a wooden float attached to it, is then thrown to a third person, who remains on the beach, and immediately the two in the canoe paddle her into the rapid stream as quickly as they can, throwing out the net all the time. When this is all out, they paddle ashore, having the end of the other tow-line made fast to the canoe. . . . All haul in together."[31] The Chinook Indians usually caught sufficient salmon with their beach seines to survive. Since the seines could fish only a small part of the river's width, they did not diminish the Columbia's enormous productivity.

Though seines were important, weirs were by far the most important and efficient means for capturing salmon. All cultural groups throughout the Northwest used them. On smaller streams, individuals or families owned weir sites, but on larger rivers, communities built and operated the weirs collectively. Because weirs blocked entire streams, they had the potential to capture an entire salmon run. But tribal customs coevolved with the use of weirs to create an effective salmon management system that precluded such problems. A weir could mean the difference between a winter of plenty or starvation for an entire village, so tribal members often constructed them with much ceremony and religious ritual. The whole process of building and using a weir might be supervised by a fish chief or shaman.

The Kepel fish dam, built by the Yurok tribe on the Klamath River in California, reveals how ritual, religion, and fishing technology were all deeply intertwined. Yurok legend attributes the original selection of the Kepel weir site—a wide, shallow river reach just above a sharp bend—to the Woge, ancient ancestors of the Yurok who tried constructing weirs at several places but through trial and error found the Kepel site to be the best. The

Yurok had a taboo against harvesting the first fish that swam upriver in summer, but once they confirmed that the salmon had returned in abundance, the shaman—really a combination of master mechanic, chief priest, and salmon manager—signaled the start of the rigorous ritual for constructing the weir. First the shaman performed a series of purifying rites: he marked the actual site of the weir, fasted, avoided looking at women, and recited prayers to the trees that would supply logs for the weir. Then workers from several villages cut the trees needed for the various parts of the weir. When all the materials were collected, the shaman gave the order to start construction.

According to Yurok custom, the weir had to be built in exactly ten days and used for exactly ten days; then it was torn down. The reasons for this self-imposed limitation on harvest are not clear, but the effect is obvious. Limiting the fishing period to ten days allowed the salmon to move upstream, where they could be harvested by other villages that had also constructed weirs. Most important, it guaranteed that some fish would reach the spawning grounds to create a new generation.[32]

While the native northwesterners possessed the skills, technology, and knowledge to fully exploit salmon in nearly all the freshwater, estuarine, and nearshore oceanic habitats, the limited glimpses we have of their prehistory show that they learned to live within the carrying capacity of their environment. The success of their salmon-based economy was in no small measure due to their belief that the fish were kindred spirits capable of engaging in personal relationships. The gift economy reinforced this relationship between people and salmon. Gifts received from kindred spirits were treated with respect, and it was this belief system that led to what anthropologist Richard Nelson calls a limiting ideology—an intentional restriction on the use of technology.[33] The destruction of the Kepel weir after ten days of fishing illustrates the practical effect of this ideology. Although the origins of many of the mechanisms used by Indians to limit resource use remain shrouded in the fog of prehistory, it is clear that the early fishing economies faced and resolved the same problems of overharvest confronting modern fisheries managers.[34]

As the gift economy and salmon management systems evolved, the Indians undoubtedly made mistakes that led to waste, overexploitation, collapse of food sources, and starvation. For example, archaeologists uncovered direct evidence of such an "error" in the midden piles of prehistoric Aleuts on Amchitka Island, Alaska. Aboriginal hunters apparently overharvested the sea otter in the Aleutian Islands, which permitted an increase in sea urchins, the otter's major prey. Abundant sea urchins then overgrazed kelp beds, causing nearshore fish populations to collapse. An analysis of animal remains in the refuse piles reveals that the Aleuts made a dramatic shift from a diet of sea otters and fish to sea urchins and other benthic herbivores.[35] A similar human-induced shift in marine community structure also occurred in California in the late nineteenth century.[36]

Knowledge of such mistakes was passed from generation to generation in the form of beliefs, rituals, and taboos that functioned to prevent overexploitation. For example, underlying the First Salmon Ceremony was the belief that the salmon king could prevent his people from entering a river, and would do so if the salmon were not treated with respect. This belief reflected past experiences, including natural failures in the salmon run and the ill effects of human overexploitation.[37]

The lessons learned through trial and error resulted in several different mechanisms to keep harvest within the limits of natural productivity. Some mechanisms, such as limiting the time of fishing, sharing of fishing sites, and diversifying the economy, have counterparts in our modern industrial economy, but the fundamental belief that animals and plants had conscious spirits was uniquely Native American.[38] To the Indians, the natural world was a community of kindred spirits with whom humans could cultivate advantageous relationships to ensure a continuous supply of food. Disrespect could turn the animal spirits against people. For example, the Puyallup Indians of Puget Sound believed that the dog salmon would take the soul of a wasteful person to its house under the sea and kill him or her.[39] The fear of retribution placed a self-imposed limit on the harvest and greatly reduced the possibility of overexploitation.[40]

Author Mary Giraudo Beck has recounted a Tlingit story illustrating the tremendous power animals were believed to have over humans who treated them with disrespect. The story began in a village where people had enjoyed several years of plentiful salmon harvest. The easy times made the people lax and caused them to neglect traditional taboos. One day, four boys took a canoe to a nearby stream to fish. The salmon were abundant, so the boys built a fire and began to heat rocks to cook their catch for dinner. Soon they got caught up in the good fishing, pulling in more and more fish, and in their excitement threw live fish onto the hot rocks. When the fish squirmed, the boys laughed. One boy even threw a frog onto the rocks just to torture it. For their acts of disrespect to the salmon and the frog, the four boys died, but the animal spirits retaliated further. Other boys died, and the whole village was threatened. When the death of the tribe's shaman finally appeased the spirits, the village was spared, but the people vowed never again to treat animals with disrespect.[41]

Contrast this story of remorse for waste and disrespect with the early days of commercial fishing at the turn of the twentieth century. During the peak of the run, fishing fleets so overwhelmed the capacity of the canneries that piles of salmon would accumulate until they began to rot. Cannery workers would then shovel the carcasses back into the river. This kind of overharvest and waste could never happen in a culture where people engaged in personal relationships with the salmon, in a culture where people regarded the salmon as an important gift that had to be treated with respect.

By the early decades of the nineteenth century, Northwest Indians had built a robust economy and a rich culture fully adapted to the local climate, food resources, landscape, and native materials. But in the early decades of the nineteenth century, change was on its way. Like the family in the story that opened this chapter, the Indians did not hear the rumble of a new culture as it crossed the continent. They failed to feel the winds of change, and when they looked up they were confronted with a new culture so powerful that it had no intention of adapting to local conditions. When the new culture with its different economy washed over the Northwest, it altered the region's ecosystems no less profoundly than the thunderous floods that poured out of ancient Lake Missoula.

CHAPTER

3 New Values for the Land and Water

January often squeezes out a few days unique to the Pacific Northwest—bright, clear skies following weeks of gray, sunless days and rain. It's the contrast that makes them so special. On one of those sunny days in January, I headed west out of Corvallis, Oregon, to do what I often do when a tough decision is at hand; I went river watching. I have long since forgotten what was troubling me, why I needed the peace and solitude of a river, but I do remember that later that day I found myself on the Alsea River, watching, thinking, letting the flow and rhythm of the river clear my head.

The Alsea's gray-green water slid quietly over a sheet of bedrock. The morning silence was complete except for the gentle rippling of the river being cut by a branch from a leaning alder. Several minutes passed before I realized I shared this spot on the river with another. I first saw her out of the corner of my eye: a light spot, just below the river's surface, moving upstream and then drifting back. As I focused in on the spot, the shape of a chinook salmon slowly emerged from the murky water. She would drift downstream and then, with an effortless flex of her caudal fin, dart back upstream over a shelf in the bedrock sheet. But this was late January, and the chinook usually finish spawning in the Alsea by the end of December. This old girl was a living example of the salmon's biological diversity.

I watched the big female and thought about what her species has endured since the arrival of Euro-Americans, with their gauntlet of nets and hooks, logging, splash dams, high dams, irrigation diversions, and polluted water. I thought about the arrogance of humans, who in 1915 attempted to improve on nature by building a dam across the Alsea River down low in the watershed, preventing all but a few salmon from reaching the upper river to spawn. Most of the fish were captured at the dam, and their eggs were sent to a hatchery for incubation. Few of the fry survived, but somehow the ancestors of this chinook had kept the thread of life alive generation after generation.

As I continued to watch, what she did surprised me and continues to puzzle me every time I think about it. She swam upstream, rolled over on her side, and beat her flanks against a solid sheet of rock in the typical redd-building motion. I wondered what instinct brought her to this spot, told her this was as far as she could go, and pushed her to beat her life out against this unyielding rock. Was this the right spot, just 100 years too late? What did this stretch of river look like 100 years ago, before the timber was cut and before the salmon were "protected" by hatcheries?

Watching this single female wasn't at all like watching the thrashing of several salmon digging normal redds. Each time she rolled over and beat her tired flanks against that rock, I felt a sense of urgency. She was trying hard to fulfill the genetic plan she had been carrying in her blood, bone, and tissue for the past four or five years. Her whole purpose now was to extend the thread of life to the next generation, the thread that would connect her offspring to 10,000 years of ancestors. That was her mission, yet each time she assaulted this scoured piece of river bottom the thread grew weaker. Her battle was a contradiction. It showed the strength of the instinctive drive for survival and, at the same time, the vulnerability of natural reproduction to the changes humans have made in rivers.

This salmon's struggle was an individual tragedy. In the big picture, her struggle didn't count for much. She was not part of the mainstream, the peak of the run. But at the same time she embodied the biological diversity that has kept the salmon alive for millennia.

Ever since Euro-Americans arrived in the Northwest, the belief that ecosystems can be simplified and controlled has guided their relationship with the salmon. In the management of watersheds and salmon, not only has diversity been ignored, but humans have waged a war against it. Historically, hatchery programs, harvest regulations, and habitat destruction have all diminished the salmon's biological diversity.

When humans try to control an ecosystem, they must decide what will live and what will die. They must determine which salmon are "viable" and which are not. Throughout most of the last century, viability has been measured solely in economic terms.[1] From an economic perspective, this female would be condemned to nonviable status. She didn't belong to the peak of the run, that part that supports the fishery, that part that is converted to cash. She was an important part of the biodiversity of the Alsea's chinook, but although she has value in the natural economy of the salmon, she has no value in the industrial economy. And the industrial economy has been the final arbiter of viability.

As I watched the chinook perform the fruitless spawning dance that would end her life, I was reminded of a story I had read about Indian ghost dancers. Through the power of their dance, they tried to bring back customs, culture, and a way of life that were slipping away.

Conflicting Economies

For thousands of years, Indians used the natural resources of the forests, plains, and rivers. Although their use of resources did not involve the massive restructuring of the landscape that followed Euro-American settlement, the Indians did make extensive use of fire to shape the vegetation and the ecology of forests and prairies. They employed fire to enhance the growth of food such as camas or to attract deer and elk to the rejuvenated vegetation following a burn.[2]

When they arrived in the Pacific Northwest, the first Euro-Americans found the land and water teeming with what appeared to be inexhaustible resources. The early settlers believed that this great natural abundance was part of an untouched wilderness, but in reality they were in a landscape that had been shaped and manipulated by human hands. That manipulation, however, was part of a sustainable relationship between the natural economy of the region's ecosystems and the gift economy of the Indians. In most areas of the Northwest, it was primarily a relationship between the Indians and one of their principal resources: the salmon.

The relationship between Native Americans and the salmon coevolved over thousands of years. Shortly after the arrival of Euro-Americans, however, a different economy, a different worldview, and a different set of rules governing the use of natural resources changed the long-standing relationship between the salmon and humans. First, pale-skinned men on floating islands wanted furs and fish in exchange for wondrous new tools. Then Indians were dying of strange diseases over which their shamans had no power. Salmon-drying racks and the potlatch were soon replaced by noisy canning machines and a market-driven cash economy. The salmon were no longer a gift from the powerful salmon king but a commodity—free wealth available to anyone willing to take it. Within less than fifty years, a new and totally foreign economy and culture, one that viewed nature as a warehouse, displaced the Indians' economy and their view of nature as a community. At first Indians tried to adapt to the situation and worked to supply the canning machines with salmon. But because subsistence fishing offered little advantage in the new cash economy, the Indians' age-old gift economy collapsed. Eventually, the native fishermen were largely replaced by the white settlers.

When Euro-Americans arrived in the Northwest, their industrial economy set the coevolutionary clock back to zero, throwing aside all that had been learned since humans harvested their first salmon at Milliken or The Dalles. Since the arrival of white men, the relationship between salmon and humans has been reevolving in the context of a new economy and a new set of rules. But the new industrial economy has not dovetailed well with the salmon's natural economy; instead it has insisted on new behaviors, new beliefs, and new technologies untuned to the natural rhythms of the Northwest's landscape. As the new economy took hold, it shifted from native fishermen and subsistence harvest to nonnative fishermen and commercial harvest, from a goal of gathering a gift of food to a goal of maximizing profits. As a consequence, the salmon had to face entirely new pressures.

Throughout the world, native subsistence fishermen have been and are being replaced by a new class of fishermen working within the industrial economy. When the maximization of profits becomes the dominant goal, diverse and labor-intensive fisheries controlled by local communities of fishers are usually replaced by expensive technology operated by a few people and governed by distant, centralized bureaucracies.[3] This change weakens the negative feedback loops between fish and fishers, making it virtually impossible for a sustainable relationship to coevolve between the fish and the industrial economy.

In the Northwest, for example, when a tribe overharvested the salmon in the rivers of their home territory, they enjoyed the benefits of extra food and possibly the prestige of a large potlatch. But at some point, if the overharvest persisted, negative feedback kicked in: that same tribe suffered a diminished salmon run and possibly starvation. With the development of the industrial fishery in the late nineteenth century, distant markets determined the size of the harvest and created a competitive atmosphere that did not reward fishermen for limiting their catch out of concern for future harvests, a situation that has been called the tragedy of the commons.[4] One fisherman's restraint out of concern for the future only left more fish for those who sought to maximize harvest, and thus profit, in the short term. The end result was that everyone tried to maximize harvest in the short term. Once the legends and taboos that regulated the Indians' use of salmon were destroyed and replaced by the myth of inexhaustible abundance, fishermen had wide-open access to the fish, with little or no constraint or regulation.

Furthermore, in the market-driven industrial economy, local fishermen still bore the cost of overharvest, but the benefits were reaped largely by entrepreneurs in London, New York, or other financial centers. Their distance from the Pacific Northwest and the mobility of their capital insulated them from the consequences of overharvest. They regarded the region solely as a colony and a source of raw material for eastern markets and industry.[5] When the canning industry exhausted the fish in one river, the distant entrepreneurs simply moved their canneries to other rivers and other salmon stocks. Some canneries were actually built on barges to facilitate such movement. The social costs—depleted salmon and unemployed fishermen—were left for local communities to absorb.

The Northwest's new industrial economy had an insatiable appetite for other resources too. While native people made good use of the region's watersheds, their impact on salmon habitat was probably negligible. In contrast, the newly arrived pioneers and entrepreneurs fully exploited all of the watersheds' economic resources through trapping, mining, logging, grazing, and irrigation. As a result, the salmon's habitat came into direct contact with the industrial economy throughout the entire length of the chain of habitats that make up its ecosystem—logging and mining in the headwaters, agriculture in the rivers' lower elevations, cities and industry in the broad alluvial plains and estuaries, and finally pollution and large-scale fishing in the ocean. A major part of the salmon's problem, then and now, is that there is not just one threat to their existence but a continuous series of threats at nearly every point in their range, throughout their entire life cycle. At every point of contact with the industrial economy, from the headwaters to the sea, the salmon have long engaged in a losing struggle for habitat.

Although the first pioneers in the Northwest thought the salmon were inexhaustible, the continuous decline of the fish over the past 100 years has made their vulnerability to industrial development too obvious to ignore. Acknowledging the problem, however, has not led to a solution. In fact, the industrial economy and its worldview so dominate our culture that our attempts to restore the salmon are inevitably shaped by its principles, and thus our efforts often conflict with the natural economy. For example, the belief that hatcheries can circumvent most of the salmon's freshwater life history and eliminate the need for healthy rivers has allowed the riverine ecosystems to become degraded. Where fish hatcheries are the primary source of reproduction, the rivers are regarded as simple conduits that carry the juvenile salmon to sea following their release from concrete ponds. In the highly developed Columbia River, even the simple channel to the sea is now too dangerous because of mainstem dams, so the hatchery juveniles are loaded into barges and shipped to the estuary. Principles of the industrial economy have also influenced stream restoration projects, which often attempt to "engineer" site-specific habitats for salmon rather than to restore the natural riverine processes that create and maintain salmon habitat throughout the entire length of their life-history pathways. These programs, at best, have done little to stop the salmon's demise and, at worst, have contributed to it.

The salmon's problem is—at its root—a clash of two economies: the industrial and the natural. The gift economy of the Indians evolved a sustainable balance with the natural economy. Eventually the industrial economy will also have to evolve a balanced relationship with the natural economy of the Pacific Northwest. As the dust bowl that ravaged the Great Plains in the 1930s clearly illustrated, there is a heavy price to pay if the needs of the ecosystems are ignored too long. To remain productive, the industrial economy of the Northwest will have to back away from a conflict with the natural economy and seek ways to achieve a balance with it.[6]

A major obstacle to achieving a balance between these two economies is the fact that they approach the world from different perspectives, work toward different objectives, and have different operating principles. The industrial economy encourages individualism, extraction, economies of scale, and simplification of natural systems, whereas the health of the natural economy depends on interdependence, renewal, dispersal of production, and diversity. But the centers of organization in the industrial economy—cities, railroads, highways, power transmission lines, farms, factories, ports, and shopping centers—are clearly visible and familiar. One does not need a degree in economics to understand how railroads, farms, and factories contribute to the productivity of the industrial economy. The centers of organization in the natural economy—floodplains, groundwater, wetlands, riparian vegetation, alluvial stream reaches, and natural events such as landslides that rearrange the structural elements of a watershed—are not readily apparent except to trained experts. And even experts do not fully understand or agree how the features and organization of watersheds contribute to a healthy, natural economy.

A simple comparison between the natural and industrial economies is the first step to understanding the roots of their conflict, and it is also the first step in searching for a way to balance the two (see Table 3.1). The industrial

Table 3.1. Characteristics of the natural and industrial economies

Industrial Economy	*Natural Economy*
The landscape is organized into political hierarchies (countries, states, counties, townships, cities, individual homeowners).	The landscape is organized into watersheds (catchments, mainstem rivers, subbasins, tributaries, creeks, lakes, and springs).
Political divisions compete and often conflict over the ownership, use, and distribution of resources. Watersheds are often fragmented by political boundaries.	Salmon and their habitat are part of a coevolving relationship. Boundaries are imposed by biophysical constraints. Watersheds are integrated wholes.
The industrial economy's infrastructure is visible and recognizable, and its function is generally understood—roads, dams, cities, factories, and so on.	The natural economy's infrastructure is only partially visible, and its function is poorly understood, even by experts.
The industrial economy is driven by fossil fuel and the need to accumulate capital.	The natural economy is driven by solar energy and the need to reproduce.
The industrial economy favors large centralized production facilities (factories/hatcheries), which lead to biological and technological monocultures.	The natural economy favors dispersed production among several small units (salmon stocks), which encourages diversity.
In biological monocultures, fertility is maintained through the use of fertilizers, pesticides, and hybrids produced outside the ecosystem.	Productivity is maintained by processes internal to the ecosystem, including the mass transfer of nutrients captured in the sea and transported to headwater streams via the salmon's bodies.
The industrial economy is linear and extractive, emphasizing production.	The natural economy is circular and renewable, encouraging reproduction.
The industrial economy creates waste—toxic waste as well as a failure to fully recycle resources.	In the natural economy, there is no waste; everything is recycled.
Watersheds are partitioned into discrete economic spheres: mining, logging, grazing, irrigation, transportation, industrial water supply and waste disposal, and so on. Each industry operates as though it is independent from the others.	To the salmon, watersheds are complex mazes of connected habitats.
Rivers are restructured, simplified, and controlled to meet human needs.	Rivers are complex and connected mazes of habitats, to which the salmon adapt by evolving complex life histories.

economy organizes the landscape into political hierarchies including states, counties, townships, cities, and property owners. The United States is divided into segments bounded by straight lines and perfect squares, regardless of the irregularities of landscape features. To the salmon, however, watersheds are the logical way to divide the earth. In the industrial economy, political subdivisions often work at cross-purposes. For example, states may claim conflicting rights over the use and distribution of resources—especially resources such as salmon and water, which move across political boundaries. In the natural economy, salmon integrate all the ecological processes of the watershed through their life histories and species-specific distributions.

The industrial economy is driven by fossil fuel, whereas the natural economy is driven by solar energy. The industrial economy is motivated by the desire to accumulate capital, but in the natural economy, survival and reproduction are the measures of success.

In the industrial economy, a salmon has value only after it is removed from the ocean, placed in a can, and converted to cash—a one-way trip that leads to an emphasis on production. In the natural economy, a salmon has value if its genes contribute to the fitness of the next generation. This leads to an emphasis on reproduction.

The industrial economy produces waste: toxic waste, at its worst, but also—less obviously—the waste involved when resources are not used. In the natural economy everything is recycled; there is no waste. Those who talk about the "waste" of old-growth trees that are not cut, the "waste" of water allowed to escape to the sea, or the "waste" of salmon when "too many" escape the fishermen's nets and spawn, are expressing values derived solely from the industrial economy.

In the industrial economy, production favors large, centralized facilities that lead to technological and biological monocultures: for example, assembly lines manufacturing thousands of identical cars, farms growing thousands of acres of the same hybrid variety of corn or wheat, and large hatcheries producing genetically uniform salmon. One big problem with industrial and biological monocultures is that they cannot adapt to environmental change. The most famous technological monoculture, the gas-guzzling automobile, could not cope with the fuel shortages of the 1970s. Its inability to respond to that changing environment opened the U.S. auto market to foreign competitors. In the case of salmon, monoculture fish released from hatcheries survive at about half the rate of their wild cousins after they enter the natural environment. In contrast, the natural economy favors dispersed reproductive units: in this case, thousands of genetically distinct salmon stocks.

In the industrial economy, the watershed is seen not as a connected whole but as a collection of discrete spheres of economic interest; for example, a timber company claims the trees, a farmer claims the water, and a factory claims the hydropower. Although the spheres usually overlap in space or time, each interest operates as though it is independent from the others and from the ecological processes that maintain the ecosystem. From the salmon's perspective, watersheds are complex mazes of connected habitats. To fully use those habitats, salmon species, races, and populations have evolved a rich biodiversity.

The differences between the industrial economy and the region's natural economy have had profound consequences for the salmon. With its singular focus on profits and efficiency, the industrial economy readily gained dominance in the Northwest. Powerful market forces drove the exploitation of watersheds and heedlessly broke the ecological bonds between the fish and their habitat.[7] Degradation of the region's watersheds and ecosystem resulted not from a particular malevolence toward the salmon but simply from the everyday assumptions and workings of the industrial economy and the worldview it engenders.

To illustrate the dominance of this perspective, consider how salmon managers measure the success of their work. Since the rise of the industrial economy, management success has been calculated strictly in economic or production terms: pounds of fish harvested, retail and wholesale value of the catch, number of licenses sold, number of fishing days, and number of fish released from hatcheries.[8] In his study of the California fisheries, Arthur McEvoy emphasized the domination of industrial economics over ecology in fisheries management: "Economic distress was the goad to public action and economic gain the gauge of its effect. . . . The real cause of depletion—social costs transmitted to the fisheries ecologically and the social forces that sustained the Hobbesian struggle of all against all in which the fishers were trapped—went entirely unaccounted."[9] Since we tend to manage only what we measure, salmon managers focused their attention on narrow economic factors and ignored the deteriorating ecosystems upon which salmon depend.

Finally, from the vantage point of the industrial economy, Euro-Americans have viewed watersheds solely in terms of economics. This perspective went hand in hand with the belief of nineteenth-century pioneers that it was their destiny and duty to control and use nature for their exclusive purposes.

In their view, progress was inextricably tied to bending nature to the will of men. And if it was the will of men to make as much money as possible, then all other considerations—such as conservation of forests for future generations and the beauty of undisturbed nature—were secondary. That view, coupled with the industrial economy, reduced all natural resources, including the salmon and even the water that flowed in the Northwest's great rivers, to the status of commodities or instruments whose only value was in their economic benefit. The commodification of resources and dominance of economic values made it hard to recognize and accept the different values of the natural economy and to see watersheds from the salmon's perspective.

The voracious appetite of the industrial economy, coupled with the belief that nature should be controlled by man, eventually led to the depletion of the salmon throughout much of the Pacific Northwest. Embedded in the story of that depletion lie lessons that must be learned if we are to find a balance between the natural and industrial economies and restore the salmon. The next three chapters examine how the industrial economy and its worldview destroyed the salmon through habitat degradation (Chapter 4), overharvest and waste (Chapter 5), and hatcheries (Chapter 6).

CHAPTER

4 The Industrial Economy Enters the Northwest

On Monday, August 12, 1805, Captain Meriwether Lewis and three of his men reached the headwaters of the Missouri River. One member of the group, Hugh McNeal, stood with one foot on each bank of the Missouri and thanked God he had survived to straddle the mighty and seemingly endless river. Lewis and his small party then crossed the Continental Divide and proceeded down along Horseshoe Bend Creek, a tributary of the Lemhi River. Later Lewis refreshed himself beside the clear, cold stream and noted in his diary, "Here I first tasted the water of the Great Columbia River."[1]

The next day the men followed a trail down the valley of the Lemhi, all the while making intermittent visual contact with Shoshone Indians. Late in the day, about sixty warriors rode directly up to the explorers. Lewis advanced unarmed and was greeted by the Shoshone with an enthusiastic display of cordiality. "Both parties now advanced," he wrote, "and [we] were caressed and besmirched with the grease and paint until I was heartily tired of the national hug."[2] After smoking a pipe and informing the Indians they hadn't eaten all day, Lewis and his men were given dried cakes of chokecherries and serviceberries, some antelope meat, and a piece of salmon. Lewis wrote, "This was the first salmon I had seen and perfectly convinced me that we were on the waters of the Pacific Ocean."[3] When Meriwether Lewis entered the Pacific Northwest, he was welcomed by a meal of Pacific salmon.

Lewis and Clark were not the first white men to enter the region. Spanish and English sailors had visited the Northwest coast several years earlier looking for the Northwest Passage, a mythical waterway that some believed stretched across the North American continent and offered an easy trade route to the East Indies and China. Captain James Cook looked for the Northwest Passage in 1779. Although he failed in his original mission, he did find something of great commercial value. While in Nakoota Sound on the west coast of Vancouver Island, Cook purchased 1,500 beaver skins so the crew could repair or replace their badly worn clothing. Later, when he sailed to China, the skins—which had cost him but a few pennies—sold for $100 each.[4] News of Cook's astonishing profit piqued the interest of American, Spanish, and English traders and started an important sea-based fur trade.

Exposure to the market economy changed the Indians' attitude toward beavers and otters. Furs were no longer a gift from the beaver to the Indian; they were a commodity sold to Europeans to satisfy distant markets. The fur trade opened the first hairline crack in the gift economy of the Northwest coast. But as long as the Europeans stayed on their ships, which they usually did, the Indians controlled the trade and minimized the intrusion on their traditional culture and economy. The Indians quickly shifted from a gift to a market economy, at least in their dealings with the Europeans, and quickly learned the art of bargaining, often pitting English against Spanish or American traders to obtain the best deal for their furs.[5]

Overland explorers Alexander McKenzie, Lewis and Clark, and David Thompson widened the crack in the region's traditional economy and started what would become a flood of change in the Northwest. McKenzie crossed the Rockies in 1789, traveling partway down the Fraser River and then cutting overland to Bella Coola, not far from Namu. Lewis and Clark came next, with their epic journey in 1805 and 1806. Then David Thompson explored the upper Columbia River and followed it all the way to the sea in 1811. By the time Thompson reached the mouth of the Columbia River, the Americans had already established a fur-trading post at Astoria. The Northwest was being squeezed by Euro-American settlement approaching from both the east and the west.

President Thomas Jefferson sent Lewis and Clark west to explore the new territory of the Louisiana Purchase and to find the "most direct and practical water communication across this continent for the purpose of commerce."[6] Not in their wildest dreams could Jefferson or his explorers have imagined the degree to which commerce would expand after the 1805–1806 expedition. Nor could they imagine how profoundly the landscape would be transformed. The waves of people that followed in the 1840s and 1850s brought massive and irreversible change to the Northwest that altered forever the relationship between people and salmon.

Fur traders and trappers were the first to arrive. Then successive waves of miners, loggers, ranchers, fishermen, and farmers entered the Northwest, each seeking free wealth from a different resource. Each wave of new people targeted a different resource, exploited it fully, and then receded, leaving behind permanent changes in the land and in the salmon's habitat. As each wave crested, the economic interests driving it grew in political power. But as each wave receded, its political power remained strong, allowing the loggers, ranchers, irrigators, and miners to hold onto practices that destroyed salmon habitat and to fight off legislation that would have reduced their impact on the salmon's home. The political legacy of this history is a body of laws, policies, and myths generated in the nineteenth century, when society accepted and government encouraged laissez-faire access to public resources. Legal scholar Charles Wilkinson has called these successive interests, with their lingering power over the land and water, the "Lords of Yesterday."[7]

For over a century, the Lords of Yesterday have used their power to delay, obfuscate, and prevent attempts to develop a balance between the Northwest's natural economy and the Euro-Americans' industrial economy. They have aggressively guarded their dominance over the use of the Northwest's watersheds. The consequence of their success is clear and unmistakable. Salmon are now extinct in 40 percent of their historic range in the Pacific Northwest. Listings and potential listings under the Endangered Species Act include salmon populations from Puget Sound to northern California and east to the Snake River in Idaho. The Lords of Yesterday have been successful in part because our understanding of habitat destruction and salmon depletion has lacked a historical context. The salmon's plight is generally treated as a recent phenomenon, as a by-product of rapid population growth following the development of the region's hydroelectric potential. The Northwest's growing population and its capacity to destroy habitat through the use of modern technology certainly do pose a serious threat to what is left of the salmon's ecosystem. But before 1930, fewer people, using less technology—but with fewer constraining regulations—effected massive changes in the salmon's habitat. Those changes started with the first men who came to the Northwest looking for furs.

Fur Trade

Land-based fur traders and trappers followed close on the heels of the first overland explorations of the Pacific Northwest. By 1824, when the Hudson's Bay Company moved its headquarters from Fort George (Astoria) to Fort

Vancouver, near the confluence of the Columbia and Willamette Rivers, the fur industry had firmly established itself in the Pacific Northwest. The company's traders built outposts throughout the Columbia Basin, and the trappers soon decimated the beavers.[8]

The depletion of the beaver, particularly in the Columbia Basin, was the result of a scorched-earth policy carried out by the Hudson's Bay Company. The company intentionally exterminated the beaver in many areas of the basin because its leaders believed that the United States would eventually control the Pacific Northwest, especially south of the Columbia. Hudson's Bay saw no point in leaving any valuable resources for its American competitors. Company executive George Simpson summarized the policy this way: "The first step that the American Government will take towards colonization is through their Indian traders and if the country becomes exhausted in fur bearing animals they can have no inducement to proceed thither. We therefore entreat that no exertions be spared to explore and trap every part of the country."[9] And they did. Within fifteen years, the beaver trade had run through the sequence of discovery, rich harvest, depletion, and local extinction.

The records of the Colville District of the Hudson's Bay Company illustrate the effect of the policy. Between 1826 and 1834, district trappers killed an average of about 3,000 beavers per year, but by 1850, after fifteen years of steady decline, the trappers could find only 438 beavers.[10] By 1843, the center of the fur industry had moved north to Vancouver Island.[11]

Even after England and the United States settled their boundary dispute and signed the Treaty of 1846, Americans continued intensive beaver trapping, and by 1900 the beaver was nearly extinct not just in the Northwest but throughout North America.[12] The destruction of salmon habitat in the rivers of the Pacific Northwest started with this near annihilation of the beaver in the last half of the nineteenth century.[13]

Beavers are nature's river engineers; however, unlike the U.S. Army Corps of Engineers, the beaver and the salmon coevolved in Northwest watersheds over thousands of years, developing mechanisms that allowed them to coexist. In fact, the beaver's activities changed the ecology of the rivers in ways that enhanced salmon habitat. Beaver dams create pools that store sediments, organic material, and nutrients, releasing them slowly to the stream. They reduce fluctuations in flows, increase dissolved oxygen in the outflowing waters, create wetlands, and modify the riparian zone, all of which stabilize the ecosystem and buffer the effects of natural disturbances such as floods and droughts.[14] Extensive beaver activity in a watershed stabilizes salmon habitat and dampens fluctuations in salmon abundance. In arid regions, water and wetlands conserved by beaver activities were particularly important to the salmon.[15]

The depletion of beavers in the 1830s took place 100 years before the first extensive surveys and inventories of salmon habitat. Though the beaver dams and their beneficial effects on stream habitat did not disappear as soon as the beavers were trapped out, by the time biologists surveyed salmon habitat, little evidence of the dams and their positive contributions remained. During the century following the Hudson's Bay Company's extermination of the beaver, the streams of the Northwest were assaulted by miners, ranchers, irrigators, and timber harvesters, whose activities further degraded the quality of salmon habitat. During the height of that assault, and before effective environmental legislation, salmon habitat sorely needed the stabilizing effect the abundant beaver had once brought to the stream ecosystems of the Pacific Northwest.

The John Day River flows through Oregon's high desert country. It is one of the few major rivers with no large dams and no fish hatcheries. Although it flows through a harsh desert environment, it contains five of the twenty-three healthiest salmon and steelhead stocks in the entire state of Oregon.[16] Many people believe that the relative health of the John Day River salmon is due to the work of fisheries biologist Errol Claire.

Claire started his career as steward of the John Day River salmon in 1959, and for the next four decades he focused on protecting habitat and nurturing wild salmon without the aid of hatcheries. When I asked Claire about the importance of beaver to salmon in the John Day, I discovered that he not only had fought the day-to-day battles over salmon habitat but also had intensively studied the ecological history of the watershed. He told me that its riparian areas once consisted of extensive galleries of cottonwoods and willows—food for beavers, which the grazing cattle have all but eliminated. "Cows and beavers don't mix," he said to emphasize the point. He then described beaver dams 100 to 200 feet long that traversed floodplains along major tributaries. The skeletons of some of these dams, now 100 or more years old, can still be found. The desert beavers that built these dams were large, weighing up to 100 pounds and reaching up to five feet in length.[17]

Claire strongly believes that beavers were an extremely critical part of the salmon's ecosystem. In the harsh environment of the desert, the beaver ponds and the riparian vegetation they promoted provided essential habitat, especially during the low flows of summer. Beaver ponds today serve as important rearing areas for juvenile steelhead, and adult steelhead use them for holding habitat before spawning. The beaver ponds are especially critical rearing areas. Any attempt to restore the health of the John Day River ecosystem must include the return of the natural engineers of salmon habitat, according to Claire.[18]

Beavers also created salmon habitat west of the Cascade Mountains. Research biologists in Oregon recently documented the benefits of beaver ponds to juvenile coho salmon in coastal rivers. Small pools are the preferred habitat of juvenile coho salmon in summer, but with the onset of high flows in winter, the juvenile salmon dramatically shift their habitat preference to beaver ponds or quiet water in side channels. Those beaver ponds were once abundant but now make up less than 9 percent of salmon habitat in Oregon's coastal rivers. The lack of protected, quiet water, which the beaver ponds once provided, during high winter flows is now a major bottleneck to overwinter survival of juvenile coho salmon in coastal rivers.[19]

Mining

When James W. Marshall found gold in John Sutter's millrace on January 19, 1848, he triggered a frantic search for the yellow metal that ultimately had grave consequences for the Pacific salmon. With shovels, dredges, and high-pressure hoses, the forty-niners turned the Sacramento River watershed inside out. They washed whole mountains into the riverbed, choking much of the salmon habitat of the upper basin with cobble and silt. The period of intensive gold mining was, like beaver trapping, relatively short lived, but the single-minded search for gold so degraded salmon habitat that the scars are still visible today in many rivers. Viewing the rivers from their own narrow economic perspective, miners took little notice of the salmon as they systematically destroyed rivers and adjacent landscapes. The destruction went unimpeded until sediment washing downstream began to erode the productivity of farms in California's Central Valley. In 1884 the Ninth Circuit Court of Appeals finally granted an injunction to stop the mining practices that were cutting into the profits of agriculture. But by then, salmon from the San Joaquin, Tuolumne, and Stanislaus Rivers had already been devastated.[20]

The first wave of forty-niners quickly worked over the stream channels and collected the readily available gold. Then, to get the gold hidden in the hills, they turned to hydraulic mining. With high-pressure water hoses, miners literally disintegrated whole hillsides and washed them into tributaries of the Sacramento River. Rivers ran thick with soil, which clogged the salmon's gills and covered the natural river bottom, smothering the incubating salmon eggs as well as the aquatic insects, such as mayflies and stoneflies, that the juvenile salmon depended on for food. So much gravel and mud washed into the rivers that by 1884, the bed of the Sacramento River had risen twenty feet at its confluence with the Feather River.[21] According to the California Division of Fish and Game, mining was one of the primary causes for the decline of salmon in the Sacramento River.[22] Damage to salmon habitat was not limited to the Sacramento, though; rivers in northern California, such as the Trinity and the Klamath, suffered similar damage.[23]

The shift to the more capital-intensive method of hydraulic mining prompted many independent prospectors to leave California and move north, where they found gold in the Rogue River of southern Oregon in 1852 and in the John Day River in 1862.[24] They also made strikes in Idaho, Washington, and British Columbia. After the initial rush of individual prospectors, mining companies followed with more destructive technology, such as hydraulic mining and dredging, which took over and persisted for decades. In the Rogue River, miners worked with hydraulic equipment and dredges until the turn of the century.

During the peak of mining activity, the Rogue and its two principal tributaries, the Illinois and Applegate Rivers, ran brick red with the heavy silt loads, which degraded habitat and reduced the abundance of the salmon.[25] In 1888 the Oregon State Board of Fish Commissioners lamented that the salmon runs into the John Day River had been greatly diminished because the river was "given over nearly entirely to mining and irrigating interests," which made the waters very muddy.[26] Fisheries managers also noted that mining had caused heavy siltation of other streams, such as the Grande Ronde River.[27]

Gold dredges physically devoured whole rivers, throwing the gravel out of the stream channels onto the banks where, in many places, it remains today. Some dredges were huge. One that operated in the John Day River was 100 feet long and three stories high. It was a self-contained dredge and placer mining operation, which could move up small rivers by continuously digging a channel large enough to float in. The mining company built the dredge near the river, where it dug a pond large enough to float itself. The dredge then began chewing up the riverbed and running the gravel through sluices to winnow out the small amount of gold. The dredge dug its way through the river channel, but it also cut new channels through streamside meadows. During hard times, ranchers sold the rights to the gold lying in gravel deposits under their meadows. When the dredges finished, however, only sterile piles of gravel remained. The land was ruined forever.[28]

Remains of gold dredge, Sumpter, Oregon, December 2, 1959. (Photo courtesy Oregon Historical Society, OrHi 11013)

Snow-covered landscape showing the impact on a stream channel from gold dredging in the Powder River Basin, late 1940s. (Photo courtesy Homer Campbell)

The effect on salmon habitat was just as devastating. Dredges swallowed the complex physical and biological structure of the stream and spit out piles of gravel. They even changed the very courses of rivers, because "where the dredge went the river followed," in the words of one article describing its effect.[29] But what the dredge left behind was actually a grotesque distortion of a living river: ugly piles of gravel and simple channels in which water flowed but few of the ecological connections that had once made it a living ecosystem survived.

The consequences of dredge mining persist to this day. For example, in Oregon's Blue Mountains, Bull Run Creek was once the center of a booming gold district. Today, the chinook salmon that find their way up to this small tributary of the John Day River must climb nearly 5,000 feet above the level of the ocean. Lower Bull Run Creek meanders through beautiful meadows, but in its upper reaches the stream suddenly becomes a straight, incised channel, all its gravel thrown into high piles on the stream bank. There are few salmon in this section of the creek. It is a barren and inhospitable reminder of the legacy of a single-minded pursuit of riches. The stream had more to offer humans than the few gold nuggets found there 100 years ago: It could have grown salmon in its waters. It could have nourished grass in the meadows to graze cattle. It could have poured its clean, cool water into the John Day River. It could have provided men and women from cities with a sense of peace and relaxation. Now it is just a reminder of lessons we should have learned.

Timber Harvest

In November 1852, James G. Swan sailed north on the brig *Oriental* from San Francisco to Shoal Water Bay on the Washington coast. One night during the trip, the small ship ran into heavy seas and strong winds. The next morning, thirty miles off the mouth of the Columbia River, Swan awoke to a strange sight. "The Columbia River, from which a huge volume of water was running," he wrote in his diary, "carry [*sic*] in its course great quantities of drift-logs, boards, chips and sawdust, with which the whole water around us was covered."[30] What Swan saw thirty miles off the coast of Oregon and Washington was, like the tip of an iceberg, an indication of something much larger that was hidden from view. The source of the sawdust that covered the sea lay deep in the green coastal hills and river valleys. If a single storm event could fill the ocean's surface with sawdust for thirty miles offshore, the small streams where the debris originated must have been covered with a suffocating layer of this mill waste. In fact, lumber mills dumped sawdust by the ton into rivers and streams throughout the Northwest. Winter storms eventually flushed the debris out to sea, but not before it smothered millions of incubating salmon eggs and miles of rearing habitat for juvenile salmon.

The first lumber mills in the Pacific Northwest were small water-powered operations, located on or very near the streams they tapped for energy. As the mills reduced giant Douglas firs to dimension lumber, they also produced huge piles of sawdust and other waste. A few mills burned their debris, but most used nearby streams to carry the sawdust out of sight. When heavy fall and winter rains drenched the Northwest, the streams rose and flushed the mill debris down into larger rivers and eventually out to sea. Sawdust covered the river bottoms, smothering incubating salmon eggs and killing the natural food webs. When the debris began to rot, it sucked oxygen out of the water. Furthermore, sawdust suspended in the stream could clog the gills of juvenile and adult salmon. In his first annual report, Washington State Fish Commissioner James Crawford described the debilitating effects of sawdust on fish: "On one of my visits this summer to Chinook Beach, I saw a fine salmon that would have weighed 25 or 30 pounds, gasping on the beach; and upon examining its gills found them clogged with sawdust, and the irritation produced from the foreign substance had caused a festering sore that had eaten away a large portion of the gills."[31]

Eventually evidence of the destructive effects of sawdust in salmon streams convinced the legislatures to act. In 1876, the legislature of the Washington Territory enacted a statute prohibiting the dumping of sawdust in rivers and streams.[32] The Oregon legislature passed a similar law in 1878.[33] But these statutes, like most early attempts to protect salmon and their habitat, were enforced haphazardly or not at all. Sawmills continued to dump sawdust in streams for another fifty years.

Salmon managers failed to enforce the laws and protect salmon habitat for several reasons. The legislatures sent conflicting directives. They enacted laws to protect habitat but in many cases failed to provide sufficient funds for enforcement. The political ambivalence probably stemmed from the conflict between the recognized need to protect habitat to ensure the persistence of the salmon industry, coupled with a laissez-faire attitude toward resource-based industries. Legislatures were reluctant to restrict the timber industry even though they realized that its practices infringed on the fishing industry.

When James Swan sailed through the sawdust-covered sea, the Northwest was in the midst of a lumber boom spurred by the population explosion in California following the discovery of gold at Sutter's Mill. In 1848 about twenty-two mills operated in western Oregon and Washington, but by 1851 the number of mills in Oregon alone had grown to over 100.[34] The increasing number of mills needed steady supplies of logs, but supplying those logs was not an easy task, even in a region covered with dense forests. Horses and oxen could not haul enough logs over the poor or nonexistent roads. As timber harvesting moved farther away from the mills, mill owners and loggers solved their transportation dilemma the same way they solved their sawdust problem. They turned to the region's rivers to carry logs from the forest to the mills.

Because the enormous old-growth logs had to be "skidded" by oxen from the forest to the nearest point on a river suitable for transport, early loggers first cut the trees close to or in riparian zones along the riverbanks. For example, by 1881, loggers had cut a one-and-a-half-mile-wide band along the banks of Washington's Hood Canal.[35]

For 3,000 to 5,000 years, coastal rivers had flowed through groves of old-growth western red cedar, Douglas fir, western hemlock, and Sitka spruce. With their massive root structures, these giants protected stream banks from erosion. Forests buffered the effects of other natural disturbances such as droughts, floods, and even fires. The trees shaded the streams, keeping them cool. Logs that fell across the streams eventually worked their way into the channel, where they trapped gravel needed by the salmon for spawning and created pools where juvenile salmon reared. The large streamside cedars, firs, and hemlocks played the most important role in stabilizing salmon habitat, and they were the first to be cut.

Once loggers cut the big trees, they hauled the logs to a river and stacked them on the banks. There they sat all summer, until autumn rains and high water finally washed them downstream to a mill. By the end of the 1880s, every river that could float logs was being used for transport.[36] Because the flows of the Northwest's rivers vary with the seasons, the mills were often without a steady supply of wood, especially during the dry summer months when flows were lowest.

Splash dams overcame the problem of seasonal flow variations. Logger Alex Polson built the first splash dam in Pacific County, Washington, in 1881, and Charles Granholm built Oregon's first splash dam on the north fork of the Coos River in 1884.[37] These log structures stopped the flow of a river and created large reservoirs. As the dam backed up water, logs were skidded from the harvesting site and dumped into the growing pool. At the appropriate time a sluice gate was opened, rapidly releasing logs and water for their trip downstream. Depending on the river and the logging operation, such "splashing" from a single dam could occur once a week or as fre-

Logs in a riverbed awaiting the spring flood, which will carry them downstream to the mill. (Photo courtesy Oregon Historical Society, OrHi 4698)

quently as every day. The logs flushing downstream often hung up in the stream and quickly jammed into massive, helter-skelter tangles, some extending upstream for as much as a mile. Loggers routinely used dynamite to break up these jams. On the Humptulips River in southwestern Washington, five boxes of powder were used each day to blow the jams loose.[38]

The rapid change in flow, the "flush" of logs, and the liberal use of dynamite killed and injured large numbers of salmon and destroyed their habitat.[39] During the seventy or more years that loggers used splash dams and log drives, rivers became dangerous places for the Pacific salmon. The giant logs—each weighing many tons—dragged along the stream bottom, crushing and gouging everything in their path, ripping large chunks of earth from the banks, and turning the river chocolate brown. The salmon had few places to hide.

Each splash of logs and water left behind a devastated river with fewer and fewer of the ecological attributes the salmon needed in their habitat. Hiding places under overhanging banks were gone. Spawning gravel had disappeared, and the danger of being killed by yet another load of logs was just a few hours or days away. Splash dams made life particularly dangerous for chinook salmon, because they spawned and reared in the mainstems—directly in the path of every log drive. Old-growth forests once shaded the rivers and kept them cool, protected stream banks from erosion, and added nutrients to the aquatic food web, while downed trees with root wads intact created pools for juvenile salmon to rear in. Now logs from those forests were killing the rivers and the salmon. Hundreds of splash dams operated throughout the network of coastal streams of Oregon and Washington and into the interior on rivers such as the John Day, the Minam, and the Grande Ronde in northeast Oregon.[40] By 1900, the destruction of mainstem salmon habitat had limited reproduction and severely reduced the number of chinook salmon in Oregon's coastal streams.[41]

Splash dams and log drives became increasingly controversial as more people settled along the coastal rivers. The periodic floods of water and logs not only degraded salmon habitat but also destroyed docks and impeded river travel, the primary means of transportation and communication for early settlers on the coastal rivers. Facing mounting criticism, timber companies leaned heavily on economic arguments to maintain their control and use of the rivers. It was cheaper to use the rivers to transport logs, they contended, and mill owners on coastal rivers needed the savings to compete with the big mills operating on Puget Sound.[42] That argument worked for seventy years. But while the timber industry gained an economic benefit through the use of splash dams, their extra profits came at a heavy cost to the salmon and to the fishing families that depended on them. The last splash dams in coastal streams didn't cease operation until 1959, and the last log drive took place in 1970 in Idaho's Clearwater River.

Aside from the physical damage to streams, splash dams left another negative legacy that lasted for decades after the log drives ended. They helped shape an erroneous vision of healthy rivers. Before the arrival of white pioneers, the coastal rivers of Oregon, Washington, and Puget Sound were tangles of logs, beaver dams, side channels, and wetlands. This condition provided productive habitat for juvenile salmon.[43] In order to use the rivers as "highways" to move their logs to the mills, loggers had to clear the stream channels of these obstructions. They blasted large rocks free of the channel and cleared the stream of natural logjams. Side channels, which are important winter habitat for the salmon, were blocked off to keep the logs in the mainstem of the river.

As a result, people came to regard cleared rivers with no large woody debris, beaver dams, or side channels as the standard for how economically healthy rivers should look. This vision even infected the thinking of fisheries biologists, to the detriment of salmon habitat. For example, in 1953 Robert Schoettler, director of the Washington Department of Fisheries, equated rivers to highways: "Just as automobiles need smooth roads to operate on," he wrote, "fish need clean, unobstructed rivers and streams. . . . "[44] In some cases, especially in the 1930s to 1950s, logjams did block salmon migration and needed to be removed. But with "clean highways" as the guiding vision, management agencies were overzealous in their removal of wood from streams. So much woody debris was removed in the 1950s and 1960s that stream habitat for salmon actually degraded.[45]

Eventually, after the trees closest to the rivers were cut, logging moved farther into the mountains, but the damage to salmon habitat did not subside. Using steam donkeys, loggers pulled logs up steep slopes to central yarding areas and then loaded them onto flatcars for transport to mills. To use this new technology effectively, loggers had to clear-cut the hillsides to eliminate all obstructions to the movement of logs.[46] Later, after the roots of the harvested trees rotted and lost their ability to hold the soil, the steep hillsides slid into streams and overloaded channels with sediments and logging debris, creating a new but still deadly set of habitat problems. Logging on the steeper slopes and the construction of logging roads eventually replaced splash dams and log drives as the major cause of habitat destruction in today's logging operations.

The early loggers did not possess the technology that could have lessened the damage to the land and rivers, nor were they particularly inclined to do so. In his novel *Woodsmen of the West,* Allerdale Grainger captured the mood of the turn-of-the-century logging industry, in which "loggers were doing desperate work, fighting against time, to put in logs and sell completed booms while the prices were so high." Few thought about the future, except to dream about rising land values and fortunes to be made. "Someone would find it worthwhile, someday," Grainger wrote, "to buy from [the logger] the stretches of forest whose sea fronts he had shattered and left in tangled wreckage. As for him, he was going to butcher his woods as he pleased. It paid."[47] This spirit of exploitation for private gain persisted even as the consequences of heedless logging became apparent.

When the first pioneers cut trees to build houses or clear spaces for gardens, the loss of those trees went unnoticed in the vastness of the Northwest's forests. The ecosystems retained their natural resiliency and productivity after those first inroads. But soon, the subsistence economy gave way to a market economy, and trees began falling at a breakneck pace to satisfy distant markets. Long before the start of the twentieth century, logging and milling practices that increased profits for the timber industry became economic costs for the fishermen and Indians to bear. Within a matter of decades, these destructive practices shattered the natural economies of forested watersheds that had been evolving for 4,000 to 5,000 years. The salmon are still paying the price for the economic efficiency of the timber industry's splash dams, log drives, and clear-cuts.

As one of the Lords of Yesterday, the timber industry used its power to assert the dominance of the industrial economy over the natural economy, and to reap short-term profits at the expense of the long-term productivity of riverine ecosystems. They defended the destruction of the natural economy by arguing that it was an economic necessity—an industrial economic necessity. Consequently, many streams were logged to the water's edge prior to the enactment of forest protection laws in Oregon (1972) and Washington (1976).[48] These attempts to protect stream habitat were major improvements, but they fell short of their objective. For example, in 1986, a task force from the Oregon legislature concluded that "current implementation of the Forest Practices Act does not adequately protect riparian fish and wildlife habitat."[49]

Today, after a century of human failure to seek a balance between the natural and industrial economies, the coho and chinook salmon and steelhead are on the verge of extinction. New protective regulations are being enacted, but more may be needed to ensure recovery. In the meantime, the Lords of Yesterday are using the same arguments of economic cost to avoid achieving a balance between the natural and industrial economies in commercial forestlands—the same arguments they used a half-century ago to continue the use of destructive splash dams.[50]

Grazing

The discovery of gold and the wave of miners and prospectors that followed generated a demand for meat, which was largely satisfied by increasing local herds throughout the Northwest. Farmers not lured to California to search for gold discovered that grass converted to beef was just as good as gold. Even after the waves of gold seekers peaked, growing urban populations in the rapidly industrializing areas of the eastern United States and Europe kept the demand for western beef high.

At first, livestock production was centered in the humid western valleys of Oregon and Washington. In those confined areas, the effects of overgrazing appeared even before the first forty-niners hit the trail for California. In

1848, the Tualatin Plains south of Portland, Oregon, showed signs of overgrazing, and by 1860, the 100,000 head of cattle owned by farmers in western Oregon and Washington had eliminated the native bunchgrasses over large areas of the interior valleys.[51] Between 1860 and 1890, farmers converted the fertile lands west of the Cascades and the Sierra Nevada to cultivated crops. As a result, cattle production, which depended on open-range grazing, was forced to move to the public grasslands of the intermountain region east of the Cascade Mountains.[52]

In the arid grasslands east of the Cascade Range and the Sierra Nevada, cattle were attracted to the riparian areas that bordered the streams and wetlands. As their numbers grew, cattle ate and trampled the streamside vegetation, and when the riparian vegetation disappeared, salmon habitat rapidly degraded. Loss of the extensive cottonwood galleries and the thick growth of willows exposed streams to the desert sun and heated the water to lethal temperatures. Stream banks eroded under the constant pounding of bovine hooves, releasing heavy sediment loads into the stream. The excess sediment then settled on the stream bottom and smothered incubating salmon eggs and aquatic food webs. Stripped of the stabilizing effect of riparian shrubs and trees, and suffering from the earlier loss of beaver dams as well, salmon streams became especially vulnerable to natural disturbances such as floods and droughts. In streams unprotected by riparian vegetation, floods ripped out banks and scoured channels, causing them to downcut. The resulting arroyos lowered water tables, making recovery of riparian zones more difficult.[53] Streams that once flowed year-round with cool, clean water dried up in summer or became a series of stagnant pools at the bottoms of raw gullies.

Sometimes it was not the cows but the cattlemen themselves who destroyed the riparian areas and wetlands. According to Herman Oliver, who wrote about his father's 1880s ranch on the John Day River, "One of the first jobs on the Clark homestead was to clear off the brush and trees. Big cottonwoods grew all along the river and the meadows were covered by wild thorn bushes, to be chopped out by hand." Aside from removing vegetation, Oliver wrote that his father "took out the big bends, straightened the channel, rip rapped the banks and made each meadow safe. He dried up the wet places. For draining, he dug by hand ditches about two feet deep and 18 inches wide."[54]

The cattle industry that developed in the Columbia Plateau prior to 1900 was not an organized enterprise in which a proprietor carefully controlled the processes that led to production and profits. Free and open access to the public domain without restriction meant that greed and nature were the only constraining factors. The early cattlemen cared little about the connection between beef production and the grassland ecosystem that made it possible, and any possible connection to the salmon was given even less thought. As pioneer W. J. Davenport recalled, "The term 'industry' . . . presupposed business methods, proper supervision and care . . . cattle raising was wholly lacking in these essentials. To be a cattle raiser in those days one had only to purchase the stock and turn the same loose to shift as best it might, without food or shelter during the winter months. If the herds managed to survive under these conditions, the owner was counted as a successful cattleman."[55]

Free grass on the wide-open public lands attracted individuals and corporations expecting to get rich. By 1890, with at least 26 million cows on the range, the West had become a large, unfenced, and unrestricted cattle pasture.[56] Huge herds of cattle led to lower prices, creating a prolonged depression in the industry beginning in the early 1870s. As the market became saturated, ranchers could not sell their beef. The herds increased, and so did overgrazing and the destruction of riparian vegetation. Cattlemen looked for alternative markets. Some processed their beef in salmon canneries during the off-season. The surpluses persisted, however, until the severe winters of 1881 and 1882 killed large numbers of cattle.[57]

During this period, overgrazing destroyed many streams and their salmon habitat. Camp Creek, a tributary to the Crooked River in central Oregon, exemplifies how dramatic changes in the land occurred within the twenty-five-year period of intensive grazing before the turn of the century.

Today, Camp Creek lies at the bottom of a deep vertical-walled gully, which cuts through an arid plain known as Price Valley. The creek flows only in the spring; the rest of the year it is a dry wash. Sagebrush dominates the valley floor, growing right up to the edge of the gully. There are no signs of the healthy ecosystem—beavers, wetlands, grasses, willows, and aspens—that Euro-Americans found when they first entered Price Valley in the early 1800s.

When trapper Peter Skene Ogden visited the Camp Creek area in 1826, the salmon still returned to the Crooked River. Ogden's diary includes a description of an Indian fishing weir near the north fork of the Crooked River, which is near the mouth of Camp Creek.

Between 1826 and 1885, thick stands of willow and aspen grew along the banks of the creek. Early visitors noted little difficulty in crossing the stream, which indicates that it had not downcut into the gully that it flows through today. Early surveyors described grassy meadows, an abundant beaver population, and well-developed wetlands in the floodplain. As late as 1876, a General Land Office survey noted that the Camp Creek floodplain provided a rich expanse of bunchgrass, wild rye, and swamp grass.

In the late 1800s the number of people living in the Camp Creek area increased. As the human population grew, so too did the livestock population. Within twenty-five years, the physical appearance of Camp Creek changed dramatically. In 1903, a surveyor described the valley as a sagebrush-covered plain cut by a deep gully 60 to 100 feet wide and 25 feet deep. The lush grass, riparian vegetation, wetlands, and beavers were already gone. The surveyor speculated in his notes that the arroyo was the result of overgrazing.[58]

Local people agreed. According to one early Price Valley resident, "There was too much stock . . . they'd eat the willows and everything and that['s] what did it. . . . There were too many people with too many cattle. Everybody had cattle."[59]

Geoff Buckley, who, as a graduate student at the University of Oregon, carefully documented the transformation of the Camp Creek drainage during this period, attributed the changes in the landscape to both overgrazing and climate change. Once cattle destroyed native vegetation in the late 1800s, the land and the stream became vulnerable to periods of high rainfall and floods, which subsequently scoured and downcut the channel.[60] Author Charles Wilkinson said, "There are Camp Creeks all across the American West—thousands of them."[61]

By the end of the nineteenth century, a number of factors—including deterioration of the range, heavy losses of cows during severe winters, and irrigation and agricultural settlement—had contributed to a decline in open-range cattle grazing. Mass starvation of cattle during the severe winters of the 1880s convinced most cattlemen to abandon their use of the open range and adopt a more conservative practice: they reared cattle on the open range only during summer, and then brought the animals into fixed holding areas to be fed hay through the winter months. This strategy meant the herds were reduced to the number that could be cared for through the winter. As the number of cattle shrank, however, the number of sheep increased. In 1879, 5,000 sheep ranged in the Yakima River Basin, but by 1899 that number had exploded to more than 250,000.[62] By 1900, sheep outnumbered cattle in most western states.[63] Small towns in the high desert, such as Shaniko, Oregon, suddenly found themselves surrounded by a sea of grazing sheep. Millions of dollars in wool sales earned Shaniko the title "Wool Capital of the World" in 1903.[64]

The movement of large flocks of sheep grazing through the mountain West prompted conservationist John Muir to label them "hoofed locusts."[65] But it wasn't only the direct effects of overgrazing that earned the sheep this scorn. Some activities of the shepherds were also destructive. To "improve" the range, shepherds set fire to the forests to remove shrubs and stimulate the growth of new forage. At times the fires inflicted significant damage on the range and stream habitat.[66] In the Yakima Basin, the fires were so extensive around the turn of the century that the U.S. Geological Survey predicted that their effects would destroy normal watershed function and reduce summer flows, threatening the supply of water for irrigation.[67]

In an effort to safeguard the natural resources of public lands, Congress passed a law that gave President Benjamin Harrison the authority to create federal forest reserves in 1891, and he did. But ranchers disputed just what the president could do with the reserves once created, especially how far he could go in regulating the commercial use of the forests. In 1894, when the Department of the Interior tried to terminate grazing in all the reserves, the army refused to patrol the forests, and western stockmen vigorously protested the exclusion of sheep and cattle. Their protests led to a change in national policy, and in 1899 the government started to issue permits to graze cattle in most of the forest reserves. Sheep were excluded, except in Oregon and Washington.[68] This was the first tentative move to control grazing on public land—land that adjoined and determined the quality of hundreds of miles of salmon habitat. Outside the forest reserves, the use of the range remained largely uncontrolled, and overgrazing continued to be a problem. The combination of overgrazing and climate change would later intensify destruction of the range.

Severe drought in the 1930s, combined with unregulated grazing, added enough additional deterioration of the public range to move Congress to further regulate its use through the Taylor Grazing Act of 1934.[69] The act created the U.S. Grazing Service, which later became the Bureau of Land Management. Although access to the range was no longer open and unrestricted, grazing regulations did little to curb the destruction of riparian zones and salmon habitat. The way the act was implemented accounted for part of the failure to protect stream habitat. Until 1950, local administration of the act was dominated by grazing interests.[70] The environmental movement of the 1960s and 1970s gradually eased the entrenched control by ranchers. In

1975, however, after fifty years of "management" of the public range by the Bureau of Land Management, only 17 percent of those lands were in good to excellent condition.[71]

In early June, before irrigators divert most of the water for their crops, there is still enough flow left in the John Day River today to allow peaceful rafting through spectacular desert canyon country. Drifting through the canyon, it's hard to imagine these hills once covered with millions of sheep and cattle. It's even harder to visualize this desert river as the first settlers saw it. The large herds of cattle and flocks of sheep are now gone, but they left behind a record of their passage through this country a century ago. The hills are covered with old trails cut into the steep slopes. These small ridges and shelves are so regular and so pervasive on all the hillsides that at first they appear to be a natural part of the landscape, but they are actually trails cut into the hills by millions of hooves. The trails that cover the walls of the John Day Canyon are a silent testament to what happened in the intermountain West a century ago. The cattle and sheep left behind tracks that will continue to remind us of their historical presence here in this dry country. What is harder to find are traces of the riparian zones, the extensive galleries of cottonwoods and willows laced with beaver ponds and wetlands. The ecosystems that once helped to nurture millions of salmon in these desert streams have long been a forgotten part of the history of the West.

Irrigation

Shortly after the turn of the century, Lester Campbell was seasonally employed by the Agricultural Experiment Station at Oregon State University. He often traveled from his home in Corvallis, in western Oregon, to the field station in Union, at the far eastern edge of the state. On his trips between Corvallis and Union, Campbell usually stopped at a small hotel in Boardman, a high desert town near the Umatilla River. At that time, the Umatilla supported strong runs of spring and fall chinook, coho, and steelhead. Years later, Campbell would vividly remember the view from that little hotel. In the spring and summer, he looked out over the nearby farmland and saw cultivated fields covered with a shimmering layer of silver where millions of salmon smolts lay drying in the desert sun.[72] The irrigation diversions in the Umatilla River intercepted young steelhead and salmon as they migrated downstream toward the sea and sent them into feeder ditches that eventually carried the fish onto the farmers' fields, where they died. By 1904, farmers were diverting so much water from the Umatilla River that at times its lower reaches ran dry. Juvenile salmon that made it past the irrigation diversions found their way to the sea blocked by a dry riverbed. As a result, few salmon made it out of the river and into the ocean. The chinook salmon disappeared from the Umatilla River not long after the last big run in 1914.[73]

From the start, the pioneers who settled in the arid high plains of eastern Washington and Oregon and Idaho wanted to convert the endless wilderness of sand, sagebrush, and rabbitbrush into a garden—a verdant sanctuary for themselves and a chosen few plants and animals.[74] To them, nature in its wild state seemed a dark and forbidding place, filled with savage animals locked in a Darwinian struggle for survival.[75] This perspective allowed the pioneers to equate irrigation with reclamation. Using irrigation, they could reclaim wasted land from the wilderness and, through human nurturing, convert it into the garden that God intended for humans. Settlers therefore regarded "reclamation" of the desert as divinely ordained and fully consistent with their mission to control and subdue nature.

The early irrigation movement made no pretense of harmony between the industrial and natural economies. Irrigation was a weapon in the war against nature, and the only acceptable outcome was the complete victory of the industrial economy over its natural counterpart. Farmers and ranchers settling the Northwest began diverting water from streams to water crops soon after they arrived. Because all the diversions were completely open, there was nothing to stop juvenile salmon from being funneled into ditches and deposited on the cultivated fields.

At first the irrigation projects were mostly private, and they were subject to land speculation, fights over water rights, and many failures.[76] In spite of those problems, irrigation spread so rapidly that by the turn of the century, farmers' claims to water exceeded the natural flows of rivers. For example, the water rights claimed in the Deschutes River above the city of Bend, Oregon, in 1914 amounted to forty times the river's flow.[77] With all of the water committed to quench the thirst of farmers' crops, there was little or none left for the salmon. Many rivers were dried up below the last water diversions, and in others, the combination of greatly reduced flows, lack of beavers, and loss of riparian shade due to logging and grazing rapidly warmed the remaining water to temperatures that either killed the salmon or made it difficult for them to reproduce and rear.

In 1890, while visiting the salmon-producing rivers in eastern Oregon, the state fish commissioners heard many complaints about the fish loss in the irrigation ditches. According to their report, in springtime "a great many young fish of all species, as they are descending the stream, are run into these ditches and carried out on the farms and left to perish." The report recommended that the legislature pass a law requiring the owners of all ditches using water from streams inhabited by fish to put a screen of fine wire netting across the intakes during the months of April, May, and June. "This will not be a hardship to anyone," they explained, since "very little water is used during those months, and the screens can be so arranged to interfere very little with the flow of water."[78]

Although states made some attempts to screen irrigation diversions, their feeble efforts met with little success. In Washington State in 1904, crude devices were installed at a few diversions to separate the salmon from the irrigation water, but they were ineffective.[79] In California, primitive screening devices were installed in many irrigation diversions before 1900, but these also failed to stop the mass killing of juvenile salmon.[80] So many fishes were lost in irrigation ditches and canals that the California Division of Fish and Game eventually tried a different tack—creating a Bureau of Fish Rescue and Reclamation in 1928 to capture stranded salmon and return them to natural waters—but this too met with little success.[81] In Oregon, the legislature finally acted on the fish commission's recommendations and in 1929 passed a law requiring the screening of irrigation ditches. When the state game commission took a recalcitrant farmer to court to enforce the law, however, the judge in the district court ruled in favor of the defendant, effectively killing the statute.

The war against the arid wilderness and the fervor with which it was pursued is clearly illustrated in the dialogue of a play called *Visions Fulfilled,* written by Alice M. Tenneson and Sue M. Lombard and performed by local residents of Yakima, Washington, in 1917. The curtain rose on a landscape in its natural state: a desert filled with "noisy demons" who pierced the quiet with their "fiendish mockery." As the plot unfolded, demons held captive the "lovely maiden, Irrigation." The Indians ignored Irrigation's pleas for help, but the pioneers heard her cries and forced their way through the "demons sneering numbers." Through hard work, the pioneers soon began to free the maiden. One player proclaimed, "Such aid has freed her arms, her body moves with grace. And tho her feet are fettered still, the place has been transformed from desert waste of sand." Because the job of completely freeing Irrigation was too big for the pioneers, they called on their Uncle Sam, who sent his daughter, Reclamation, to finally free Irrigation from the demons of the desert. The pageant ends with the lines "Let no oppression bear upon the weak—and let us not, . . . forget to value life the most."[82] But the pioneers valued only those lives they chose to bring into their cultivated garden.

Many Euro-Americans believed the western deserts were a temporary condition—a dormant garden just waiting to be reclaimed. Real estate agents working for the railroads and other speculators convinced many prospective settlers that rain would follow their plows, that the very acts of human settlement and pursuit of agriculture would so intimidate Nature that she would bring water to the desert.[83] They soon learned it would take more direct action on their part.

Farmers claimed most of the natural summer flows for irrigation, but that still left dry land that could be reclaimed if they could store the high winter and spring flows and release that water later in the year to irrigate more crops. Rearranging the flows of large rivers to meet human needs was beyond the capacity of private capital. Funding, building, and operating massive irrigation projects would take the backing of the federal government. But before the federal resources could be applied to the control of natural river flows, "the most important single piece of legislation in the history of the west" had to be enacted.[84] With pressure from Nevada's Senator Francis Newlands, Congress approved the National Reclamation Act in 1902, bringing into existence a federal irrigation agency: the Bureau of Reclamation.[85]

Reclamation of the western deserts through massive federal projects appealed to turn-of-the-century progressive reformers such as Theodore Roosevelt and Gifford Pinchot. They believed that rivers should be managed by experts who could use science, technology, and centralized planning to make the most efficient use of natural resources. In their view, it was best to use river water for agriculture instead of "wasting" it by allowing it to flow to the sea, instead of allowing it to nurture the wild salmon. It's ironic that the word "conservation" originated from the Progressives' vision of storing water in reservoirs so it could be used to irrigate crops during the summer growing season. To Progressives, the key to conservation was *efficient use* of all natural resources.[86] In fact, the Progressives' conservation movement did not seek to end the massive exploitation of natural resources; rather, it sought to replace laissez-faire exploitation with well-organized and planned exploitation.[87] In 1900 the term "conservation" had a different meaning than it does today.

The Bureau of Reclamation applied the full power and resources of the federal government to rearrange the seasonal flow patterns of western rivers for irrigation, but the agency had little understanding of the effects of its "conservation" practices on the salmon. It built water storage dams without any provision to allow adult or juvenile fish to pass, and did not screen irrigation canals to prevent the salmon from being deposited on farmers' fields to dry in the sun.

As late as 1931, salmon managers and fishermen complained that the Bureau of Reclamation showed careless disregard for state laws intended to protect the salmon. Although Henry O'Malley, the U.S. commissioner of fisheries, believed that "the federal government in its consideration of dams and irrigation projects should conform to the laws of the state in which they operate," the Washington Department of Fisheries and Game pointed out that the Bureau of Reclamation ignored state laws requiring fishways over dams.[88] With its zeal for maximum efficiency, the bureau contended that agriculture made far better use of water than fish did. This attitude was summed up by the bureau's general counsel, B. E. Stoutemyer, in a letter about irrigation in the Yakima River: "As there are about 100,000 people living in the Yakima Valley, all dependent on irrigation, I do not believe any serious argument could be made that the water should be taken from the farms and orchards to improve fishing conditions."[89]

As irrigated agriculture spread through the Northwest, massive numbers of juvenile salmon were killed in scenes like the one witnessed by Lester Campbell, who described the farmlands covered with silvery salmon. Looking back with the clarity of hindsight, it seems incredible that silver fields like those near Boardman, Oregon, could be found throughout a large part of the salmon's range. Although salmon managers knew about the irrigation problem and understood the impact it had on the fish, they made little effort to actually document the number of juveniles lost.[90] One biologist did try to estimate the number of young salmon killed by irrigation diversions in the Yakima River. In 1916, federal fisheries biologist Frank Bryant studied a subsample of 200 irrigated acres in the basin. After the fields had been watered, he walked over the entire 200 acres and counted the dead fish. Ninety percent were juvenile salmon, and there were, on average, twenty of them on each acre of land. Assuming that twenty fish were killed on each acre of irrigated land at each watering, Bryant then extended his subsample to the entire irrigated acreage in the watershed and concluded that 4,500,000 juveniles were killed in the Yakima Basin with each watering.[91]

In 1927, biologists estimated that 20 million young salmon were still being killed each year by unscreened irrigation diversions in the Yakima River alone.[92] In spite of these massive losses, there was little serious or effective action to prevent the diversion of the salmon onto agricultural lands until the 1930s—over eighty years after the first irrigation ditches began sending juveniles to their deaths. At that time, the federal government began experimenting with different mechanical devices to prevent the young fish from entering irrigation ditches.[93]

In 1949, the Lower Columbia River Fisheries Development Program was initiated to compensate for the effects of massive water development projects on the salmon. As part of the program, the federal government funded the screening of irrigation diversions, but thirty-seven years later that program was still identifying unscreened diversions in the Columbia Basin.[94] As recently as 1996, the National Research Council, in its study of Pacific salmon, reported that fewer than 1,000 of the 55,000 water diversions in Oregon were screened, and 3,240 were listed as a high priority for screening. Furthermore, a survey of pumping sites on the Oregon shore of the Columbia River in 1994 found that 80 percent failed to comply with measures for protecting juvenile salmon.[95]

But the screening of irrigation ditches corrects only one part of the problem. The water is still diverted, drying up streams or causing them to overheat. The lack of sufficient water in streams remains a problem that has largely been ignored.

Dams

Most people associate dams and the problems they create for the salmon with the large mainstem dams constructed on rivers such as the Columbia, Snake, Skagit, Rogue, Trinity, and Sacramento between the 1930s and 1970s. Even before the 1930s, however, hundreds of smaller dams were built for municipal water supplies, stock watering, irrigation, placer mining, and power generation. Like their large counterparts, these small dams also prevented the salmon from reaching spawning areas, flooded upstream habitat, and degraded salmon habitat downstream by altering flow patterns. In the Columbia River Basin alone, thirty-two major dams were built on tributary streams before 1930. Many of those dams towered over fifty feet in height.[96]

The problems that dams posed for the salmon were well known when the pioneers arrived in the Northwest. In New England, where the manufacturing industry depended heavily on hydropower and dams, the Atlantic salmon had already been eliminated from many streams by the middle of the nineteenth century.[97] As a result, the Northwest's lawmakers enacted legislation to protect the salmon from dams almost as soon as settlement and development began in the region. As early as 1848, Oregonians included a clause in their territorial constitution mandating that "rivers and streams . . . in which the salmon are found or to which they resort shall not be obstructed by dams or otherwise, unless such dams or obstructions are so constructed as to allow salmon to pass freely up and down such rivers and streams."[98] In 1890, Washington State's first legislature also passed a law requiring fish passage at all dams.[99] But these laws—like most laws intended to protect salmon habitat—were poorly enforced.

In southwest Oregon's Rogue River Basin, for example, several hundred dams were built on tributaries within the range of salmon migration. The dams largely supplied water for hydraulic mining and irrigation. Prior to 1920, many of those structures had no provision for the passage of the salmon. Public pressure and attacks orchestrated by lower Rogue cannery operator R. D. Hume finally led to the construction of fish ladders at the most destructive dams.[100]

Around the turn of the century, two dams built across the main channel of the Rogue River near Grants Pass, Oregon, did not provide passage for the salmon. One was a 12-foot-high log dam built by the Grants Pass Power Supply Company in 1890. A few salmon could pass the structure at certain flows, but most of the run was blocked. For half a mile below the dam the river was crowded with fish throughout the summer. A 1905 flood destroyed the dam, but it was replaced by a six-foot structure that also lacked a fish ladder. That dam gradually deteriorated until, by 1940, it no longer posed a problem.[101]

The second dam was so blatantly destructive, and law enforcement so ineffective, that private citizens took action on their own. Ament Dam was a 25-foot-high structure built three miles above Grants Pass in 1905. At first the dam had no fish ladder, and no salmon could pass upstream. In 1906, a ladder was built in the center of the dam, but it was ineffective. Angered by the complete blockage of the salmon run, someone floated a boatload of dynamite into the dam and damaged it beyond repair.[102]

In the 1930s, federal planners laid the groundwork for a forty-year dam-building frenzy in the salmon's home waters. But long before the massive program began, smaller impoundments had already destroyed significant amounts of salmon habitat in tributary streams. For example, in 1932 the Oregon Fish Commission prepared "an industrial, vari-colored map" of the Columbia River Basin, based on data collected over a fifteen-year period. According to the commission, the map revealed that "approximately 50% of the most important productive area within the basin has [*sic*] been lost to the [salmon] industry by the construction of dams for irrigation and power, thus isolating spawning areas."[103] Those dams destroyed great salmon-producing streams, especially in the arid interior areas. The salmon in the high desert rivers, such as the Boise, Malheur, Owyhee, Payette, Weiser, Burnt, and Powder, were destroyed in the early decades of this century.[104] Dams on the Bruneau and Owyhee Rivers blocked chinook salmon that had once migrated from the ocean all the way to the state of Nevada.[105]

In 1932, when the Oregon Fish Commission made its assessment, the first systematic habitat surveys were still six years away, so the 50 percent loss probably did not include damage from logging, grazing, and mining. Between 1840 and 1930, in the short ninety-year period (twenty-two chinook salmon generations) that followed the first major migration of pioneers into the Northwest, the region's rivers and salmon habitat underwent significant transformation and degradation. The waves of settlers divided the watersheds among themselves according to their economic interests. The fur trappers took the beavers; the miners took the gold and gravel; the loggers took the trees and the riverbanks; the ranchers took the grasslands and riparian zones; the irrigators took the water; and hydroelectric dams took the river's energy and vitality. Each group took their piece of the ecosystem with little or no regulation by the government and with little or no concern for the costs imposed on others. In the process, they pulled apart, fragmented, and destroyed the salmon's home.

To fully appreciate the consequences of this history of habitat degradation, it's necessary to return to the salmon's evolutionary legacy, first touched on near the end of Chapter 1. As described in that chapter, the landscape of the Pacific Northwest is incredibly diverse, comprising mountains, deserts, flat valleys, coastal foothills, scablands, and lava flows. Rivers of all sizes wind their way through this complex mosaic. The rivers, through their natural, seasonal flow patterns, are continuously rearranging the river channels, rebuilding and maintaining the basic structure of salmon habitat. The interaction between land and water produces a diverse array of habitats, reaching from the lowlands to headwaters in a continuum of connected places where the salmon can live. Superimposed over this is a patchwork of climates—coastal, inland, rain shadows on the eastern side of mountains, wet areas on the western slopes. Within this crazy quilt are thousands of microhabitats, each with its own challenges to the salmon's survival.

Viewed from the salmon's perspective, this patchwork of habitats is a riverscape of salmon-friendly areas, where survival is high, and other areas, where survival is low. The survival peaks and troughs of the riverscape are always changing, much like the ever-changing peaks and troughs of the sea. For example, chinook salmon eggs buried in the clean gravel of a headwater stream may be more likely to survive than eggs deposited in the silt-laden gravel of the lower river. Later in the year, juvenile chinook salmon in the lower river and estuary may grow bigger and be more likely to survive than the fish that remain in the cold headwater streams. On the other hand, in some years the lower river may be too warm for the salmon to survive through the summer. The salmon respond to the survival peaks and troughs in the riverscape through their life histories—the way they live in and use the riverscape through all the stages of their lives. Life-history diversity is the salmon's solution to the problems of survival in this diverse and ever-changing mix of habitats.

In undisturbed rivers, each salmon population is composed of a bundle of several life histories, or several alternative survival strategies. Unlike the salmon raised in a hatchery environment with its feedlot regime, the salmon in a natural population in a healthy river do not all do the same thing in the same place at the same time. They follow different pathways through the time and space dimensions of their habitat. As the riverscape changes due to natural disturbances (fires, floods, droughts, and so on), some of the salmon's life histories are in survival peaks, while others drop into troughs. This diverse array of life histories diminishes the risk of catastrophic mortality and the loss of an entire population in a naturally changing environment. Historically, even though habitat changed in response to natural disturbances and climate fluctuations, the salmon's multiple survival strategies allowed them to remain productive and a reliable base for the Indians' economy—a system that sustained itself for several thousand years.

Then white settlers began rearranging the rivers of the Northwest, and one by one those life-history pathways became death traps for fish trying to follow them. Irrigation dried up the lower reaches of rivers, cutting off salmon migration through the summer months. Grazing eliminated riparian vegetation, raising stream temperatures and reducing rearing areas in streams. Logging destroyed the natural habitat structure of forested streams. Hatcheries stripped away diversity while molding the salmon to a uniform sameness that fit the facilities' factorylike operations. Dams eliminated spawning and rearing areas and migration corridors. In many rivers of the Northwest, salmon spawning and juvenile rearing is confined to small areas in the headwaters. Migration is restricted to a narrow window in spring. Life-history diversity, the complex array of pathways through the riverscape, has been severely diminished. The effect of this loss of diversity becomes most evident during periods of natural stress, such as prolonged drought or changing ocean productivities.

Today fisheries managers produce giant schools of domesticated salmon that are programmed to migrate to sea and return to spawn at the same time. These herds of uniform salmon released from hatcheries have been spawned by the vision of a controlled river and ecosystem. They are a manifestation of our attempt to bend the salmon to our worldview. The recovery of the Pacific salmon will be thwarted until at least some of the natural pathways through the riverscape are restored, until we give life to the ghosts of those salmon life histories that were once present in healthy rivers.

Chapter 5 Free Wealth

The first wave of Euro-Americans came to the Pacific Northwest with a narrow and specific purpose, and that was fur. But while furs were their primary objective, necessity quickly introduced them to the salmon, and it was never long before a trapper ate his first meal of the silver fish. In the early years, before farming at the larger posts added diversity to the traders' diets, many of the lonely men collecting furs in the isolated outposts of the Hudson's Bay Company subsisted solely on a monotonous diet of dried salmon. Most of them did not look upon the silver fish that filled the rivers as a gift, nor did they regard them as an important commercial commodity. After a winter of little to eat but dried salmon, the trappers considered them a curse. When consumed in large quantities over extended periods, the salmon's rich flesh was a powerful laxative. Thomas Dears, a clerk in an outpost on the Thompson River, complained about this situation in a letter to a friend: "You Know that I am a slender person what would you say if you saw my emaciated Body now I am every morning when dressing in danger of Slipping through my Breeks and falling into my Boots. many a night I go to bed hungry and craving something better than this horrid dried Salmon we are obliged to live upon. it is quite Medicinal this very morning one of my men in attending on the calls of nature evacuated to the distance of six feet. this is a reall fact, and is almost incredible and often are we troubled this way."[1]

Euro-Americans bought salmon from the Indians mostly for their personal consumption, and the amount was generally insignificant.[2] At some specific sites, however, the traders consumed large numbers of the fish. For example, fur traders at Fort Okanagan ate 18,411 dried and 739 fresh salmon in 1826. At another post on Stuart Lake, the residents consumed 12,000 salmon. Such heavy reliance on salmon at the early outposts meant that in years when the runs were poor, trappers had to suffer through winters of near starvation.[3]

As intensive trapping rapidly depleted the beaver populations, a few traders, including John McLoughlin and George Simpson of the Hudson's Bay Company, began to look beyond the fur trade and envision a regional economy based on the export of salmon and agricultural products. McLoughlin and Simpson believed salmon were potentially more important than farming for the long-term economic success of their company in the Pacific Northwest.[4]

Other visitors to the region also recognized the commercial possibilities of the massive salmon runs. Captain John Dominis of Boston sailed the brig *Owyhee* into the Columbia River in 1829 and became the first American to cure and ship Pacific salmon to the East Coast. The preparation of salted salmon was simple, if not appetizing. Heads were removed and the salmon split lengthwise. The fillets were cleaned and stacked in a large cask filled with a solution of salt and water strong enough to float a raw egg. When a brown scum floated to the top, the brine was poured off and replaced with fresh solution. This process was repeated several times; then the salmon and fresh brine were sealed in the cask.[5] The salted salmon would usually remain preserved for several weeks, long enough for the ocean voyage to its destination.

Dominis salted and packed his salmon into fifty-three empty rum hogsheads, which sold on the Boston wholesale market for $742. Despite his initial success, Captain Dominis did not return for more salmon in part because of difficulties he had with British officials of the Hudson's Bay Company. Moreover, he was irritated by the $106 import tax assessed by the U.S. government because it considered the lower Columbia River a foreign country. But Dominis most likely abandoned the salmon business because his pickled fish tasted so bad that it retailed rather poorly.[6] The bad taste of the processed fish was a fundamental stumbling block for the early salmon-export business.

Nevertheless, another Bostonian, Captain Nathaniel Wyeth, came to the Columbia River to trade for furs and to export salted salmon. While he traveled overland, Wyeth sent his supply of barrels and salt on the ship *Sultana*. When he arrived at Fort Vancouver in 1834, he learned that the *Sultana* had been lost at sea, so he sent for more supplies. In the meantime, he built a fort on Sauvie Island and started a commercial salmon fishery near the confluence of the Willamette and Columbia Rivers. Because Indians were the best fishermen, Wyeth planned to hire them to catch the salmon, which he would then salt and sell in Hawaii and Boston. But the Hudson's Bay Company had other ideas. Accustomed to their monopoly on trade in the Northwest, the company had no intention of surrendering the salmon fishery to an American. McLoughlin raised the price he paid the Indians for their salmon until Wyeth decided he could no longer compete. After a year, he closed his operation and left the Columbia with only a partial load of salted fish.[7]

The Hudson's Bay Company marketed the salmon through its network of company stores. In London, its subsidiary firm, Pelly, Simpson and Company, helped develop markets for the salted fish in Europe, but the volume of exported salmon remained small. Because salt and barrels were often in short supply and because the cured salmon product was of poor quality, the Hudson's Bay Company could not expand its fish export business to make full use of the huge numbers of fish that returned to the rivers each year. Eventually, poor quality reduced the market, and the company's export of salted salmon to England failed.[8] Yet even with these problems, ships leaving the Columbia customarily carried a few barrels of salted salmon as part of their cargo. When the British and Americans signed the Treaty of 1846, setting the international boundary at 49° latitude, the Hudson's Bay Company withdrew from the lower Columbia River, leaving the fishery to the Indians and the American settlers.

When emigrants poured into the Northwest in the 1840s and 1850s, headed for the fertile Willamette Valley of Oregon or the California goldfields, American entrepreneurs once again tried to establish a commercial salmon fishery. In the 1850s, sixty boats fished the Sacramento River between Sutter's Fort and Suisun Bay.[9] By 1853, salted salmon harvested from the Sacramento River had made its way through San Francisco markets to Australia. But labor unrest among the fishermen, the urge to prospect for gold, and shortages of salt all hampered the early California salmon fishery.[10]

American entrepreneurs also tried again to establish a commercial salmon fishery farther north on the Columbia River. Local residents caught the salmon in gillnets, then salted and shipped them to Hawaii or the East Coast. In 1861, H. N. Rice and Jotham Reed packed 600 barrels, each containing about 200 pounds of salted salmon, and sold them for $12 apiece. Four years later, they packed and sold 2,000 barrels of salmon, but growing competition had forced the price down to $6 a barrel.[11] By the 1860s, a commercial salmon industry was poised to emerge on the Columbia River, but the unpredictable and often poor quality of the salted salmon product kept the profits and the markets small. Each year, free wealth poured into the rivers of the Pacific Northwest as the salmon returned from their ocean feeding grounds to spawn. To capture this wealth, fishermen had to figure out how to convert the salmon into a marketable commodity.

Birth and Growth of the Salmon-Canning Industry

The early white fishermen had to confront the same problem faced by the Indians 3,000 years earlier. The salmon were easy to capture in large numbers for the few weeks or months each year when they migrated up the rivers to spawn, but the harvested salmon spoiled within days if left untreated. Salted or smoked salmon lacked consistent quality, and the poor flavor did not generate a market large enough to consume the available supply. Entrepreneurs needed a way to preserve the salmon so it remained tasty even after long periods of storage. The early commercial salmon fishermen were not aware that the solution to this key problem already existed. In fact, it had been around for about a half-century.

In 1809, French biochemist Nicholas Appert entered a contest to devise a way to preserve wholesome food for Napoleon Bonaparte's scurvy-ridden army. He won the 12,000-franc prize for his relatively simple but effective idea. He put fresh or cooked meat, fish, or vegetables in bottles, sealed them, and then heated them in boiling water. Food treated in this way retained its flavor and could be stored for several months. Although originally intended for Napoleon's army, canning eventually revolutionized the entire food-processing industry.[12]

Nicholas Appert supplied the idea, but William Hume supplied the energy, imagination, and courage to bring it to the Northwest and launch the salmon-canning industry. Hume was among the thousands of Americans who migrated across the continent in 1849 to look for gold in the streams and mountains of California, but his background and his purpose were different from that of his fellow forty-niners. Before leaving Augusta, Maine, Hume had fished for the Atlantic salmon. Unlike the other immigrants, he did not intend to look for gold in the gravel of California's streams. He came with a homemade gillnet to harvest the silver that swam up the Sacramento River. After several seasons of fishing and selling salmon on the fresh-fish market, Hume returned to Maine in 1856 and told his brothers stories about the millions of salmon in the rivers of the West Coast and how easy it was to harvest them. Brothers George and John were convinced and returned with William to California. It was during this second trip to California that the Humes saw the connection between Nicholas Appert's invention and the Pacific salmon.

Because each can was constructed and sealed by hand, tinsmiths were an essential part of the early canning industry. The Humes knew how to fish and catch salmon, but they lacked this key skill needed to launch a salmon-canning enterprise. Fortunately, one of George Hume's classmates worked as a tinsmith, and he actually had some experience canning lobster and Atlantic salmon. George returned to Maine in 1863 and persuaded his friend, Andrew Hapgood, to meet the Humes in California the next spring. The youngest Hume brother, Robert Deniston, also joined the company as a "sub" at a small salary. As a boy, he had watched two of Augusta's wealthiest men argue in public over how to divide a salmon from the Kennebec River; salmon had become so scarce that only the wealthy could afford them, and even then there weren't enough.[13] Now he was anxious to see for himself the millions of salmon that crowded the western rivers. The next spring, on a barge anchored across the river from K Street in Sacramento, the firm of Hapgood, Hume and Company began packing chinook salmon into Andrew Hapgood's handmade cans.[14]

Thirty-nine years later, R. D. Hume, by then a major figure in the salmon-canning industry, reminisced about that first year. They canned 2,000 cases of salmon, but the difficulties were so great that he wondered why they hadn't given up. All of the cans had to be soldered by hand with Hapgood's crude equipment, a process that routinely produced defective seals, and there was no way to check for leaks. Several cans exploded in the cooker, and about half of the salmon they packed were spoiled before they left the barge because of leaky cans. Those obstacles were big, but they also had the bigger problem of convincing merchants in San Francisco to buy their new product. By the end of the first season, all these problems brought Hapgood and the Humes to the point of despair, such that "a few hundred dollars would have purchased all their interests in the business." Just when they were about to give up, a San Francisco businessman paid the shipping charges and found a market for the canned salmon at $5 a case. The industry survived.[15]

Their euphoria did not last long. Although Hapgood believed he could solve the problem of leaky cans, there was one essential factor that the Hume brothers could not control: the supply of salmon. And by 1864, the supply of salmon in the Sacramento River was already dwindling fast. Irrigation, railroad construction, and hydraulic mining were rapidly degrading the salmon's habitat and limiting the supply of fish.

Despite a disappointing run in 1865, Hapgood, Hume and Company managed to can another 2,000 cases of chinook salmon on their makeshift barge, but the Humes understood the consequences of the river's deteriorating habitat. Three generations of Humes had fished for Atlantic salmon on the Kennebec River in Maine, and during that time they had witnessed the collapse of the fishery from pollution and impassable dams. To the Humes, the grim situation on the Sacramento River must have seemed all too familiar. But the brothers remained optimistic. After all, this was the West. There was room to move, and there were always other rivers to fish. In 1866, George and William Hume went north to the Columbia River, where millions of salmon still returned to spawn.

They selected a site at Eagle Cliff, about 40 miles from the sea, on the Washington side of the river, built a cannery, and moved all their equipment from California. In the first year they produced 4,000 cases, each containing forty-eight one-pound cans.[16] To appreciate the magnitude of the change that new canning technology brought to the salmon fishery, consider the rapidly changing value of the catch. In 1860 the entire salmon fishery of Oregon was worth $13,450.[17] In 1866 the value of the Humes' 4,000 cases of canned salmon was $32,000, and fourteen years later the value of the salmon from the Columbia River had increased a hundredfold, to $3,257,790.[18] The Humes had finally struck it rich, and their discovery started another freewheeling rush of fortune seekers—this time to mine the salmon from the rivers.

Entrepreneurs who looked at the Northwest's pristine rivers, which each year attracted millions of prime Pacific salmon, literally saw free wealth swim from the ocean right into their nets. All they had to do was convert the salmon into a commodity and ultimately into cash. But to do that, they still had to solve four basic problems. They needed a technique to make the salmon a tastier product with a long storage life. They had to develop markets larger than those originally established by the Hudson's Bay Company. They had to streamline fish processing with efficient technology. And finally, they needed capital to develop and buy the canning machinery.

The Humes' pioneering work solved the first problem. Salmon from their crude, handmade cans tasted far better than salted or dried fish. Markets were well on their way to expanding, especially in the industrialized cities of the eastern United States and England, where growing numbers of factory workers created a strong demand for cheap, wholesome food. In England, for example, factory workers bought canned salmon for half the price of fresh meat. Canned salmon became so common in England that the British army made it part of its standard ration.[19] The efficient technology needed to mechanize salmon canning was already being developed on the East Coast and in Europe, where the mass processing of fish, meats, fruits, and vegetables was well under way.[20] Still, the West needed capital.

Fortunately for the cannery entrepreneurs, the United States was riding a wave of economic growth spurred by the Industrial Revolution. As French political scientist Émile Boutmy saw it, the most striking feature of America at the time was not its heralded democracy but its economy. Boutmy described the country as a huge commercial company with the purpose of discovering, cultivating, and capitalizing the large expanse of its territory.[21] America was in the mood to invest in the exploitation of its natural resources. Capital had a way of finding potential profit, even when it was but a mere spark of an idea kept flickering in the western wilderness by four struggling fishermen from Maine. The Humes easily found investors with money that would give them more access to the free wealth that swam up the Northwest's rivers.

With the infusion of capital, the salmon-canning industry entered a period of rapid growth. Within a few decades, the entire region was engaged. When the Humes moved to Eagle Cliff on the lower Columbia, they took the only cannery on the Sacramento River with them. In 1874, however, an increase in the salmon runs stimulated new investment in canneries. The pack on the Sacramento River reached its peak in 1882, when nineteen canneries produced 200,000 cases of salmon.

After their first successful year of canning Columbia River chinook salmon, the Hume brothers and Hapgood obtained financial backing, and each man built his own cannery. The huge profits and wide-open fishery attracted many others to the new industry. By 1883, thirty-nine canneries operated on the Columbia River. That same year, the harvest of the prime spring and summer runs of chinook salmon peaked. When the spring chinook run started to show signs of depletion, cannery operators began to process what they considered the inferior fall-run chinook. Eventually other species such as sockeye, coho, and chum salmon also supplied the growing number of canneries with fish (Figure 5.1).

Canneries also spread to other rivers throughout the Northwest. For example, in 1876 the youngest Hume, Robert, sold his cannery on the Columbia and moved south to the Rogue River. That same year, the firm of

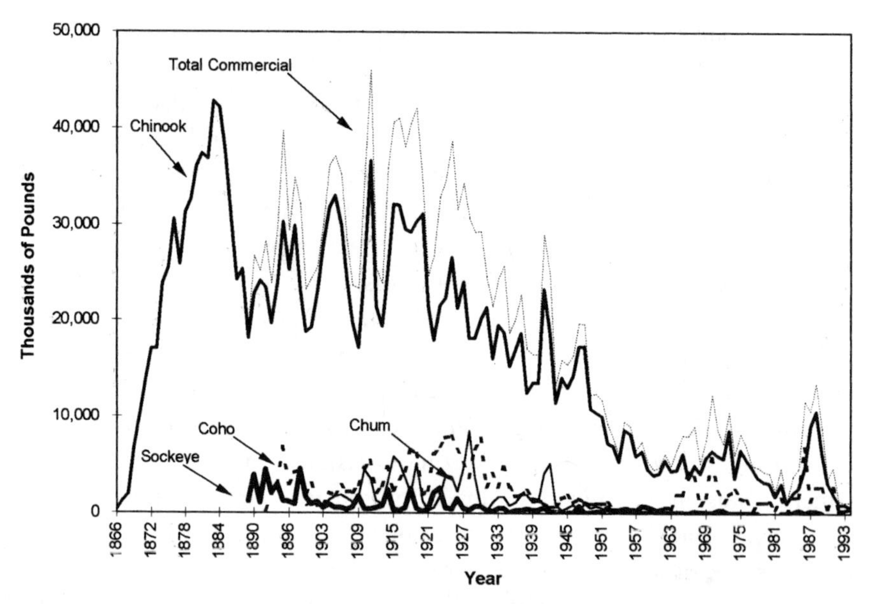

Figure 5.1. Catch of chinook, coho, chum, and sockeye salmon in the Columbia River, 1866–1994. (Sources: Beiningen 1976, Oregon Department of Fish and Wildlife and Washington Department of Fish and Wildlife 1995)

Jackson, Myers and Company headed north to Puget Sound, where they built a cannery north of Seattle at Mukilteo.[22] Like a giant octopus with its body setting on the Columbia River, the canning industry extended its tentacles to large and small rivers in the most remote corners of the Pacific Northwest. By the late 1880s, most of the coastal streams of Oregon, Washington, and California with salmon populations large enough to support a cannery were being mined (Table 5.1). Even in the smaller streams, salmon were harvested and shipped to the nearest cannery.

At the same time, the canning industry was expanding in Canada as well. James Syme canned a few cases of salmon from the Fraser River on an experimental basis as early as 1867. He exported a dozen cans to England, then closed his operation. More small canneries appeared again in 1870 and remained in continuous operation. As with the growth of the industry in America, once canneries secured a foothold on the Fraser, they spread rapidly to other rivers. By 1900 there were seventy-eight canneries operating in British Columbia.[23]

On the Columbia River, the shift to harvesting and canning inferior species only postponed the inevitable. The combined cannery pack of all salmon

Table 5.1. Date of construction of first cannery in selected areas in Oregon, Washington, and California

Location	*First Cannery*
CALIFORNIA	
Sacramento River	1864
Smith River	1878
Klamath River	1888
Noyo River	1917
OREGON	
Columbia River	1866
Rogue River	1877
Siuslaw River	1878
Umpqua River	1878
Coquille River	1883
Tillamook Bay	1886
Alsea River	1886
Nehalem River	1887
Nestucca River	1887
Coos River	1887
Yaquina River	1887
Siletz River	1896
WASHINGTON	
Puget Sound	1877
Grays Harbor	1883
Willapa Bay	1884
Queets River	1905
Quinault River	1911
Soleduck River	1912
Hoh River	1917

Source: Cobb 1930.

species in the Columbia River peaked in 1895 at 635,000 cases. By then, the salmon were under assault not only from the intensive fishery but also from the rapid deterioration of their habitat in the vast network of the Columbia's tributary streams. Even though the fish were still abundant, it was becoming clear to many observers that the canning industry would not survive unless the states took action to protect the salmon.[24] The Columbia River retained its place at the top of the industry for only a short time. In 1897, British Columbia's pack of salmon exceeded one million cases. Then Puget Sound canneries packed nearly one million cases in 1899. That same year, Alaska's canneries exceeded one million cases for the first time.

Table 5.2. Peak cannery packs of all species of Pacific salmon at selected locations south of the Fraser River

Area	*Peak Year*	*Cases*
Sacramento River	1882	200,000
Columbia River	1895	634,696
Willapa Bay	1902	39,492
Coastal Oregon rivers	1911	138,146
Grays Harbor	1911	75,941
Klamath River	1912	18,000
Puget Sound	1913	2,583,463
Coastal Washington rivers	1915	31,735

Source: Cobb 1930.

Within thirty years, the salmon fishery that came to life on the Sacramento River and grew to a full-fledged industry on the Columbia yielded supremacy to the rich waters of Alaska and British Columbia, where its center has remained ever since. By 1915, the salmon harvests in all of the streams south of the Fraser River had reached and passed their peaks (Table 5.2). There would never be more than thirty-nine canneries on the Columbia, and after 1887 the number of operating canneries began a slow, steady decline.[25] The last cannery on the Sacramento River closed in 1919 after packing only 3,125 cases of fish. In less than fifty years (twelve chinook salmon generations), nature's warehouse had been looted; its free wealth—the natural productivity of the Pacific salmon—had been devoured in the insatiable cannery lines. The salmon-canning industry south of British Columbia had already sown the seeds of its eventual demise.

The Salmon Canners

On a blustery March day in 1916 in Seattle's Volunteer Park, sixteen men and one woman posed for a picture in front of a statue of William Seward, the man responsible for the purchase of Alaska and its rich salmon rivers. Looking at the photograph more than eighty years later, it's clear that the men dressed in dark suits, overcoats, and fedoras were leaders—the cream of American capital and its industrial economy. In the middle of the gathering stood a large stack of cans that would be sent to President Wilson and his Cabinet on the occasion of the "Fourth Annual Canned Salmon Day." The stack of canned salmon symbolized what men who possessed the virtues of "thrift and industry" could do when given unrestrained access to the free wealth of the western wilderness.[26]

The dark-suited men represented major railroads, food processing, and fishing corporations. Under their influence, the return of the salmon was no longer a local event celebrated by an Indian tribe or an isolated community of Finnish or Norwegian fishermen. Canned Salmon Day replaced the Indians' First Salmon Ceremony and, consistent with the expanding markets, it was celebrated in distant and unlikely places such as Minneapolis, Cincinnati, Los Angeles, Boston, Denver, and New York. That March day in 1916, the canning industry was poised, waiting for the first salmon of the spring run to return to Northwest rivers. Fishermen, Chinese butchers, can packers, railroad engineers, and bankers were waiting, anticipating, and wondering whether this would be a good or bad year, in much the same way that native peoples had waited, anticipated, and wondered for thousands of years.

The gillnetter, canner, railroad executive, and banker were not going to treat the salmon as a gift, however, nor would they give the first or the last fish of the season any respect. They would not sing songs of supplication to the salmon, but they did take part in a ceremony and sang a song of another kind. Miss Joy Jenott, the lone woman in the picture, jumped out of a large red, white, and blue replica of a salmon can and sang "The Tale of the Salmon," which included lyrics such as "You better watch out, Mr. Salmon, or the can'ry man will get you."[27] In a speech to the assembled crowd, Judge F. V. Brown of the Great Northern Railroad summarized how the industrial economy viewed the salmon: canned salmon were "highly esteemed by the railroad," but his colleagues had no use for live salmon because "they furnished their own transportation."[28] Salmon were no longer citizens in the same community with humans. Instead they were reduced to a commodity that had value only when put in a can and converted to cash.

Not only did cannery entrepreneurs see salmon as commodities, but with their gambling spirits, they also regarded the fish as mere chits in what historian Diane Newell has called "a grown man's game."[29] In the early salmon-canning industry, grown men tried to outmaneuver each other in a highly speculative, unregulated game where fortunes were made and lost.

After R. D. Hume moved his cannery to the Rogue River in 1876, he took steps to protect his investment from the kind of wide-open fishery he'd left behind on the Columbia River. He purchased all the tidelands in the estuary and quickly established a monopoly over all the salmon fishing on the Rogue. But Hume was not satisfied to merely consolidate his economic and political power on the southern Oregon coast. He couldn't resist playing the salmon canner's game, and so he decided to challenge one of the indus-

Salmon fishing on the Rogue River near Grants Pass, Oregon. (Photo courtesy Oregon Historical Society, OrHi 4052)

try's giants, the Alaska Packers Association (APA). Formed in 1893, the APA was created by the merger of several independent canners to control prices and to consolidate production into fewer plants with lower overheads. The APA was doing the same thing Hume was doing on the Rogue River, but on a larger scale. The association controlled 72 percent of all salmon packed in Alaska.[30]

In 1893, the litigious Hume sued the APA over the use of a brand label. Although the matter was formally settled out of court, it apparently remained unresolved, so that same year Hume sent a canning ship, the *Ferris S. Thompson,* to Alaska to challenge the APA's monopoly. His Alaskan venture failed, but the APA retaliated by sending its steamer *Thistle* to the Rogue to attack Hume's monopoly. By selling trade goods, renting nets to fishermen, and salting salmon, the *Thistle*'s captain proclaimed that the Rogue was open to anyone who wanted to fish.

On October 19, 1893, several months after the APA arrived on the Rogue, a late-night fire totally destroyed Hume's cannery. The *Gold Beach Gazette,* Hume's newspaper, blamed the APA, but others suspected that

Hume had set the fire himself to collect insurance. The fight spread from the rivers of Alaska and Oregon to the courts and even to the U.S. Congress. The APA easily kept its pressure on the Rogue fishery, and Hume, failing to win in the courts, had to borrow money to maintain the *Ferris S. Thompson* in Alaskan waters. Finally, on March 20, 1895, the two antagonists suddenly settled their dispute, and each pulled out of the other's territory.[31]

The APA couldn't possibly have had a financial interest in the salmon from a small river in southern Oregon. It was a game, a chess match on a regional scale, in which men tried to "cork" each other both in boardrooms and on wild rivers.[32] Underlying the game, holding up all the industry superstructure, were the salmon—abundant, free for the taking, and seemingly inexhaustible, able to absorb man's abuses and still return in the millions to their home rivers each year. Though these dramas were commonly called fish wars, the fights were not really over fish. They were over power and the control of resources—any resource. When the salmon were gone, irrigation water, hydropower, public grasslands, or trees would take their place. The games men played while collecting free wealth from the Northwest didn't require the salmon. In fact, once it became clear that protecting salmon habitat was a hindrance to the industrial economy, the fish and the canning industry were quickly relegated to a low priority. But that part of the story was still forty years in the future.

The cannery operators took little time to think in philosophical terms about the salmon. Whereas the Indians' world focused on the seasonal cycles of the earth and the rhythm of the salmon's life history, the canners' world focused on the machinations of the markets. On a daily basis, the canners' chief concern was the difference between what it cost to put the salmon in a can and the price they received for it on the market. To maximize that difference, the production lines had to operate at maximum efficiency, especially when a large number of companies competed for market share.

At the heart of each cannery's operation was the absolute need to maintain a continuous supply of fish. Because fish were cheap, it was better to keep a surplus supply on hand than to risk running out and idling 200 to 400 workers on the line. Fishermen were willing to oblige; they caught every fish they could in order to earn enough income over the short migration period to support themselves for the entire year. The result was a huge waste of salmon. Surplus salmon were dumped on the cannery floor, where many spoiled before being shoveled back into the river. Once the canneries met their limits, fishermen who couldn't sell their catch merely dumped the fish overboard. In the early years of the fishery, only the prime spring and summer chinook salmon on the Columbia River and the sockeye salmon on the Fraser River were processed in the canneries. All other species of salmon that were caught were discarded.[33]

Mont Hawthorne was a young man in 1883 when he arrived in Astoria, Oregon, to learn the canning business. He landed in Astoria in the middle of one of the largest salmon runs in the history of the Columbia River. Salmon were everywhere, so his first task was simply to learn how to cope with the odor coming from the large piles of rotting salmon and cannery scraps that littered the riverbank. Hawthorne got so riled up at the tremendous waste of fish that he wanted to do something about it. In later years, he recalled a discussion with a government man who had come to Astoria to build hatcheries. The government man thought it was a shame to waste the fish, but there was nothing he could do because "he was sent out to preserve the salmon by building hatcheries." Hawthorne remembered throwing about 500 large salmon back to the river every other night.[34] What he witnessed was typical of the canneries operating throughout the Northwest.

Waste was pervasive and terrific. Some of the canneries operating on the Fraser River packed only salmon bellies and threw away the backs. During the years of the largest dominant runs of sockeye salmon, Fraser River canneries packed 25 to 35 million fish out of a total catch ranging from 35 to 50 million fish. The difference was largely waste.[35]

With the advent of the industrial fishery, fishermen and cannerymen attacked the huge runs of salmon like sharks on a feeding frenzy. Thousands, possibly millions, of salmon were wasted by overzealous fishermen who continued to fish after the canneries stopped buying, knowing full well they would have to throw away their catch. Abandoned traps continued to capture the salmon, filling up with dead fish until the webbing rotted away.[36] Cannerymen caught up in the frenzy sometimes processed fish that had lain on the loading docks for several days, even though the fish had become infested with maggots. Congress did not enact federal regulations to help ensure the quality of canned salmon until 1906.

To the Northwest Indians, the salmon was a gift that had to be treated with respect. But to the white men and their industrialized economy, the fish were inexhaustible raw materials to be fed into their machines. The two attitudes toward the salmon could not have been more different. Although a few, like R. D. Hume, spoke out against waste and for greater conservation, most of the men bent on profit rarely thought about the salmon at all—at least not until declining catches began to threaten profits.

The Fishermen

Although they built their warehouses and cannery lines at the water's edge, the cannerymen were detached from the rivers that flowed past their doors. Their interests connected them more closely to eastern cities with large markets and financial institutions—the source of profits and capital. The cannerymen's lives and livelihoods rose and fell on decisions made by men in New York, London, and Chicago—men who had little knowledge of the Northwest, its rivers, its people, or the salmon their investments were exploiting.

But the fishermen were different. Although their economic fate was also ultimately tied to the eastern industrialists, the fishermen also had strong ties to the river—the river whose salmon could feed them and their families, the river whose angry moods could kill them. The fisherman, river, salmon, boat, and net merged into a way of life. As Sir Walter Scott said, "It's no fish ye're buying—it's men's lives."[37]

Nearly all the early fishermen used heavy linen gillnets about 250 to 1,500 feet long and fished them at night, when the meshes of the net were invisible to the salmon. Every evening, while the salmon were running, they maneuvered a fleet of small boats with distinctive "butterfly" sails highlighted in the setting sun. The fleet sailed into "fish dark"—the time when the salmon began to move as the river changed through a spectrum of colors from bright blue to gold and finally black.[38] Once out on the river, the fishermen battled wind and rain and cursed the large tree trunks, buried in the river bottom, that snagged and tore nets. Fishermen knew firsthand that the river was moody. On one day it could destroy them in a violent rage, while on the next it could be so peaceful that it seemed to caress the boat. The river was an extension of their lives, a living thing that they had to know intimately to survive. Those who didn't paid a high price. Each year during the 1880s, the Columbia River killed twenty to sixty fishermen. On a single day following the end of a fishing strike, when they rushed out to the river right into the jaws of a storm, twenty fishermen died. Coastwide, as many as 250 fishermen lost their lives each year.[39]

The fishermen belonged to what historian Carlos Schwantes has called "the wage worker's frontier." Because the sparsely populated Northwest had no local markets of any size for the volume of commodities it produced, timber, fish, and minerals had to be shipped east. Fishermen, miners, and loggers were always at the mercy of those who controlled the markets, kept wages low, and kept workers insecure and just above poverty.[40] Although

Salmon-fishing fleet (butterfly sails), Astoria, Oregon, circa 1911. (Photo courtesy Oregon Historical Society, OrHi 12436)

salmon were abundant in the early years of the fishery, fishermen earned little for their labor. According to A. L. Flesher, who fished Yaquina Bay on the central Oregon coast, "You couldn't make anything at 2 and 3 cents a pound. There was no way. You would have had to catch tons and tons in order to make anything. . . . We had a whole acre of garden and that's the way we survived."[41]

It was a hard life, but there were times, as Edward Beard remembered, when the fishermen and the river seemed to be in harmony. Born in 1895 in Young's Bay on the lower Columbia River, Beard spent his working life in the salmon industry. "There was no more beautiful sight," he recalled, "than to see a fleet of 3 or 4 hundred gillnetters spread their sails and sail out of Astoria on a Sunday evening near sunset to seek what they considered a favorable spot to cast their nets and hope for good fishing."[42]

Salmon fishermen in the late 1800s were enigmatic, however, and so are difficult to characterize with generalities. Fishing and cannery work provided seasonal employment to immigrants from Europe and China, and many were unruly drifters who drank and fought their way through the salmon runs. In 1880, about 90 percent of the labor force employed in the Oregon fishing industry were foreign born, and about half of those were Chinese cannery workers.[43] In time, transient fishermen were replaced by permanent residents who fished the salmon runs and worked in logging camps during the off-season. This later group formed close-knit, industrious communities of largely Scandinavian descent. They summed up the fear, hard work, uncertainty, and joy of the fisherman's life with the saying "Beginning is always difficult, work is our joy, and industry overcomes bad luck."[44]

In the freewheeling, unregulated fishery, competition was intense. In the absence of effective government authority, salmon fishermen fought among themselves over who could fish and what type of gear could be used. In time, a set of unwritten rules established a semblance of order. Ethnic communities often laid claim to specific fisheries or, in the case of the salmon, certain gillnet drifts. For example, Greek fishermen had their own gillnet drift, marked off with flags, in the lower Sacramento River; and they strictly enforced their ban against its use by other ethnic groups.[45] On the Columbia River, Swedes and Finns dominated the prime fishing grounds. Without the aid of government fisheries regulations, the fishermen enforced their own customs, sometimes with fatal results.

Chinese fishermen were totally excluded from the salmon fishery. In a 1887 report on the Sacramento River fishery, Stanford University professors David Starr Jordan and Charles Gilbert noted that all the salmon fishermen were white. According to Jordan and Gilbert, there was no law regulating the nationality of the fishermen, but everyone knew that any attempt by the Chinese to fish for salmon would meet "summary and probably fatal retaliation."[46] Chinese fishermen in northern California harvested shrimp, squid, and abalone, but even in those fisheries, they were harassed and eventually excluded by statute. George Hume brought Chinese laborers to the Columbia River in 1872, but their participation was restricted to cannery work because unwritten rules did not allow them to fish. In 1880, a group of Chinese men rebelled and attempted to fish for salmon on the Columbia River. Pieces of their boats and nets were found on the beach, and local fishermen did not hide the fact that they were drowned.[47]

The Chinese were not the only ethnic group excluded from fishing. Anyone from outside the group controlling a local fishery or a specific location could be considered a foreigner, and that was enough to prevent his participation in the fishery. Foreigners were often characterized as greedy men who opposed conservation and thus had to be stopped to save the fishery.

Xenophobia worked against Chinese fishermen on San Francisco Bay and on the Columbia River. It was also used against the Austrians who tried to catch salmon with purse seines on Puget Sound and the lower Columbia

River in 1914. The fact that World War I had made the United States and Austria enemies added to the list of grievances against them. The native peoples, to whom everyone else was foreign, also found themselves on the outside of the new fishing communities. Although they had fished the salmon for thousands of years, they were now treated as if they were the aliens and blamed for the depletion of the salmon runs.[48]

Indian Fishermen Displaced

Among the diverse array of natural resources that made up the Indians' economy, the salmon were unique because they had commercial possibilities in the white man's economy. The salmon fishery might have been the key to the Indians' entry into the market economy, but that didn't happen, at least not during the early decades of the salmon-canning industry.

As the first white settlers entered the Northwest, the Indians continued to fish in their usual and accustomed ways, but they also began to harvest a surplus—more fish than they needed for subsistence—which they traded for commodities such as coffee, molasses, and iron implements. Because the native peoples possessed superior knowledge about the habits of salmon as well as harvest technology specifically adapted to catch salmon, they could participate on the edges of the market economy and still retain most of their culture. But once the canneries became operational, profits and the demand for fish increased dramatically, which in turn attracted to the fishery more Euro-Americans, who gradually displaced the Indians. A few Indians worked in the canneries, but in the end, with few exceptions, they lost their fisheries.

Many of the treaties between Native Americans and the U.S. government gave the Indians the equivalent of a property right to half the salmon. In spite of those treaties, however, Indians were continuously forced out of the fisheries throughout the Northwest. Euro-Americans displaced more than the fishing villages and the fishermen, though: they replaced a worldview and an economy that had persisted for several thousand years with a worldview that emphasized economic efficiency, short-term profits, and the massive use of technology. In the late nineteenth and early twentieth centuries, these two ways of thinking, and the economies derived from them, were not compatible. Perhaps the most important reason for the incompatibility was this: The unregulated industrial economy used the entire watershed and, as a result, destroyed the ecosystems that were the basis for the salmon's abundance. It destroyed the natural underpinnings of the native people's gift economy.

Rivers were a focal point in the culture and economy of Northwest Indians; rivers were the source of life. Before the Euro-Americans arrived, the river was a great table spread with gifts of food from the salmon chief—gifts that everyone shared.[49] The industrial economy denied the Indians a place at this table and then divided it into independent economic spheres. Irrigators, shippers, loggers, cattle and sheep ranchers, cannerymen, and fishermen all claimed not their place at the table, but their *piece* of the table. Rivers were broken into a mosaic of property owned by individuals, each pursuing his or her own legitimate economic interests without regard for the others and without interference by the government. Only later, after the industrial economy had destroyed the ecological balance of the rivers and conflict over diminishing resources had heightened, would the government try belatedly to mend the fragmented ecosystem.

The Indians' right to fish was guaranteed by treaties, but their property rights to the fish conflicted with the property rights of the Euro-Americans who now owned the land through which the rivers flowed. When the Indians tried to follow their seasonal round of fishing, hunting, and gathering, they were confronted with fences and KEEP OUT signs. Although the courts, especially the higher courts, upheld the Indians' right to fish as well as their access to the fishery, the lower courts and local government agencies made it clear that the native fishermen could harvest salmon only according to the terms dictated by the new culture. To enter the commercial fishery, Indians had to do more than fish. They had to learn a whole new set of rules based on the values and customs of those in power.

Although the agent for the Makah Indians, Henry Webster, suggested in 1865 that the government help Northwest Indians to maintain their fisheries, the prevailing government policy sought to relocate the Indians to reservations, where they were encouraged to take up agriculture and leave behind their relationship with the salmon.[50] The real objective was to get the Indians out of the way of the white settlers, who believed they could make better and more productive use of the natural resources of the Northwest.

In some areas, like the goldfields of California, Indian fishermen were pushed aside as the newcomers literally tore the yellow metal from the landscape. Where the Indians did not get out of the way fast enough, they were killed. The native peoples of California who depended on the salmon were the most vulnerable. Not only did they have to compete with men like the Hume brothers for fish, but other parts of the new economy, such as mining, logging, and grazing, devastated salmon habitat and further reduced the quantity and quality of the Indians' traditional food.

The Yurok, Hupa, and Karok tribes of the Klamath-Trinity River system were an exception to the general pattern of native displacement in California. Unlike the other tribes in northern California, the lower Klamath tribes did not fully exploit the rivers' chinook salmon prior to contact with the white immigrants. The unharvested surplus acted as a buffer against the ill effects of white settlement on the ecology of the basin. Moreover, the physical isolation and the rugged landscape of the Klamath watershed slowed white encroachment, helping the Indians retain their fishery longer than most other tribes.[51]

The organization and culture of the lower Klamath River tribes also helped the Indians to retain their fishery. The Yurok, Hupa, and Karok tribal cultures emphasized the individual or family group rather than the community or tribe.[52] Their emphasis on the individual was similar to the rugged individualism that characterized the white pioneer. To a large degree, the Klamath Indians were preadapted to the values of the new culture. This trait enabled them to survive white settlement better than neighboring tribes, whose way of life sharply contrasted with that of the invading culture.[53]

The Klamath River Indians started selling salmon to businessmen from Crescent City in 1876. Then John Bornhoff built a cannery at the mouth of the Klamath River in 1886. Because the cannery was on the Klamath reservation, Bornoff first had to secure permission from the Indian agent and then gave most of the fishing and canning jobs to tribal people. Even the army stepped in to evict squatters on contested land and protect the Indians' fishery.

Then, in 1887, R. D. Hume decided to add the Klamath to his Rogue River fishing and canning operation. He was refused permission to lease or buy land but entered the river anyway, brandishing a new Henry rifle and threatening the small army detachment. The soldiers arrested him, but in the trial he won the right to conduct a fishery. By then, the run of chinook salmon on the Klamath had been so depleted by the effects of gold mining that Hume could not make a decent profit. He eventually sold out to Bornhoff and left the river.[54] The Indians retained control of the fishery, at least temporarily.

What happened to the Yakama Indians was more typical, however. Before white settlement in the Yakima River watershed, native people had harvested 160,000 out of the 500,000 salmon that returned to the river each year to spawn. By 1900, the Indians harvested 20,000 salmon out of a 50,000-fish run—a run of only one-tenth its former abundance. The catch continued to decline until the 1930s, when it dropped to between 1,000 and 1,500 fish, less than 1 percent of the old harvest.[55]

The intensive fishery in the lower Columbia River accounted for part of the decline of the salmon run into the Yakima River, but the development of irrigated agriculture also had a large effect. The Yakama Indians also lost access to the river and their former fishing sites; as pioneer settlement of the middle Columbia Basin progressed, the Yakamas found their usual fishing sites on the other side of the white man's fences. By 1888 all the land near The Dalles, where Indians had fished for 10,000 years, had passed into private hands. Barbed wire and locked gates quickly put an end to thousands of years of tradition. Although the rights to fish were guaranteed by treaties, those rights proved to be hollow. When the Yakamas found a place to fish, pioneers accused them of taking salmon from the nurseries and causing the decline of the fishery.[56] To add insult to injury, giant fishwheels were placed in the best traditional dipnetting spots. The fishwheels were essentially automated dipnets. So it's not surprising that white men often watched the Indians dipnet salmon and then built their fish-catching machines in those same spots.[57]

On a summer evening in 1889, as cool breezes from the river began to temper the day's heat, special agent Henry Marchant observed a group of hungry Yakama Indians. From a spot that was probably one of their old fishing sites, the Yakamas watched as white fishermen loaded 38,000 pounds of salmon into railroad cars. By 1919, the Yakama Indians were reduced to accepting discarded salmon carcasses at fish hatcheries.[58] The salmon king still sent his subjects up the river to feed humans, but the Yakamas no longer had a place at the table.

The Yakama Indians were outmaneuvered in their attempts to fish by all the white man's institutions except one, the courts. In a series of rulings before 1919, higher courts declared that the Yakamas had the right to occupy and fish from their usual and accustomed places. But the courts that upheld Yakama fishing rights left open the question of regulating the Indian fishery. For example, in *United States v. Winans* (1905), the U.S. Supreme Court firmly established the Yakamas' right to fish in usual and accustomed places. But on the use of that right, the court gave supremacy to the state as follows: "It was within the competency of the Nation to secure to the Indians such a remnant of the great rights they possessed as 'taking fish at all usual and accustomed places.' Nor does it restrain the State unreasonably, if at all, in the regulation of the right."[59]

The Yakamas' right to actually fish at their usual and accustomed places, unfettered by discriminatory regulations, would continue to be contested until 1969, when Judge Robert Belloni's decisions in *Sohappy v. Smith* and

United States v. Oregon reestablished Indians' right to fish and led to their role as co-managers of the salmon together with state and federal agencies.

Controlling the Harvest

In his State of the Union message for 1908, President Theodore Roosevelt called the nation's attention to several weighty issues, including the financial standing of the nation, the reorganization of several government bureaus, the inadequacy of American public health laws, statehood for New Mexico and Arizona, and the deplorable state of fisheries management in interstate and international waters. He called particular attention to the Columbia River and Puget Sound salmon fisheries. Because the various state legislatures seemed incapable of reaching agreement on regulations that protected both the fish and the fishermen, Roosevelt threatened to federalize management. He believed "problems which in the seesaw of conflicting legislatures are absolutely unsolvable are easy enough for the Congress to solve."[60] Oregon's master fish warden, H. C. McAllister, lamented the possibility of federal control of the salmon fisheries on the Columbia River "because it would deprive this state of thousands of dollars annually, and probably be the death of the industry."[61] But the Portland *Oregonian* took a different view. In an editorial, the paper said it would be better to take management responsibility away from the states than to let the salmon be destroyed.[62]

Roosevelt never did follow through on his threat, but ninety years later, after the salmon had declined to the brink of extinction, the federal government did in effect gain some control over salmon management, following the listing of several stocks as threatened or endangered under the federal Endangered Species Act. State officials are now lamenting the real erosion of their authority over salmon management—authority they clearly failed to use wisely for more than a century.

Theodore Roosevelt was a leading advocate of conservation and the protection of fish and wildlife, but in his State of the Union speech, he was not breaking new ground for the conservation movement. Instead, he voiced concern about a problem that had been recognized for several decades—the nation's diminishing fish and wildlife. In the late nineteenth century, America's rapidly industrializing economy was clearly changing the landscape, destroying habitat, and noticeably affecting the abundance of fish and game.

Visitors to America recognized what was happening to the native flora and fauna. Following his trip to America, the British visitor Jeremy Bentham urgently recommended a complete inventory of American wildlife because "progress of a civilized population, whilst changing generally the face of nature, is obliterating many of the evidences of a former state of things." Bentham noted that the larger wild animals were disappearing and that "myriads of the lower orders of animal life, as well as of plants, disappear with the destruction of forests, the drainage of swamps." In Europe, he wrote, biologists could not reconstruct what the natural ecosystems were like because there was no record of what had been lost, "but in North America the change is going forward as it were close under the eye of the observer."[63]

The growing concern over the status of wildlife populations in America led to the growth of private clubs and organizations devoted to wildlife conservation. The number of such organizations increased dramatically in the late 1800s so that by 1878, a total of 342 organizations devoted to fish culture, field shooting, and/or fishing were advocating for improved fish and wildlife conservation. Four national publications, *American Sportsman, Forest and Stream, Field and Stream,* and *American Angler,* all established between 1871 and 1881, chronicled the sportsman's struggle against the loss of fish and game. In 1875 the *American Sportsman* reminded its readers of America's approaching centennial and asked, "Shall we boast that where the deer, the buffalo, the salmon and feathered game . . . were once plentiful . . . we may now tramp for many a long summer day and not find a specimen!" Pressure from organized sportsmen caused many states to create agencies to protect, manage, and conserve fish and wildlife.[64] By 1882, thirty-one states had established state fish commissions.[65] California's legislature established the Board of Fish Commissioners in 1870; Oregon's State Fish Commission was created in 1878; and the Washington Fish Commission came into existence in 1890.

Even after the states created fish commissions, the legislatures retained the authority to regulate the commercial fisheries. In Oregon, for example, the fish commission focused its efforts on the culture of the commercially important chinook salmon, while the legislature continued to regulate fishing seasons.[66] The Washington legislature did not grant regulatory authority to the state Department of Fisheries until 1921; in Oregon the same authority was not granted until 1941.[67]

The Washington legislature made one of the first attempts to control the salmon fishery in 1859, when it prohibited fishing by nonresidents in parts of the Columbia River. In 1866, the same year the Hume brothers opened their cannery, the Washington state legislature ruled that salmon-fishing gear could not extend more than two-thirds of the way across a stream. Gillnets, the primary fishing gear, remained unregulated. The salmon-canning industry and its intensive fishery were ten years old before either Oregon or Washington enacted the first, modest attempt to regulate the harvest. In 1877 the legislature of the Washington Territory closed the salmon fishery during the months of March, April, August, and September. The next year Oregon closed the fishery during the same months except April.[68] So in April, half the river was closed to commercial fishing (the Washington side) and half was open (the Oregon side). These conflicting regulations initiated four decades of haphazard harvest regulation on the Columbia River. Year after year, the two lower river states enacted competing regulations, which led to confusion and resulted in little effective enforcement of the statutes—the condition that prompted Roosevelt's threat in 1908.

Outside the Columbia River, the regulation of harvest was minimal, and it consisted primarily of fixed fishing seasons, short weekend closures, prohibitions against harvesting spawning salmon, and some regulation of gear. The legislatures' first attempts to regulate fisheries were generally the product of intuition and political pressure from the industry, with little or no science applied. In 1878, the Oregon legislature enacted a one-day fishing closure—sunset Saturday to sunset Sunday—in that state's coastal streams. In 1891, a fixed fishing season—April 1 to November 15—was enacted. For the next several decades, the opening and closing dates were modified. Some areas of the streams, usually in the upper watershed or tributaries, were closed, and gear restrictions were established. Washington State took a similar approach to regulating the salmon fishery on Puget Sound. In British Columbia, the first specific regulations, enacted in 1878, stopped the canners from dumping fish waste back into the river and closed fishing for thirty-six hours each weekend. Fisheries in Alaska were controlled by the federal government; Congress waited until 1889 to begin regulating the salmon fisheries there, and did not appropriate funds to enforce the new law until 1892. In its first attempt to regulate and conserve the Alaska fishery, Congress addressed only the most egregious form of fishing. It outlawed the use of dams or weirs that stretched across entire streams and prevented the salmon from reaching their spawning grounds.[69]

These first feeble attempts to curtail the harvest and allow some salmon to reach their spawning grounds were, like the statutes protecting habitat, either not enforced or completely overwhelmed by the growing number of fishermen and canneries. Under the best conditions, salmon managers would have had a hard time trying to restrict a fishery that had been freewheeling and uncontrolled for a decade or more. The fact that cannery operators were often wealthy and politically powerful members of the community made enforcing regulations even more difficult.

In some cases, though, salmon managers did attempt to enforce the early harvest regulations. Jack London's book *Tales of the Fish Patrol* describes the often harrowing attempts to enforce fishing regulations in the lower Sacramento River and San Francisco Bay. The task of enforcement usually overwhelmed the limited resources that legislatures allotted for salmon management. In 1887, the Oregon State Board of Fish Commissioners complained that although they were just three individuals, the legislature expected them to patrol 200 miles of the Columbia River to enforce closed seasons. To make matters worse, the fishery operated largely at night.

Beyond these difficulties, salmon managers sympathized with the fishermen and were just plain reluctant to enforce the laws. They appreciated the laws' short-term economic impact on the fishermen and often gave that more weight than the long-term health of the salmon and the industry. For example, six years after the first seasonal closures were enacted by Oregon and Washington, the Oregon Fish Commissioners expressed the belief that if they had enforced the closure—which they hadn't—investments by the canneries amounting to "$200,000 would have been rendered worthless." Further, they asserted, enforcement would have imposed an even greater hardship on the fishermen than on the canneries, since the fishermen "have their all in their fishing gear and that being their only means of support for their families." In addition, they believed the punishment of a $500 fine or one year's imprisonment placed too heavy a burden on the fishermen.[70] Enforcement of harvest regulations in Washington's Puget Sound was also negligible.[71]

In many respects, the rapidly growing fishery simply overpowered the weak regulations. Inventions such as the "iron chink" butchering machine, the sanitary can, and the double seamer, which eliminated hand soldering, greatly increased the industry's output and created a huge demand for fish to feed the cannery lines. For example, in British Columbia by 1877, production from an average cannery ranged from 240 to 450 cases a day with a crew of 130 to 300 workers, but by 1883 average daily production rose to 1,000 cases a day with crews of only 120 to 140 persons.[72] Further south, on Puget Sound, the average production per cannery during the 1880s was about 9,000 cases per year. By the 1890s, the annual production per cannery had jumped to 24,000 cases. At the same time the number of canneries had grown from four in 1888 to nineteen in 1900. Improved efficiency as well as more canneries stimulated demand for more salmon to keep the lines running.

The fishermen responded to the increasing demand for fish by adding more gear. On the Columbia River, the number of gillnets increased from 900 in 1880 to 2,200 in 1894, traps increased from 20 in 1881 to 378 in 1895, and fishwheels went from one in 1882 to fifty-seven in 1895.[73] The harvest was so intense that canneries just twenty miles upriver had to send fishermen to the mouth to ensure a supply of fish.[74] The river contained so many gillnets by 1887 that the U.S. Army Corps of Engineers actually concluded they were slowing the river and causing significant shoaling. The corps' observation indicates the intensity of the fishery, although the shoaling was perhaps more likely the result of grazing, mining, and beaver depletion, which destabilized streams and sent large amounts of sediment downstream to settle out in the lower river.

The early harvest regulations were supposed to increase the number of salmon reaching the spawning grounds, but it was not until 1895 that a fishery expert—Marshall McDonald from the U.S. Commission of Fish and Fisheries—studied the condition of the spawning stocks of salmon to evaluate the effectiveness of the existing regulations. Based on the reduced number of salmon reaching the spawning grounds, he concluded that overharvest was evident and predicted that abundance would decline. His conclusion was not "a matter of wonder," he stated, but "it is, indeed, a matter of surprise that any salmon have been able to elude the labyrinth of nets which bar their course to the Upper Columbia."[75] Forty-five years later, biologist Willis Rich studied the counts of salmon passing over fish ladders at the newly completed Bonneville Dam. He concluded that the fishing regulations up to that time had done little to reduce the intensity of fishing or to increase escapement (the number of breeding salmon returning to the spawning areas).[76]

Gear Wars

The growing intensity of the salmon fishery meant that fewer and fewer salmon were being divided among more and more fishermen. Competition among fishermen intensified and erupted into battles, especially among users of different types of gear. Each group of fishermen—gillnetters, trapmen, fishwheel operators, and seiners—tried to use their political power to outmaneuver the others to obtain legislation that would favor their way of fishing and give them more salmon. Some salmon managers—such as Henry Van Dusen, Oregon's master fish warden—favored one side in a gear conflict and deliberately increased the intensity of the gear wars.

Van Dusen lived in Astoria, the center of the lower river gillnet fishery, where he reluctantly enforced fishing regulations applying to his friends and neighbors. Part of Van Dusen's reluctance to carry out the laws stemmed from the confusion created by the Oregon and Washington legislatures. They had enacted fishing statutes that opened or closed the fishery on different dates or imposed different restrictions on the kind of gear used. For example, in 1905 Oregon closed its spring season on March 1, whereas Washington closed its season on March 15. Oregon closed its fall season on August 10, and Washington closed its season on August 15.[77] In addition to the different regulations, Oregon and Washington disagreed as to where the state boundaries in the river channel were.

Attempts to reconcile the differences between Oregon's and Washington's fisheries regulations failed, however, largely because Van Dusen kept his hometown fishermen stirred up against any bill that did not give preferential treatment to the lower river. In fact, critics accused him of catering to the interests of the Astoria fishermen.[78] Many people, including President Theodore Roosevelt, recognized that something had to be done to effectively curtail overharvest before the industry destroyed itself. But Warden Van Dusen's support for the lower river gillnetters and his political maneuvering threw the legislative process into gridlock. His "pestiferous" stirring up of an "anarchist" following eventually cost him his job.[79]

President Roosevelt's recognition of the salmon crisis followed Oregonians' expression of their deeply felt concern over the issue. Through their votes on two ballot initiatives in 1908, they inadvertently forced the states of Washington and Oregon to cooperate on the regulation of the salmon harvest. The process began when gillnetters from the lower Columbia River presented an initiative petition to prohibit all fishing for salmon except with hook and line upstream of the Sandy River (119 miles from the Columbia's mouth). The petition would have eliminated commercial fishing in the upper river. Commercial fishermen above the Sandy River countered with a petition of their own, one that outlawed net fishing at night. Since gillnets were the major method of harvest in the lower river and were effective only at night, the second petition would have closed the lower river fishery except for the traps near the mouth of the river at Baker's Bay (Figure 5.2).

When the two petitions were put before the concerned public in a general election, they approved both, closing most of the Columbia River fishery and throwing the salmon industry into chaos. H. C. McAllister, the new master fish warden for Oregon, was sensitive to the criticism that his predecessor had been soft on the commercial fishermen, so he proceeded to vig-

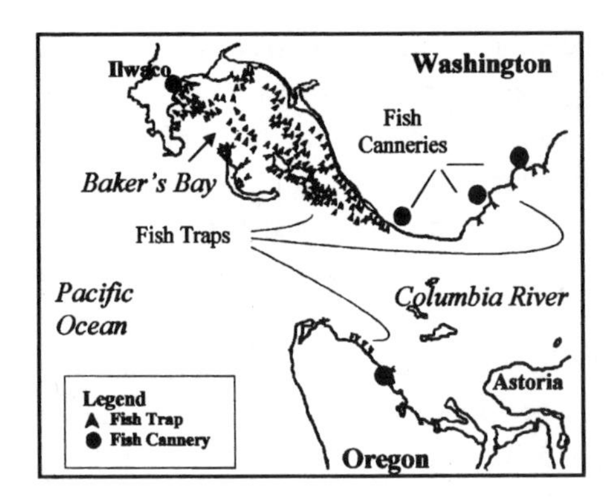

Figure 5.2. Concentration of fish traps (triangular markings) and fish canneries (circular markings) on the lower Columbia River and Baker's Bay in 1887. (After Jones 1887)

orously implement the people's will. He hired special deputies and began arresting fishermen on both sides of the river, regardless of whether they held an Oregon or Washington fishing license. The state of Washington immediately asked the federal court for an injunction, which was granted, but the injunction only restrained McAllister from enforcing the law on the Washington side of the river. He thought it unfair that Washington fishermen could continue to fish while Oregon fishermen had to sit idle or risk arrest, so he convinced the federal court to extend the injunction to the entire river. At this point President Roosevelt elevated the conflict to the national level with his State of the Union address.

Pushed to the brink of disaster and threatened with losing control of a major industry, the legislatures of Oregon and Washington finally worked together in 1909 to craft a single set of regulations for the Columbia River salmon fishery. The states enacted concurrent regulations until 1918, when Congress approved an interstate compact between Oregon and Washington to regulate the harvest of salmon on the Columbia River.

Although the lower Columbia River fishery was under a unified set of regulations, the competition and conflict among trapmen, gillnetters, fishwheel operators, and seiners continued to boil. Henry Van Dusen believed that the Oregon legislature had sold out the lower river fishermen for the benefit of the upper river interests. In April 1909, after the Oregon and Washington Joint Legislative Committee had drafted its unified fishing regulations, Van Dusen wrote a threatening letter to the committee chairman, Senator I. H. Bingham: "Our fishermen down this way are very much in hopes that they will have a chance one of these days to take a crack at you for the stand you took on this matter."[80] Senator Bingham replied, "Censure of our work emanating from a man whose official career is as replete with official imbecility and political mendacity as is yours, certainly has no terror for me. Lay on, McDuff."[81]

Although the states now managed the fishery with uniform regulations, they were still unwilling to regulate an industry that had been largely uncontrolled since 1866. And as one federal biologist, Claudius Wallich, explained, "If the fish are in the can we cannot expect to find them at the spawning grounds."[82] The unified regulation scheme resolved a political conflict between Oregon and Washington, but the biological problem of too few spawning salmon remained largely untouched.[83]

Oregonians have used the initiative petition twenty-three times to decide salmon management issues, and ten of those initiatives were adopted by popular vote. Several of them resolved many of the "gear wars" between different types of fishermen. By the 1920s, the conflict between fishermen had evolved primarily into a fight between fixed and mobile fishing—gillnets and seines on one side and traps and fishwheels on the other. The gillnetters eventually won the political battle for supremacy on the Columbia River. Through initiative petitions, Oregon eliminated fishwheels in 1927 and then all remaining fixed fishing gear on the Columbia River in 1948. Washington prohibited the use of traps, fishwheels, set nets, and seines in 1935. Voters approved the gear restrictions because they recognized the need to reduce the harvest and increase the number of salmon returning to the spawning grounds. But this did not happen. The will of the people was not enough to shake the management institutions out of their complacency, so overharvest continued. The ballot measures only redistributed the catch to the remaining fishermen.[84] The gillnetters had won but, as we shall see, their victory was short lived. Uncontrolled ocean trolling, which intercepted the salmon before they reached fresh water, would eventually eclipse the river fishermen.

Other ballot measures closed commercial fishing in certain areas. For example, by 1956, after several ballot measures, the people voted to close commercial salmon fishing in all of Oregon's coastal streams.

The Fishery Moves to the Ocean

In 1898, F. J. Larkin moved to Portland from San Francisco and brought with him an idea that would revolutionize the salmon fishery. Larkin wanted to replace the picturesque butterfly sails that powered the Columbia River gillnet boats with gasoline engines. His idea quickly took hold. Nine years later, half of the boats operating out of Astoria were equipped with engines, and the number was rapidly increasing.[85] Power from gasoline engines increased the effectiveness of the traditional gillnetter, but it also gave rise to a new and highly mobile way to catch salmon—the troll fishery. Trollers hunted salmon over large areas of the ocean. Each troller could hang as many as 140 lines with a lure or bait to attract feeding salmon.[86] This highly effective technique dramatically changed the harvest patterns and created a set of biological and international problems that have persisted to this day.

The troll fishery off the mouth of the Columbia River grew from a few boats around the turn of the century to 500 boats in 1915 to 1,500 boats in 1919.[87] Prior to the development of nylon webbing, fishermen constructed gillnets from heavy linen fibers. When World War I cut off the supply of flax, which increased the cost of linen gillnets, many gillnetters switched to trolling.[88] By 1920, there may have been as many as 2,000 trollers fishing in the sea off the Columbia River.[89]

The harvest of salmon on the open ocean raised a new question: Whose fish were the trollers catching? Suddenly it became very important to determine if salmon undertook long ocean migrations, where the salmon from specific rivers spent their oceanic life, and who was catching them. The Canadians were the first to try to answer those questions. They caught and tagged chinook salmon off the west coast of Vancouver Island from 1925 to 1930 and then tracked the recovery of those tagged fish to rivers throughout the Northwest. Their study revealed that salmon migrated long distances in the ocean and that the troll fishery intercepted fishes from several different rivers. For example, significant numbers of chinook salmon tagged off Vancouver Island were later recovered on the Columbia River.[90] The troll fishery would elevate the "gear wars" from fights between fishermen using different kinds of gear within a river to battles between fishermen from different states and different countries.

Early on, salmon managers recognized the problems that the troll fishery would create. As early as 1921, the California Fish and Game Commission noted the growth of sea fishing for salmon and recommended that it be carefully investigated. By 1923, California biologists recognized that the salmon could not survive both the river and ocean fisheries, so they recommended eliminating the ocean fisheries.[91] Professor E. V. Smith from the University of Washington completed a study of the ocean fisheries in 1920 and concluded that they harvested immature salmon before they reached their full size and therefore represented a huge loss of the fishery's potential. Smith recommended that the harvest of immature salmon be stopped.[92] Edward Blake of the Washington State Fisheries Board went so far as to consider the ocean fishermen enemies of the salmon. As he explained, "A man who goes out to the feeding grounds of the salmon and injects his hook into a four- or five- or six-pound salmon that eventually if permitted . . . would have weighed about twenty-five pounds, jerk[s] that fish off his hook and let[s] it die out there is an enemy of the salmon."[93]

With all this opposition to trolling, the states of Oregon and Washington banned ocean fishing within three miles of their coasts, but they could not regulate fishing outside three miles, nor could they regulate fishermen who intercepted their fish in other states or in British Columbia. The growing troll fleet, which could fish beyond the three-mile limit of state jurisdictions, went largely uncontrolled until 1949, when a tri-state compact (Oregon, Washington, and California) brought the Pacific States Marine Fisheries Commission into existence.

The ocean troll fishery, despite opposition from many salmon managers, academics, and policy experts at the time of its emergence, continued to grow until it had the political clout to survive and eventually replace gillnetting as the principal means of harvest. Over the years, all of the problems predicted by the early critics of the troll fishery have materialized—interception fisheries that created interstate and international conflict, the harvest of immature fish and fisheries in mixtures of strong and weak stocks, and more. The management of ocean fisheries has grown into a convoluted system fraught with intractable problems that persist year after year while the salmon continue to decline.

The Hatchery Fix

The early decades of the canning industry in the Pacific Northwest can be described as nothing less than an all-out attack on salmon. Believing that the resource was inexhaustible, many fishermen and cannerymen harvested far more fish than could be processed or sold. Millions of fish, whose carcasses rotted unnecessarily, never made it to their spawning grounds to reproduce. In the face of this profligate waste, salmon managers were reluctant to enforce the laws that restricted fishing, even though they knew full well the consequences of poor stewardship. The complete collapse of the Atlantic salmon fisheries had already revealed the consequences of failure to prudently regulate fishing and protect habitat. Still, the enforcement of regulations was uncomfortable. It was difficult to tell fishermen and cannerymen what they could and could not do—especially when they stood to make large profits by mining the free wealth of nature.

In 1875, with the salmon-canning industry just nine years old and growing very rapidly, businessmen and politicians looked for advice on how to avoid subjecting the Pacific salmon to the same fate as their Atlantic cousins. Soon after Congress appointed the Smithsonian scientist Spencer Baird to head up the newly created U.S. Commission on Fish and Fisheries, the Oregon legislature sought his advice. In response, Baird wrote a letter identifying three major threats to the survival of the salmon industry in the Pacific Northwest: excessive fishing, especially when salmon were in nursery streams and in the act of spawning; dams that blocked the migration of salmon to their spawning grounds; and changes in the physical habitat of the streams. These three problems identified by Baird over 120 years ago are the same threats still facing the salmon in the late twentieth century.

Although Baird thought that restrictions placed on the fishery might allow enough salmon to reach their spawning grounds to maintain the supply of fish, he did not believe such regulations could actually be enforced. Guided by the technological optimism of the day, Baird offered an alternative, more appealing approach—an approach that he believed guaranteed a continued supply of fish *without* the need to regulate the fishery. Baird's new approach was fish culture.

Three years earlier, the U.S. Commission on Fish and Fisheries had built a small hatching station on the Sacramento River, which showed that the eggs of the Pacific salmon could be fertilized and hatched successfully in large numbers. Based on this limited and incomplete experiment, Baird made a dangerous and gigantic scientific leap. He unhesitatingly recommended that "instead of the passage of protective laws, which cannot be enforced except at very great expense and with much ill feeling, measures be taken, either by the joint efforts of the States and Territories interested or by the United States, for the immediate establishment of a hatching establishment on the Columbia River, and the initiation during the present year of the method of artificial hatching of these fish."[94]

Baird offered the salmon canners an industrialist's dream: unrestricted access to a public resource that could earn them millions of dollars, along with the promise of an endless supply of fish, for the trifling cost of $15,000 to $20,000 a year.

Baird's letter of advice is one of the most significant documents in the history of salmon management. First, it aptly identified the three major threats to salmon in the Northwest. But, more important, it marked the conception of an unfounded belief that has persisted even in the face of evidence challenging its validity. Baird's letter gave birth to the myth that artificial propagation could maintain or enhance salmon abundance, no matter how many fish were caught. Salmon managers, with no better evidence of hatchery success, extended Baird's promise further: hatcheries not only could make up for excessive harvest but would compensate for habitat destruction as well. This myth has continued to guide salmon management for the past 120 years, and its effect on salmon cannot be overestimated.

As might be expected, given a choice between artificial propagation and restricted harvests, the industry seized the hatchery alternative. In 1887, when proposed legislation threatened to impose minor restrictions on the fishery, 169 cannery companies or individuals signed a petition against the bill. The cannerymen pledged to raise $25,000 to build and operate a hatchery as an alternative to restrictive regulation.[95]

Thus, with absolutely no scientific evidence to support his recommendation, Spencer Baird launched a myth and a program that would have profound effects on the salmon.

CHAPTER 6

Cultivate the Waters

In 1883, while fishermen on the Columbia River hauled in an all-time record catch of chinook salmon, fisheries experts from many nations gathered in London to discuss the status of their industry at the International Fisheries Exposition. Both London and Astoria looked to the future with optimism, for different reasons. In Astoria it was the record catch of chinook salmon. In London it was the introduction of two new technologies that made the future seem brighter. First, the 1883 exhibition boasted an unprecedented use of electricity. New dynamo-electric machines generated enough electricity to light the entire great exhibition hall, the largest area ever to be illuminated from a single power plant. The exhibition organizers were so proud of this feat that they devoted an entire volume of the written proceedings to a description of the electrical system.[1] These new dynamos had important connections to fisheries. Fishing boats supplied with electricity would be safer, more efficient, and less labor intensive.

The exhibition's second reason for optimism was another form of technology—the artificial propagation of fishes in hatcheries. Samuel Wilmont (the father of Canadian fish culture) explained to the delegates that the exhibition's purpose was to search for ways to counteract the effects of human greed and avarice on the abundance of world fisheries.[2] Wilmont believed hatcheries filled that need. He pointed out that Canada led the world in fish culture and stood second to no other country in terms of the number of artificially propagated fish it produced each year. Then Professor George Brown Goode of the U.S. Fish Commission picked up the hatchery banner and carried it even further, proclaiming that the Sacramento and Columbia Rivers were more productive than they had ever been due to the release of artificially propagated salmon. According to Goode, the salmon industry of the Pacific coast was "thoroughly under the control of fish culture."[3] How he reached that conclusion, especially for the Columbia River, is difficult to understand since at the time, the only hatchery on the Columbia had been closed for two years.

These two new technologies gave fisheries experts reason to look to the approaching twentieth century with high expectations. No one at the exhibition could foresee a time in the mid-twentieth century when these same two technologies—electricity and artificial propagation—would converge on the mother river of the Pacific salmon. No one could foresee that large dynamos placed in concrete dams would completely block and severely hamper the migration of salmon, and artificial propagation would be chosen as the primary means of counteracting the deadly effects of the dams. No one could foresee that hatcheries would fail miserably, that the technologies of electricity and artificial propagation would ultimately bring the Columbia River salmon to ruin.

An Old Idea Whose Time Had Come

The culture of fishes was not a new idea. Ancient Chinese were culturing carp 2,500 years before Professor Goode made his outlandish claims of success. Written records of carp culture in Bohemia go back 800 to 900 years. By the late sixteenth century, Bohemia and Moravia were cultivating fish in 450,000 acres of developed ponds. Jan Dubravia, Bohemia's bishop of Olomuc, published a five-volume treatise on fish culture in 1547. But although fish culture has a very long history, the propagation of salmon is relatively new, dating back only about 230 years.

The first full account of salmon culture came from German nobleman Steven Ludwig Jacobi, who published a monograph about his experiments on the artificial propagation of the Atlantic salmon in 1763. Jacobi and his sons fertilized the eggs of salmon and trout and incubated them in wooden boxes placed in small streams. Then they reared the hatched fry in ponds on their estate. In recognition of his pioneering work, King George III of England awarded Jacobi a lifetime pension.[4] Despite the award, Jacobi's work attracted little attention. His carefully described techniques for fertilizing salmon eggs and incubating them under semiartificial conditions circulated among a small group of scientists and a few wealthy landowners who wanted to keep their private ponds and lakes fully stocked. In the still-emerging postfeudal economy, the commercial potential of Jacobi's ideas lay dormant and unrecognized.

Several individuals working independently rediscovered salmon culture between 1815 and 1840, but this time the technology emerged in a more favorable social environment. During the first half of the nineteenth century, a scientific renaissance was sweeping over Europe and America. In its wake, science and technology were linked to commerce, producing an optimistic new vision for the future. Science would not only unlock nature's rich stores for commercial gain, but it would also give people the power to dominate nature and bring it into their total service.[5]

It wasn't the scientists, however, who made the connection between fish culture and commerce; their work was too focused and their purposes too narrow. For example, M. Coste, a French scientist, used fish culture to further his studies of embryology, while John Shaw in Scotland artificially propagated the Atlantic salmon in 1837 to test his theories about their life history. He believed that salmon parr converted to silver smolts, then migrated to sea and eventually returned to the river Nith as full-grown fish. Although this life cycle is considered conventional wisdom today, his ideas were considered radical in the 1830s.

While the scientists conducted their experiments, two uneducated but practical French fishermen, Joseph Remy and M. Gehin, independently discovered how to artificially propagate salmon. Seeking to do something about the declining numbers of salmon and their shrinking income, on a November night in 1841 they crept to the edge of a river and, under a full moon, watched salmon in the act of spawning. Once they figured out how the salmon spawned naturally, they quickly developed a practical method of spawning the fish artificially and, in their mind, improving on nature. Like most fishermen, they were good at improvising gear, and after some trial and error, they reinvented Jacobi's hatching boxes.

When the two fishermen publicized their invention, having immediately recognized its commercial possibilities, many scientists rightfully pointed to Jacobi's work and dismissed the Frenchmen's claim that they had discovered how to propagate salmon. In response, supporters of Remy and Gehin criticized the scientists (who may well have been jealous of the attention given Remy and Gehin) for their inability to give the fishermen credit.

Scientists also came under fire for their failure to recognize the value of the discovery. As one critic put it, "Those learned bodies could not put together three simple ideas: fecundation (artificial fertilization), artificial hatching, and production, and draw from them, for the interests of humanity, the natural results which must strike the sense of any intelligent person; . . . it needed two simple fishermen to verify their ideas and spread them before the world."[6] The tension between unfounded optimism and scientific skepticism marked the birth of modern fish culture, and it has continued to mark the discourse surrounding fish culture ever since.[7]

While Remy and Gehin worked out the methods, refined the technology, and pointed out the commercial value of fish culture, their discovery was not merely a practical, economic endeavor. Salmon culture fulfilled higher human goals. The fishermen readily grasped this fact when they drew a parallel between their discovery and agriculture. "Does not man sow his entire field with the single sort of grain he wishes to cultivate?" they asked. "What, then, is there to hinder us from stocking bountifully our streams with fish, by aiding the process of hatching, and protecting the young from destruction by their innumerable enemies?"[8]

The comparison to agriculture worked well. In both cases there were seeds to sow, hatching boxes or plots of ground to tend, and food to harvest. But the analogy went beyond concrete similarities. At the time, agriculture represented both the triumph of civilized societies and the promise of the future. By extending the familiar values of agriculture to fish culture, the fishermen invested their technology with a powerful image of abundance obtained from human ingenuity and labor—an image that has proved enduring and unshakable. Comparing a fish hatchery to a farm fixed the ideas of progress associated with agriculture onto artificial propagation. This had two important long-term consequences. Fish culturists transferred the success of agriculture to hatcheries without the critical intermediate step of actually evaluating and verifying that success. In some cases, hatchery programs for Pacific salmon have yet to be adequately evaluated, even though they were initiated more than 100 years ago.[9] Second, fish culture's close association with agriculture caused the science of fisheries management and biology to diverge from the science of ecology.[10] As a result, institutional thinking focused on an agricultural paradigm and ignored its negative effects on the ecological processes that sustained the salmon. Today, more than a century later, we still know very little about the impacts that hatcheries have on the ecosystems they release their fish into, and furthermore, there is little research under way to identify and understand those impacts.

Remy and Gehin were not the only ones to recognize the broader social significance of fish culture. In a pamphlet describing their discovery, C. E. P. Godenier explained the higher human purposes of fish culture. "Man reigns on the earth supreme: he bends the sun to his use for the plants he requires;

domestic animals submit to his will and produce at his pleasure; he commands the waters to transport him with fearful speed, and soon perhaps the air will be conquered. All nature seems to obey his laws. The fish alone have escaped his dominion, but not his nets. Now Mr. Gehin has discovered the secret of their reproduction, and placed in our grasp the means to enliven our rivers and watercourses, as we cover our fields with corn, hemp and flax, as we multiply our flocks, our domestic fowls, our silk worms."[11] Agriculture and fish culture not only permitted people to increase their food production but also enabled people to assert their dominion over nature.

From the very beginning, artificial propagation of salmon had the services of good publicists, such as Godenier, who effectively embellished its success and pointed out its higher purposes. Due in part to the enthusiasm of their supporters and the personal interest of Napoleon III, Remy and Gehin were rewarded with government pensions, and each was given an interest in a tobacco factory.

With growing popular support for salmon culture, French scientist M. Coste gained considerable government backing to expand scientific studies of artificial propagation. Most important, he secured funds to construct the world's first commercial-scale salmon hatchery at Huningue, on the upper Rhine River. Coste believed Huningue was the grandest enterprise of the century, with the potential to produce an unlimited supply of good food for everyone.[12] Visitors from across the globe soon came to Huningue to learn techniques and copy the technology. The hatchery distributed 20 million fertilized eggs all over the world. Coste called his creation a "piscifactory," or fish factory—an appropriate name, since he intended to mass-produce fish in the same way that factories manufactured other goods. The piscifactory reflected the promise and popularity of technology in the rapidly industrializing economy of Europe. Once the hatchery took on the image of a fish factory, it was easy to extend that vision to the fish—salmon became products rather than living animals.

Not everyone accepted artificial propagation wholeheartedly in the early years of salmon culture. One Scottish critic, writing under the pen name "Salmo," harangued hatchery proponents as "men of tanks and incubators . . . philosophers of conjecture and romance . . . and feeble drivellers, who have voted [the salmon] incompetent to discharge the functions which constitute the chief end and object of her existence." Instead, Salmo proposed an approach to protecting fishes in their natural habitat based on the belief that Providence had "formed every animal perfect in its kind, and . . . given it the instincts and the capacity to accomplish worthily and effectively the purpose for which it was created."[13] Salmo's attack on fish culture indicates that critics, who abound today, were also present at the birth of modern hatchery programs.

Hatcheries in the United States

By the time Remy and Gehin rediscovered artificial propagation in 1841, industrialization was rapidly destroying salmon fisheries in the northeastern United States. For example, in the summer of 1839, when Henry David Thoreau and his brother John took their trip down the Concord and Merrimack Rivers, the town of Lowell, Massachusetts, boasted twenty-eight cotton mills with 150,000 spindles and 5,000 looms, all driven by the force of the Merrimack's water. Throughout the Northeast, rivers were being altered to meet the needs of industry. Aside from building dams to harness power for their factories, new industries also diverted water and channelized rivers to transport raw materials and products. The factories dumped their putrid wastes directly into the nearest waterway. Rivers were being so rapidly transformed and their ecology changed on so massive a scale that, in spite of feeble attempts to ensure fish passage at dams, the Atlantic salmon in many streams were doomed.[14]

With salmon numbers declining in the wake of industrialization, New England fishermen began to look hopefully at the new technology that the French were promoting at Huningue. Fish culture not only restored fish abundance but was profitable. Pioneering American fish culturist Stephen Ainsworth claimed that an investment of $47,100 in a trout hatchery would, at the end of four years, return $468,750, for a profit of $421,650.[15] Fish culture would be lucrative as long as demand stayed high and the natural supply was kept low—a condition that depended, ironically, upon the destruction of natural fish habitat by industry.

Yet even with the potential for high profits, could private enterprise be expected to restock salmon in northeastern rivers? Stocking trout into ponds or streams on private estates might work, but restocking rivers with a migratory fish was another matter entirely. Why would entrepreneurs invest in a salmon hatchery, only to see their cultured fish captured by fishermen farther downstream or in the estuary? Like the French at Huningue, the Americans would have to look to government to shoulder a major portion of the costs of restoring depleted salmon runs. Accordingly, by 1857, the legislatures of Connecticut, Massachusetts, and Vermont had begun looking into the feasibility of fish culture.[16]

Vermont chose George Perkins Marsh to study the need for hatcheries. As author of the book *Man and Nature,* Marsh had been one of the first to warn the world about the dire consequences of the industrial alteration of the earth. While he certainly did not condone the degradation of habitat, Marsh did not believe it was possible to stop the destruction of rivers caused by industrialization or to effectively restrict the harvest of fish. In his report to the Vermont legislature, Marsh recommended that the state encourage private fish culture efforts to restore depleted fisheries through financial grants to individuals. Marsh also found a social justification for restocking Vermont's rivers with artificially propagated fish. Without enough fish and wildlife, Marsh worried that men would give up hunting and fishing and lose their virile, robust character. More fish would encourage more fishing and thereby prevent American males from becoming "duller, . . . more effeminate, and less bold and spirited."[17]

When Marsh presented his findings in 1857, Vermont and other New England states were reluctant to engage in fish culture. They didn't like the idea of government meddling in what had been the domain of private enterprise. But by 1864, with fisheries dwindling, the New Hampshire legislature took action and appointed the first state fish commission in the nation. Still, the new commission waited two years before trying artificial propagation. Because the new technology required viable eggs, which were in short supply in New Hampshire, the commission turned to New Brunswick, Canada, where the salmon still returned to the rivers in large numbers.

The first attempt to import eggs from Canada in 1866 largely failed. The commission managed to release only a few hundred fry into the headwaters of the Connecticut and Merrimack Rivers. Two years later, New Hampshire obtained 70,000 Canadian eggs, but only 10,000 hatched. Frustrated by the unreliable supply of eggs, the state hired a former minister, Livingston Stone, to build a permanent egg-collecting facility on the Mirimachi River in New Brunswick. When Stone began mining the Mirimachi's eggs and shipping them to New Hampshire, many Canadians voiced their opposition to what they regarded as the theft of their salmon seed by Americans. The hostility of the locals grew, compelling Stone to abandon the project. The Canadians recognized, however, that they controlled a supply of salmon eggs that the Americans desperately needed. Responding to the market's invisible hand, Canada agreed to sell the eggs at the outrageous price of $45 per 1,000.[18]

To avoid dependence on the high-priced Canadian eggs, the states of Maine, Massachusetts, and Connecticut formed a joint venture to collect eggs from Maine's Penobscot River. Live salmon were purchased from fishermen and artificially spawned; the fertilized eggs were then distributed to the three states in proportion to their investment in the enterprise.[19] The consortium began operation under the direction of Charles Atkins, and in the first year, the states obtained 72,000 eggs for $18.09 per 1,000.

Although the northeastern states recognized the need to build fishways over dams, to regulate the discharge of pollutants, and to restrict the harvest, they made little progress on these contentious fronts. As a result, fish culture and restocking became an increasingly appealing alternative.[20] At the time, it seemed logical to solve the problems created by industrialization with the use of more technology. If industrial factories caused depleted salmon runs, then fish factories would restock them. Americans were beginning to mass-produce more material goods; why not mass-produce more fish for restocking depleted rivers? Hatcheries offered the perfect solution: unlimited fish and minimal confrontation with the industrialists, fishermen, and farmers. However, restocking the rapidly growing number of depleted rivers required capital investment and acquisition of salmon eggs on a large scale, both beyond the resources of the individual states.

On December 20, 1870, a group of fish culturists met at a skating rink in New York City and agreed to form the American Fish Culturists Association. The original goal of the association was to promote the private financial interests of its members, but Livingston Stone later recalled that the group's mission quickly changed to promoting the public good. In 1884, the association's name was changed to the American Fisheries Society to reflect that broader mission.[21] The American Fish Culturists Association and the northeastern states shared a common goal—to restock the rivers with salmon—so they combined their forces to approach the only institution capable of underwriting the task, the U.S. Congress.

Events in New York and Washington, D.C., were beginning to converge in a way that would make federal help possible. A few months after the New York meeting, President Ulysses S. Grant asked Spencer Baird to head up a federal bureau created to address problems confronting the nation's dwindling fisheries. The new U.S. Fish Commission gave the federal government an agency capable of carrying out the restocking program that the states and fish culturists wanted, and they wasted little time before acting. The Vermont Fish Commissioner wrote to his distant relative in the U.S. Congress, Senator George Edmonds, and asked for help in securing funds for two federal hatcheries for anadromous fishes. He proposed a simple plan: the federal government's fish commission would collect salmon eggs and then distribute them to the states. Although the supply of Atlantic salmon eggs was inadequate to the gigantic task of restocking the northeastern rivers, in the

West the rivers still teemed with high-quality salmon and a endless supply of eggs. Once the eastern states received the eggs, they would carry out their own programs.

When Senator Edmonds introduced the hatchery proposal to Congress in 1871, his colleagues were skeptical about funding a program that delivered benefits to only one region. To garner an appropriation for the federal salmon hatcheries, U.S. Fish Commissioner Spencer Baird had to extend the benefits of fish culture to other regions. Not only would the northeastern rivers benefit from the federal largesse, but the fish commission would also stock shad in the Mississippi and other rivers. With the help of influential supporters such as Senator Edmonds, Baird finally maneuvered a $15,000 budget for fish culture through Congress.

Soon thereafter, Baird met with New England state fish commissioners and members of the American Fish Culturists Association in Boston to launch the first national hatchery program. Livingston Stone argued that the Pacific salmon were the only fishes that could produce eggs in the quantities needed to restore eastern rivers. Baird agreed, but the group was provincial when it came to their Atlantic salmon. Some thought the Pacific salmon produced soft, white meat as poor in quality as that of catfish and suckers. Besides, many believed that the Pacific salmon didn't take a fly, unlike the Atlantic salmon.[22]

Eventually, with Baird's steering, the group agreed to give a portion of the federal funds to Charles Atkins's Atlantic salmon hatchery on the Penobscot River, which would be supplemented with eggs from the Huningue hatchery in France. Another part of the budget went toward the collection of eggs from the Pacific salmon in the Sacramento River. Because the Sacramento flowed through an area where temperatures often exceeded 100°F, Baird believed these eggs should go to the warmer rivers of the mid-Atlantic region such as the Hudson, Delaware, Susquehanna, Potomac, and James. Even rivers in the South, including the Mississippi, would receive Pacific salmon eggs. Finally, the remaining funds would be spent to develop a shad hatchery. Thus, the first plan to restock America's rivers with artificially propagated fish was essentially political, an attempt to please nearly everyone.

While Spencer Baird is best known for his many contributions as a natural scientist, he was also a skilled politician, adept at manipulating Congress, especially when it came to securing funds for his programs. To that end, he ultimately used hatcheries to spread "gifts" of fish planted in specific congressional districts to leverage support for his objectives. This use of hatcheries did not always coincide with the highest principles of science, and it exposed a different side of Baird. For example, in response to criticism that some eggs were not distributed wisely, Baird said, "It does not make much difference what [is done] with the salmon eggs. The object is to introduce them into as many states as possible and have credit with Congress accordingly. If they are there, they are there, and we can so swear, and that is the end of it."[23]

From the very start, Baird was compelled to bargain for his agency's budget. As a result, politics exerted more influence over the development of the young fish culture program than science.

Fish Culture Moves West

On July 6, 1872, Spencer Baird directed Livingston Stone to proceed to California with $5,000 of the new congressional appropriation to arrange a way to obtain large numbers of eggs of the best varieties of Salmonidae and other food fishes on the western coast. When Stone arrived in San Francisco, he had trouble finding out where the salmon spawned in the Sacramento River. Even members of the California Fish Commission didn't know. Finally, Stone located a railroad engineer who could remember seeing Indians spearing salmon near the junction of the McCloud and Pitt Rivers, so he took the train to the end of its line at Red Bluff, still fifty miles short of his destination. Locals told him the salmon on the McCloud River were spawning right then. He quickly gathered some supplies, took the stage to within a mile of the river, and went on foot from there until he found Indians fishing for salmon. The next day—September 1, 1872—Stone started construction of the first Pacific salmon hatchery.[24] He immediately wrote Baird to report that the salmon in the Sacramento River did in fact take a fly.[25]

Stone's no-nonsense attitude made him an excellent person for the job. He had little patience for the state fish and game commissioners, who frequently offered to take him fishing or hunting. He disliked their casual attitude toward fish culture and, in a letter to Baird, complained, "They seem to think I came out here to have a jolly time of it, and my idea of steady laborious work [do] not appear to comprehend."[26]

Building the first salmon hatchery in the West presented Stone with many obstacles, adventures, and dangers. His hatchery site was twenty-five miles from the nearest town, fifty miles from a railroad station, and fifty miles from the nearest sawmill. Wagons hauling supplies over the narrow mountain roads barely made twenty miles a day. With the frontier's lack of law and order, Stone often relied on his fists to discourage poachers from taking fish from the river below the hatchery. To top it off, Stone and his crew had to work in temperatures that frequently reached 100°F or higher.

Despite its difficulties, though, the hatchery work offered many rewards, especially for someone like Stone, who genuinely cared about the fish. He later recalled the enchantment of those first successful spawnings of chinook salmon. "The thrill of excitement that tingled to our fingers' ends when we first saw the little black speck in the unhatched embryo, which told us that our egg was alive. It was one of the dearest sights on earth to us then. . . . We can hardly believe that such a commonplace, matter-of-fact affair could ever have stirred our feelings and our imagination as it did once, when the sight and sensation were both new, and the world of promise before us was untried and unknown."[27] Stone's dedication to fish culture went beyond the goal of restocking rivers. It had deep roots in a personal bond between man and animal.

With the McCloud River hatchery up and running, the U.S. Fish Commission began shipping chinook salmon eggs to eastern rivers via specially designed railroad cars. On their return trip the cars often carried the eggs of eastern fishes back to California rivers. Once when Stone rode the train east with a batch of salmon eggs, a railroad bridge collapsed, plunging him and his assistants into the river, where they barely escaped with their lives.[28]

Stone shipped his salmon eggs to rivers in Maryland, Virginia, Wisconsin, and other states. In the first eleven years of operation, the hatchery stripped the McCloud River of 30 million salmon eggs and sent them to rivers in the eastern United States and to New Zealand, Hawaii, Holland, Russia, Denmark, Germany, France, and Australia. Moreover, the eggs he shipped cost the government only fifty cents per 1,000—about one-hundredth the price previously paid for Canadian eggs.

Before long, Stone encountered a familiar problem. Like the Canadians who had opposed his egg mining on the Mirimachi River, Californians began to question the export of their chinook salmon eggs. In response, Stone tried to strike a bargain to trade Pacific salmon for shad. On February 24, 1873, he wrote Baird, "I find that I am unable to give a satisfactory answer to the question, 'What is the U.S. Fish Commission doing for California?' I would suggest if the appropriations warranted that the U.S. help the state in its effort to introduce new varieties of fish from the states. . . . It seems now to the people here as if my mission here were to take every salmon egg and give nothing in return. A small turn extended in some undertaking which the state is actually engaged in, as for instance the introduction of shad, I think would have a good effect."[29] Eventually Stone and Baird engineered a satisfactory deal, and Californians readily traded chinook salmon for shad as part of the wholesale restocking of freshwater ecosystems. Once again, politics influenced the shape of the federal hatchery program.

With the hatchery on the McCloud River firmly established, Spencer Baird sent Livingston Stone to the Columbia River, where a group of cannery owners seeking to avoid government regulation of the salmon fishery had formed the Oregon and Washington Fish Propagating Company. These businessmen raised $20,000 to build and operate a hatchery, but they lacked expertise in fish culture. Baird did not want to subsidize the salmon industry of the Columbia River with a federal hatchery, but he was willing to provide Stone's expertise to businessmen willing to spend their own money. In fact, the cannery operators were following Baird's earlier advice that hatcheries were the preferred alternative to harvest regulation.

Stone arrived in May 1877 and chose a site on the lower Clackamas River, in Oregon. By the middle of September, the hatchery was built and 200,000 chinook salmon eggs were incubating. Then, on September 15—afterward called "Black Friday"—a sudden flood wiped out all the eggs. Stone immediately had 200,000 eggs shipped from the McCloud River hatchery to Clackamas.[30] The transfer of eggs between the Sacramento and Columbia Rivers was a significant act: almost as soon as two salmon hatcheries were operating in the Northwest, fish culturists began transferring eggs between them. The transfer of eggs among hatcheries in order to keep all the ponds filled to capacity was a particularly insidious practice that became common and is still practiced today.

We now know that the life histories of the salmon are tuned to the environmental cues in their home stream. The time and place of spawning, the time of emergence from the gravel, juvenile rearing distribution, and the time of migration to the sea are all outward expressions of the relationship between the salmon and their home stream. For example, on the Fraser River, biologists have identified at least forty-five stocks of sockeye salmon. Each stock returns to its home stream to spawn, but the timing of sockeye spawning on the Fraser varies among the stocks by as much as five months. This variability has an important ecological function: Water temperatures in the different spawning areas differ, and the temperature determines how long the salmon eggs must incubate in the gravel before hatching. The juvenile sockeye migrate to rearing lakes shortly after hatching; to maximize survival, their arrival in the lake must coincide with the precise time that food becomes available. The various sockeye stocks spawn over a five-month period, but the peak in juvenile emergence in those same rivers takes place over a span of a few weeks. Salmon spawning is synchronized to the local environment so the juveniles will emerge at the appropriate time.[31] If this synchrony were disrupted by the transfer of fish among streams, juveniles would emerge too early or too late to take advantage of the seasonal peak in food abundance.

The transfer of fish and eggs among hatcheries in different watersheds put the salmon in foreign waters, often hundreds of miles from their home stream. The fry found themselves in rivers where their life histories, their strategies for survival in their home stream, were out of sync with the new environmental cues. But fish culturists at the time didn't understand that the salmon were adapted to the environment of their home streams. Nor did they understand that adaptation has a key prerequisite: the population must experience a degree of reproductive isolation. That is, spawning must be largely limited to members of the same stock.

Mass transfers of salmon between rivers disrupted thousands of years of reproductive isolation and destroyed the adaptive relationship between the salmon and their home stream. The newly hatched fry, deposited in rivers distant from their natal stream, had to face a new set of survival challenges that were not part of their evolutionary legacy. The advantages of local adaptation were lost, and few of the fry released into foreign streams survived.[32]

As hatchery programs grew, transfers of salmon eggs between rivers became more common. In 1909, Oregon constructed Central Hatchery (later renamed Bonneville) on Tanner Creek in the lower Columbia River. The name "Central" underscored the hatchery's primary mission: to serve as a clearinghouse for salmon eggs brought from other hatcheries. Some eggs were transferred from the upper river, hatched at Central Hatchery, and the fry released directly into the Columbia, under the theory that releasing fry closer to the ocean would circumvent high mortality in the river during migration to the sea. But most eggs from Central Hatchery were shipped to other rivers throughout the Northwest (Figure 6.1) rather than sent back to their home streams. For example, after chinook salmon eggs taken from the Kalama River in Washington in 1925 were hatched at Central, the fry were released in the Alsea River on the central Oregon coast. Sockeye salmon from Yes Bay, Alaska, were hatched at Central and then released into the Columbia at Tanner Creek. While these transfers among rivers kept the fish culturists busy, they produced few adult salmon.

For millions of years, the salmon and their habitats had evolved into a

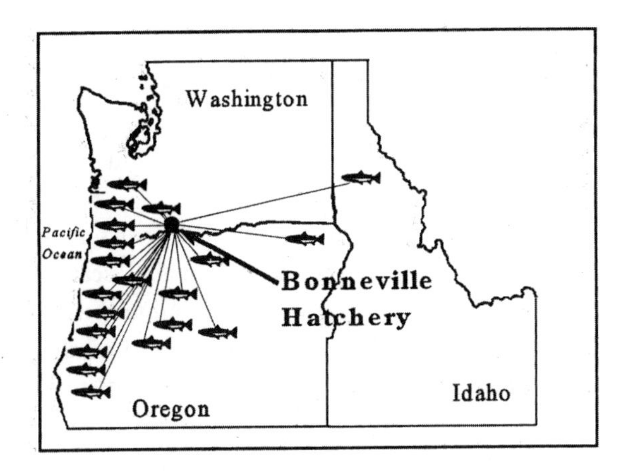

Figure 6.1. Transfers of chinook salmon from Bonneville Hatchery to other locations in Washington, Oregon, and Idaho, 1909–1950. Each line can represent multiple plants of salmon. (Source: Wallis 1964)

complex tapestry of individual populations adapted to the habitats of their home streams. Now the relocation of non-native fish species into western rivers, together with the transfers of the salmon among rivers within the region, was unintentionally ripping that tapestry to shreds. The early transfers of the salmon between rivers could be blamed on a lack of knowledge. But why didn't the fish culturists seek the knowledge they needed, especially since they did recognize that need? Their ignorance of the salmon's life history and ecology attests to the fact that the original objectives of the federal hatchery program were not implemented.

In 1884, the U.S. Commission of Fish and Fisheries published its overall objectives for the new hatchery program. First, it recognized the need to gather information and "to arrive at a thorough knowledge of the life history from beginning to end of every species of economic value, the histories of the animals and the plants upon which they feed or upon which their food is nourished, the histories of their enemies and friends, and the friends and foes of their enemies and friends as well as the currents, temperatures and other physical phenomena of the waters in relation to migration, reproduction and growth."

Second, the commission intended "to apply this knowledge in such a practical manner that every form of fish shall be at least as thoroughly under control as now the salmon, the shad, the alewife, the carp and the whitefish."[33]

The commission's first objective recognized the need to understand the biology and ecology of propagated fish throughout their life cycle and in relation to the physical environment. The biological studies called for were not carried out. In 1879 Livingston Stone had asked the commission to assign a trained biologist to the McCloud River hatchery. A biologist could have initiated studies of the salmon's life history, but Baird turned down the request. Had Stone's request been granted, it might have established a precedent and created a different approach and direction in the hatchery program. Quite possibly, our knowledge of the salmon and their status might have been very different than it is today.[34]

One hundred and twenty years later, a committee of independent scientists studying the decline of the Pacific salmon for the National Research Council concluded that the hatchery program had not been subjected to long-term evaluation and that, consequently, the success or failure of the program had never been documented.[35] In the 1880s, the fish commission could have initiated studies of the life history of the fish its scientists were propagating and the ecology of the watershed those fish were released into, but it didn't. That might have given ecological science more influence over the development and use of hatcheries. The commission chose instead to emphasize control over production and closely align fish culture with agricultural science. We still operate hatcheries and manage rivers as though they were specialized farms rather than complex ecosystems.

Acclimatization—Playing God with Ecosystems

Hatcheries fit nicely into the nineteenth-century view that ecosystems were warehouses of commodities that existed solely for human use and benefit. The earth itself was seen as a giant factory "furnished by the great engineer." People had the mission of subduing the wild elements, of ridding the land of animals and plants that were useless to humans, and of bringing the whole system under efficient control.[36] The fundamental goals of dominating, controlling, and manipulating nature for human use were deeply embedded in western culture. Hatcheries provided the perfect vehicle for ordering and controlling the aquatic realm.

Hatcheries and the distribution of Pacific salmon eggs worldwide were just a part of the larger mass transfer of favored animals and plants throughout the world. Governments and private organizations known as "acclimatization societies" industriously moved flora and fauna from one country to another, from one hemisphere to another, from one region to another, in order to "improve" upon nature. For example, carp were introduced to the waters of the United States from Germany, and thirty species of fish as well as the American lobster, the eastern oyster, and the soft-shell clam were transplanted from the eastern United States into western waters by 1895.[37] By playing God with ecosystems and rearranging aquatic communities, "acclimatization" proponents could ensure that rivers contained the species they considered most beneficial or, in newly settled lands, those species that reminded the white pioneers of home.

In his account of the attempts to transplant the salmon into Australia, Arthur Nicols described the significance that people accorded the redistribution of the world's mammals, birds, and fish species. "He who succeeded in making two blades of grass grow where but one grew before has been canonized as the greatest benefactor of mankind," Nicols wrote, "but surely he who achieves the more difficult task of . . . peopling a barren river with a noble species of fish, should not pass unnoticed by his contemporaries and those who enter into the enjoyment of his labour."[38] According to Nicols, successful "acclimatization" was an accomplishment more momentous than that of pioneering agriculture. Of course, he didn't recognize that those "barren" rivers were functioning ecosystems inhabited by native species adapted to the local conditions.

The popularity of both hatcheries and "acclimatization" reflected the prevailing scientific and social view that waters should be treated like farms and managed on an agricultural model. Even the prestigious journal *Science* praised the U.S. Fish Commission's acclimatization projects.[39] Rivers needed to be tilled by the hand of man to suit his needs. Only those fish favored by humans, those deemed capable of efficiently producing commodities, should be allowed to inhabit rivers. At the time, natural resource management was a mixture of religion, science, and technology. Actively manipulating the fauna of rivers was simply another way to carry out the providential plan of dominating and ordering the natural world.

The pervasive aim of controlling nature was not just a general attitude that guided the rearranging of the nation's land and waters. It permeated the very details of the biologists' thinking about fish. For example, the salmon that returned to the rivers each year carried millions of eggs in excess of the number needed to maintain their population. In an 1884 article, Livingston Stone explained why the salmon carried this huge surplus of eggs to the spawning grounds: "Nature, perhaps more aptly speaking, Providence, in the case of fish, . . . produces great quantities of seed that nature does not utilize or need. It looks like a vast store that has been provided for nature, to hold in reserve against the time when the increased population of the earth should need it and the sagacity of man should utilize it. At all events nature has never utilized this reserve, and man finds it already here to meet his wants."[40] Stone believed the excess salmon eggs were put there intentionally for future human use, for the time when technology would permit people to control the eggs' fate in hatcheries.

The perception that the human mission was to control nature, combined with the agricultural view of watersheds, eliminated any thought of natural limits to production. Just as a garden tended by humans can produce more commodities than the same plot of land left in its wild state, rivers would produce greater abundance if tended by people. To cultivate the rivers and convert the huge surplus of eggs described by Stone to adult salmon, people had only to protect those eggs from their natural predators. The "fish only need our help in one way. At breeding times their eggs are mostly destroyed by numerous enemies and but few are hatched. By artificial means, at trifling cost, nearly all the eggs can be saved and vast numbers of young produced," an 1890 fish commission report asserted.[41] Hatcheries gave fish culturists the power to break the natural limits on salmon production, or so they believed.

For thousands of years the salmon populations responded to the seasonal cycles and rhythms of their ecosystems. Now the salmon's production system would be built and directed by humans, whose vision of the watersheds and the salmon was severely flawed.

Salmon Without Rivers

One of the most troubling consequences of this flawed vision was that it diverted salmon managers' attention from the root causes of the salmon's decline. As a result, significant problems such as habitat destruction and overharvest were consistently ignored. Agency budgets and staff energy were devoted to artificial propagation instead of habitat protection. This tradeoff between habitat and hatcheries is illustrated by the way salmon managers dealt with the effects of logging and dams.

For example, even though the Washington legislature had prohibited the dumping of sawdust into streams in 1876, twenty-four years later the state fish commissioner complained that the devastating practice was still a major problem in the waters of Puget Sound. Then he acknowledged, "Inquiry has frequently been made by many persons why we have not instituted more prosecutions for the violations of the fish and game laws. We have been compelled to admit that these infractions are frequent and well known to this department." He went on to lament the lack of funds: "It must be readily seen that with the limited amount of deputy assistance and expense money at our disposal . . . , it is impossible for us to even attempt anything like a systematic prosecution of these violations of the law." He then offered what he considered a preferred alternative to enforcement and disclosed his highest priority: "I have considered that the most important work in connection with the Fish Department was the establishment and proper maintenance of fish hatcheries."[42]

Fisheries managers in Oregon took a similar approach.[43] The priority was artificial propagation rather than protection of habitat: circumventing the problems rather than solving them. Underlying this approach to management was the assumption that hatcheries could maintain salmon production without healthy rivers. Rigorous enforcement of the laws against sawdust dumping would wait until the 1920s, forty-five years after the statutes were enacted.[44]

Splash dams and log drives were two of the most effective destroyers of salmon habitat ever devised by humans. To compensate for the loss of habitat blocked by the splash dams or destroyed by the log drives, salmon managers again turned to fish culture.[45] To collect salmon eggs for their hatcheries, fish culturists placed racks, or wooden weirs, across the streams. The racks completely blocked the migration of the adult salmon. Prevented from reaching their spawning grounds, the salmon concentrated in pools immediately below the racks, where hatcherymen easily captured them and stripped them of their eggs.

The timber industry so dominated the use of rivers that it could not even accommodate hatcheries built to compensate for the habitat it destroyed. A wooden rack was no match for a large "splash" of water and logs. The log drives destroyed hatchery racks, causing A. C. Little, Washington State's commissioner of fisheries, to suggest that the operation of splash dams should be prohibited during the time that hatcheries were collecting their brood fish.[46] Some hatcheries reached an accommodation with the loggers. For example, fish culturists delayed placing their rack and collecting eggs at the federal hatchery on Washington's White Salmon River until the last run of logs for the season was finished.[47] Even where splash dams didn't interfere with egg collection, logging on a large scale in watersheds altered water quality to the point that some hatcheries had to be closed.[48] Instead of compensating for habitat destruction, these early hatcheries merely mined eggs from the wild salmon populations while contributing little or nothing to their survival. Hatcheries were just another stress the salmon had to contend with.

Salmon managers also believed their hatcheries could compensate for the effects of hundreds of dams that were being built in streams throughout the Northwest around the turn of the century.

On the north slopes of the Olympic Peninsula in Washington, dams on the Elwha River—like dams throughout the region—put Euro-Americans' need for energy to drive their industrial economy ahead of the salmon's need for a freshwater home.

Although the same tragic struggle between dams and fish occurred in hundreds of canyons throughout the Pacific Northwest, the story of the Elwha River has greater significance for two reasons. First, the Elwha solidified the precedent of using artificial propagation to compensate for the effects of dams; and second, the Elwha may mark the first time dams were removed from a river in order to restore the Pacific salmon. The Elwha River was a productive salmon-producing watershed that supported one of the largest runs of salmon and steelhead on the Olympic Peninsula.[49] Although all the species of Pacific salmon were native to the Elwha River, the spring chinook run was of special importance.

The story of the Elwha River spring chinook salmon begins back before the arrival of Euro-Americans and is probably older than the earliest humans on the Olympic Peninsula. Before dams blocked their migration, the spring chinook salmon made their way above the narrow canyons and the wild water at Goblins Gate to spawn in the highest reaches of the river. The narrow canyons and big rapids acted as a biological filter, which selectively admitted only the largest and strongest spring chinook to the spawning grounds. Over thousands of years, the run evolved into a race of giants, with individual fish commonly weighing over 75 pounds and many over 100 pounds.[50] Over thousands of years these large spring chinook salmon evolved to survive in the rugged habitat of the Elwha River.

In 1890, Thomas Aldwell bought a homestead claim for $300 and took possession of a tract of land at the bottom of the Elwha River canyon. Four years later he had a vision of a dam stretching across his homestead from one canyon wall to the other. Aldwell wanted to take the same power and energy that had produced the large race of chinook salmon and use it to power factories in the nearby city of Port Angeles. Seven years later, he had secured enough capital to build a dam to realize his dream.[51] The only question that remained was this: Would Aldwell's dam coexist with the spring chinook? But that question hardly entered Aldwell's plans and calculations for the river.

Although Washington State statutes prohibited the construction of dams that blocked salmon migrations, the Washington State fish commissioner had already opened a loophole in the law before the Elwha Dam began construction. In 1906, the Jim Creek Light and Power Company wanted to dam Jim Creek. The proposed dam was sixty feet high, and John Riseland, the state fish commissioner, was worried that a fish ladder on a structure that

Elwha River canyon, near where the Elwha Dam was later built, 1910. (Photo courtesy Olympic National Park)

First dam built in the lower Elwha River canyon. No date. (Photo courtesy Daishowa/Fort James Corporation)

high would not be effective. When he offered an alternative—a hatchery in lieu of a fish ladder—he both struck a deal and opened a loophole in the law, all on his own. The legislature did not sanction the policy of building hatcheries in lieu of fish ladders until 1915.[52]

Aldwell began construction of his dam in 1911, and by September of that year it had become obvious that his construction was blocking the salmon run. The local warden, J. W. Pike, wrote Riseland, "I have personally searched the Elwha River & tributaries above the dam and have been unable to find a single salmon. . . . There appear to be thousands of salmon at the foot of the dam where they are jumping continuously."[53]

Aldwell stalled on the construction of the fishways, in part because the bottom of his dam blew out during construction and had to be rebuilt. By 1913, with the run still blocked, the new Washington State fish commissioner, Leslie Darwin, found a way to resolve the impasse. He used Riseland's loophole and went a step further, merging the interests of the Washington Department of Fisheries with Aldwell's vision of a dammed river. To get around the letter if not the spirit of the 1890 law, Commissioner Darwin pointed out that the state allowed barriers to salmon migration if those barriers were for the purpose of collecting eggs for artificial propagation of salmon. Then, through baffling logic, he equated Aldwell's 100-foot-high dam to a large hatchery rack, giving legal cover to an otherwise illegal dam. Darwin proposed to Aldwell, "The only possible way to avoid this [building a fish ladder] would be for the state to select at the foot of your dam and to make use of your dam as its own obstruction for the purpose of taking fish for spawning purposes."[54]

Aldwell agreed in 1914, and the hatchery went into operation in 1915.[55] Like most of the early hatcheries, the Elwha hatchery was ineffective; in 1922, a year after Darwin retired, it was abandoned. Thus the dam that would be a hatchery rack became just a dam without a fish ladder. This was the fate of many of the hatcheries authorized by Darwin in lieu of ladders.

Washington Department of Fisheries personnel returned to the Elwha River eight years later and found some of the chinook salmon still surviving in the five miles of free-flowing river below the dam. The men captured those fish, stripped them of their eggs, and shipped them fifteen miles, to a hatchery on the Dungeness River. As author Bruce Brown explained, "The Department of Fisheries tended to approach the Elwha situation like the death of a distant relative: it was all somewhat sad, but there was nothing they could do about the outcome."[56]

It's an easy drive from Port Angeles to the upper dam on the Elwha

River.[57] When most people arrive, their attention is immediately captured by the narrow sheer-walled canyon below the dam and then by the peaceful blue of Lake Mills, nestled in a basket of firs behind the dam. But I always skip over these features and focus my attention upstream—past the dam and the reservoir and into the hills and canyons of the Olympic National Park. I think of the river flowing through those green hills and canyons. It is still wild and nearly pristine, having been protected for over fifty years by the Olympic National Park. Because it has been protected, the river above the dams retains its natural attributes in an undisturbed condition, except one—the giant silver fish.

Today a few spring chinook return to the Elwha River below the lower dam. The genes of those few surviving fish still hold the memory of the wild river crashing through canyons and flowing through mountain meadows heavy with the scent of cedar and fir. Locked away in those genes is the memory of survival in a rugged and beautiful landscape. The river and the fish have been separated for eighty years. But if the river is ever released from the grip of the Elwha Dams, the spring chinook will have little trouble recolonizing their former habitat. The studies and impact assessments have been completed, and it seems possible and feasible to restore the Elwha. All we need now is more political courage and a new vision for the salmon.

Aldwell's dam set a critical precedent that linked dams and hatcheries, and that linkage formed a partnership that was forged deep inside the heart of the industrial economy. That partnership remains the ultimate expression of human domination over nature. Dams transformed the flowing rivers into lakes and gave humans the power to control the river's energy. Hatcheries built to mitigate the effects of dams gave humans control over the salmon. The merger allowed the industrial development of the water by substituting hatcheries for wild fish and reservoirs for free-flowing rivers.

Hatcheries offered another advantage that would become more significant after 1930. They became a major source of income for salmon management agencies and, more important and tragic for the salmon, that income did not depend on performance, or the ability of salmon managers to maintain salmon abundance. Recently the U.S. General Accounting Office (GAO) tallied the cost of efforts to maintain salmon in the Columbia River. Of the $1.8 billion identified by the GAO, 40 percent was spent on hatcheries compared to 5 percent on habitat restoration. Even though three panels of independent scientists have concluded that salmon hatcheries have failed to fulfill their purpose, salmon managers tenaciously cling to those programs.[58] For half a century, the income generated by hatcheries depended only on the law of gravity. As long as water flowed downhill and turned turbines, it generated income for salmon management agencies, regardless of whether the hatcheries were achieving their objectives of mitigating the loss of salmon.

While the Lords of Yesterday changed the landscape and the rivers of the Pacific Northwest, salmon managers tried to compensate for the massive destruction of habitat by building hatcheries. A few people recognized that hatcheries would not be enough, and they concluded that the only way to protect the salmon's natural economy was to separate it completely and protect it from the destructive potential of the industrial economy. The only way of doing this, in their view, was to create salmon refuges or sanctuaries.

Salmon Refuges—The Road Not Taken

The first person to propose setting aside rivers as salmon sanctuaries was Livingston Stone, the former minister from New Hampshire who had built the first Pacific salmon hatchery, in northern California, for the U.S. Fish Commission. In 1889 he traveled from California to Alaska with a group of scientists to investigate the state's fisheries and look for opportunities to introduce artificial propagation. When he returned, Stone thought about what he had seen, comparing the Alaska salmon and their habitat with the conditions in the rest of the Pacific Northwest. He concluded that the nation needed to set aside rivers exclusively for salmon. There was no other way to prevent the rapid development in the Pacific Northwest from destroying habitat and the salmon. In a talk to the American Fisheries Society, he presented his idea for a salmon park or sanctuary.

Stone admitted that it might sound foolish to call for salmon refuges because the fish were still abundant in Oregon, Washington, and Idaho in 1892, but he reminded his audience that the buffalo were also once abundant. If Yellowstone National Park had not been created in time to protect the last few animals, he concluded, the buffalo would have become extinct.

Pointing to the fate of the Atlantic salmon, Stone argued that the Pacific salmon were caught between the greed of fishermen and the destructive forces of development. He said, "It was the mills, the dams, the steamboats, the manufacturers injurious to the water, and similar causes, which first [made] the stream more and more uninhabitable for the [Atlantic] salmon, finally exterminating them altogether."[59] Stone admitted that not even the hatcheries could overcome the destructive effects of the growing population and advancing civilization.

The massive runs of salmon in Alaska's pristine rivers impressed Stone so much that he recommended designating the Uganik River on Afognak Island (near Kodiak Island) as the first salmon park. He was an effective salesman: President Benjamin Harrison granted his request, setting aside Afognak Island and its river as the first salmon sanctuary on Christmas Eve 1892. The proclamation protected the river from development and closed commercial and subsistence fisheries in the river and surrounding marine waters.[60]

Although Stone argued effectively for salmon parks, he was not ready to give up on the belief that hatcheries provided the salmon with more protection than wild rivers did. Even at the Afognak sanctuary, fish culturists still attempted to improve on nature by building a massive salmon hatchery capable of incubating 78 million eggs. The sanctuary and its fisheries restrictions were put into place not to maintain healthy, natural salmon production, but to ensure a supply of eggs to the hatchery. In his campaign for the establishment of salmon parks, Stone apparently did not detect the contradiction in his strategy. If hatcheries were more effective and productive than wild rivers, as fish culturists claimed, then why did hatcheries need protected wild rivers to maintain their supply of eggs?

Unfortunately, the fishing restrictions around Afognak Island were gradually lifted, and passage of the White Act in 1923 abolished all special fishing regulations intended to protect the refuge.[61] In the short space of thirty years, the nation's only salmon sanctuary reverted to simply a hatchery. In 1934 it closed for economic reasons, and its buildings were turned over to the U.S. Navy for use as a recreation center. A good idea had fallen victim to a narrow vision and false assumptions about nature.

In the same year that President Harrison established the Afognak sanctuary, the U.S. Fish Commission enlisted the aid of the U.S. Navy to help fight the depletion of salmon. Commander J. J. Brice was ordered to survey military and other government reservations to identify potential hatchery sites. His job was to "formulate a plan to restore salmon in their original numbers." Consistent with the prevailing approach to salmon management in the late 1800s, Brice recommended construction of four central hatcheries and twenty auxiliary stations from California to Alaska. He made one recommendation that deviated from the status quo, however. Brice advised setting aside the Klamath River of northern California as a salmon preserve.[62] This recommendation was never implemented.

In 1911, Henry Ward addressed the annual meeting of the American Fisheries Society and reminded his audience that refuges and game parks were being set aside for mammals and birds in increasing numbers. Ward believed fishes deserved similar treatment and reminded his colleagues that native fishes were disappearing at a rapid rate, especially from the eastern rivers.[63] He recommended that river reaches or even entire watersheds be set aside for the protection of native fish fauna. His plan for preserving rivers and watersheds fell on deaf ears.

Prolonged drought, changing ocean conditions, excessive harvest, and continuing habitat degradation sent the Pacific salmon into sharp decline in the 1920s and 1930s. This decline stimulated new interest in salmon refuges. In his annual report for 1932, Washington State Supervisor of Fisheries Charles Pollock called for "definite legislation . . . to set aside certain watersheds as permanent fish sanctuaries to guarantee both commercial and recreational fisheries of this state in the future."[64]

Oregon also considered creating salmon sanctuaries. In 1928, the voters of Oregon were asked to decide if the Rogue and McKenzie Rivers and their tributaries, along with portions of the Deschutes Basin and the lower Umpqua River, should be set aside as fish sanctuaries. The petitions would have stopped all water development in those basins and declared that those rivers be maintained, "so far as is still possible, in the natural condition and free of encroachments by commercial interests," and that the use of those rivers "for food and game fish propagation and for recreational purposes, shall be and is hereby declared a beneficial use of the waters."[65] All the measures were defeated by a ratio of about two to one.[66] But salmon sanctuaries did not die with this defeat at the polls; the idea was resurrected yet again ten years later.

In 1938, Governor Charles Martin of Oregon asked the Oregon State Planning Board to study the Columbia River's declining salmon fishery and report back to him. Among other recommendations, the board stated that there was a need to study the feasibility of establishing salmon preserves—tributaries set aside exclusively for salmon spawning.[67] There seems to be no evidence that the study was ever carried out; if it was, the proposed salmon sanctuaries were never implemented.

In 1959, Ross Leffler, the assistant secretary of the interior, made a last-minute plea to declare the Snake River a salmon sanctuary instead of proceeding with dams planned for hydroelectric development. He failed, and today the Snake River chinook and sockeye salmon are threatened with extinction.

Salmon refuges are still the road not taken.[68] Hatcheries are still the preferred alternative.

The Political Tool

Several years after the U.S. Fish Commission began planting the eggs of the Pacific salmon in eastern rivers, there were still no clear signs of success. A few adult salmon had been spotted from time to time, but no obvious "runs" of the Pacific salmon had become established. Hatchery advocates remained optimistic that artificial propagation would eliminate the need to regulate the harvest of fish. The U.S. Fish Commission operated under the policy that it was "better to expend a small amount of public money in making fish so abundant that they can be caught without restriction, . . . rather than to expend a much larger amount in preventing the people from catching the few that still remain after generations of impoverishment."[69] Accordingly, during its first twelve years of existence, the commission spent 75 to 85 percent of its budget on artificial propagation.[70]

With such a large commitment to fish culture, everyone involved wanted to find signs of success. Though Livingston Stone shipped most of the eggs collected at the McCloud hatchery to the East Coast, he also released a few thousand fry back into the McCloud River every year. In 1878, when the federal and state fish commissions noticed an increase in the pack of canned salmon on the Sacramento River, they attributed it to Stone's hatchery.[71] In fact, the increase was short-lived and most likely reflected a burst of fishing effort after eighteen new salmon canneries opened between 1876 and 1881. Fish culturists anxious for any indication of success based their conclusions more on wishful thinking than any kind of real evaluation.

The claim that artificial propagation caused the increase in the Sacramento River salmon harvest reveals a troubling inconsistency in the scientific approach of the U.S. Fish Commission. In most areas of their research, the commission's scientists were scrupulous. For example, their biological surveys of the Great Lakes reflected high professional standards. But when it came to hatcheries, the scientists' skepticism and careful evaluation of evidence fell victim to blind optimism and political expediency. While blind optimism led them to premature and false conclusions about the success of their program, political expediency—the need to protect 85 percent of the commission's budget—made that success absolutely necessary.

Ultimately, the Sacramento cannery pack declined after 1883 despite massive increases in the hatchery program.[72] In fact, the growth of the hatchery program, especially after 1895, seems to have accelerated the decline in harvest (Figure 6.2).

Replenishing America's rivers with millions of fry from federal hatcheries to make the fish so abundant that there would be no need for harvest regu-

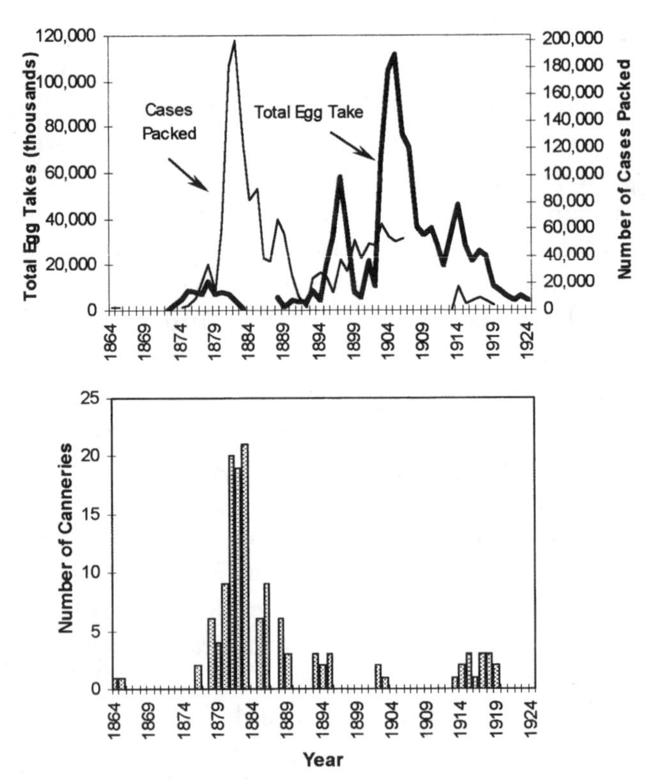

Figure 6.2. The pack of canned Pacific salmon and the total egg take for all hatcheries on the Sacramento River, 1864–1924 (upper figure). The number of canneries operating on the Sacramento River (lower figure). (Source: Clark 1929)

lations was an appealing strategy. However, this goal of fish culture got mixed up at the outset with unrealistic fantasies and politics. To further his scientific program and to protect or increase the budget of the U.S. Fish Commission, Spencer Baird used the gift of hatchery allotments to congressional districts. As Baird explained, "Some of our distribution of fish is made not so much for re-production and propagation by the parties [but] as . . . an obligation to some congressmen and individuals. So long as we can enter the matter as attended to, that is all we insist upon. Of course if fish were scarce, we could not even afford to waste even a small percentage, but . . . having an abundance we can afford to throw away a few for political or other reasons."[73]

Once the hatchery program became a political tool, it gained immunity from close scrutiny and scientific evaluation. Success and optimism had to be maintained in order to retain the value of the hatchery gifts and ultimately to perpetuate the commission's whole hatchery program. In the end, the managers deceived themselves, and the wild salmon runs paid a terribly high price.

In 1883, just eleven years after the first Pacific salmon hatchery had begun mining eggs from the wild salmon of the Sacramento River, George Brown Goode of the U.S. Fish Commission declared hatcheries to be a complete success. (As noted in the beginning of this chapter, Goode made this unsubstantiated claim at the World Fisheries Exposition.) When other scientists questioned his evidence, Goode supported his claim with a curious logic: "It was not likely that the American Congress, or the Canadian government, would for a period of ten or twelve years keep on making annual appropriations for fish culture if they were not satisfied that it was not only a success from a scientific stand-point, but a success from a commercial point of view."[74]

A program that had begun with blind optimism and quickly become a political tool had now gone full circle. Questionable releases of hatchery fish were used to gain political support and funds for the U.S. Fish Commission's program; that political support was then used, even in the scientific community, as a measure of the program's success. Because of their ideological roots and their early politicization, hatcheries became so entrenched as the primary means of managing salmon that it would be fifty years before they came under even the most limited scientific scrutiny.

Crack in the Foundation

Just one year after his claim of success, Goode was compelled to announce that all attempts to restock the Pacific salmon in the rivers of the eastern United States had failed.[75] After a decade of extensive stocking, Pacific salmon had failed to colonize the eastern rivers. Faced with this obvious failure, hatchery proponents scrambled to find examples of success. Of course, the hatchery on the Sacramento River had apparently boosted harvests. And the shad and striped bass introduced in the West were supporting successful commercial fisheries.[76] The failure of one experiment—the stocking of eastern rivers with salmon—was not enough to deflate the boundless enthusiasm for massive hatchery programs.

It was not until 1893 that Baird's successor, Marshall McDonald, cast the first shadow of doubt on the efficacy of hatcheries. In a speech to the World Fisheries Congress, McDonald said, "I am disposed to think that in this country we have relied too exclusively upon artificial propagation as a sole and adequate means of maintenance of our fisheries. The artificial impregnation and hatching of fish ova and the planting of fry have been conducted on a stupendous scale. We have been disposed to measure results by quantity rather than by quality, to estimate our triumphs by volume rather than potentiality. We have paid too little attention to the necessary conditions to be fulfilled in order to give the largest return for a given expenditure of money."[77]

McDonald unmasked three important flaws in the hatchery myth. For one, hatcheries by themselves could not overcome a lack of resource stewardship. Second, the size of the program—for example, the number of fry released—was not an appropriate measure of the success or failure of a hatchery. And finally, the condition of the river into which the fish were released would ultimately determine the success or failure of the artificial propagation. McDonald was not the only skeptic about the ability to produce salmon without rivers.

That same year, at an American Fisheries Society meeting in Chicago, society president Herschel Whitaker raised similar doubts about the practice of fish culture. Whitaker told fish culturists to focus more on replenishing fishes in their native waters and less on transplanting fish to foreign rivers and lakes. He challenged them to seek the cooperation of scientists to solve biological problems through careful experimentation. Most importantly, he told them that it was not enough to hatch large numbers of eggs. Fish culturists had to be ecologists, he contended; they had to understand the biology of the fish and the relationship between the fish and their habitat. Two years later Whitaker went further and told fish culturists that they had a *moral* obligation to avoid wasting public money by planting fry in waters where there was no hope of success because of inadequate harvest regulation.[78]

McDonald and Whitaker were pioneer fish culturists and were certainly not ready to give up on hatcheries. They retained a firm belief in hatcheries' potential to supply the nation's rivers, lakes, and bays with desirable food fishes. But both men raised important questions about the effectiveness of hatcheries and about the role of scientific investigation in perfecting the art of fish culture. Their remarks opened a long-standing debate—but the debate produced little change. In spite of some progress, McDonald's three criticisms of hatchery programs are still valid today.

Even as its successes were being strongly called into question in the eastern states, fish culture continued to gain in popularity in the West. As it happened, wild salmon were still abundant in the same Pacific Northwest rivers where large numbers of artificially propagated fry had been released. Since wild adult salmon could not be distinguished from hatchery-raised adults, the presence of large numbers of wild salmon essentially masked the failures of the hatcheries. Proponents of artificial propagation pointed to the abundant (wild) salmon and claimed success for their hatcheries.

In 1893, Senator Adlai E. Stevenson directed the U.S. Fish Commission to determine why salmon migrating up the Columbia River failed to go up the Clark Fork River and into Montana's Flathead Lake. He also asked the commission to investigate the suitability of a federal hatchery in the state of Washington. Consistent with his criticisms of fish culture, Commissioner McDonald told the U.S. Senate that any recommendation regarding the need for a new hatchery could not be made without an evaluation of the conditions that existed in the entire watershed. He said it would be futile to consider building another hatchery in the Columbia Basin until the states of Oregon and Washington adopted regulations that restrained fishing, so that sufficient numbers of salmon could reach the headwaters to spawn.

The states still maintained their faith in the ability of artificial propagation to overcome any degree of salmon depletion due to fishing, however. Henry Van Dusen, Oregon's master fish warden from 1901 to 1908, remarked in his last year in office that he was amazed the chinook salmon on the Columbia River were able to withstand the "onslaught" of commercial fishing. He could not account for their continued survival except through the benefits provided by hatcheries. In spite of McDonald's misgivings, and in the absence of any creditable study of their effectiveness, the benefits of artificial propagation were taken on faith.[79]

Western fish and game commissions came to see fish culture as the only way to override the onslaught of overharvest and habitat degradation. Hatcheries offered a ready alternative to strict regulations and an easy way to avoid confrontation with irrigators, timber companies, and polluters. For this reason, the criticisms of fish culture, which came largely from eastern biologists, had little impact on western hatchery programs. By 1930, seventy-three hatcheries were propagating Pacific salmon in rivers from Alaska to California. Striving primarily for quantity, these early hatcheries released 13.3 billion fry and fingerlings into rivers of the Pacific coast by 1928.[80]

At the same time that hatchery proponents in the West were falsely claiming that their programs were a success, fish-culture practices such as egg mining and stock transfers were undermining the wild salmon's ability to reproduce. Hatcheries not only failed to boost the number of fish but also actually harmed the salmon by mining billions of eggs from wild populations. The first hatcheries were designed to collect and incubate the maximum number of eggs with the greatest efficiency. In many cases, that meant all the eggs in the entire salmon run into a tributary were collected and brought into the hatchery. A rack or weir placed across a river blocked the migration and concentrated all the salmon trying to reach their spawning grounds. Adult fish held up below a rack were easily captured and artificially spawned. The same biological trait that made the salmon accessible to Indians at The Dalles on the Columbia River and at Milliken on the Fraser River 10,000 years ago also delivered them directly to the fish culturists.

The destructive impact of racking was tremendous. For example, the Oregon Fish Commission built a hatchery on the lower Alsea River in 1915. To ensure an adequate supply of eggs, the state constructed a permanent dam and rack, which blocked the entire run to the upper river from 1916 to 1929. Eventually, upstream residents, outraged because the weir cut off their supply of wild fish, sent a load of dynamite downriver and blew up the rack.[81]

The commercial-fishery records for coho salmon in Alsea Bay clearly reveal the effect of the hatchery and its rack. Before the rack was constructed, allowing fish culturists to take all the salmon eggs into the hatchery, the annual catch of coho salmon in Alsea Bay averaged 391,000 pounds. While the rack was in place, the catch averaged 158,000 pounds. Finally, after local residents destroyed the dam, the catch rebounded to an average of 340,000 pounds.[82] The hatchery did not even compensate for the effects of its own rack, let alone "enhance" the salmon run into the Alsea River.

All over the Northwest, this cycle repeated itself. Hatcheries were constructed, the streams racked, and the eggs mined from wild salmon populations until too few salmon returned to make continued operation worthwhile. Then the site was abandoned or eggs were transferred in from other streams to keep the hatchery full of fish.[83] Racking, which prevented the natural reproduction of salmon in specific streams, combined with the mass transfer of salmon eggs between streams to reduce salmon abundance over large areas. False assumptions and a lack of critical evaluation ultimately caused fish culturists to contribute to the destruction of the very thing they sought to improve.

The Search for Improvement

For several decades after Livingston Stone arrived in California and built the first hatchery, fish culturists released most of the artificially propagated salmon back to the river shortly after they hatched. After young salmon hatch, they derive nourishment from a yolk sack that hangs from an opening in their underside. The young fish are particularly vulnerable to predation while lugging around the clumsy yolk sack, so under natural circumstances they remain hidden in gravel, near the redd where they incubated as an egg. After a few weeks, the yolk is absorbed and the underside of the fry heals. The young salmon must then leave the gravel to find food in or on the surface of the stream.

Early fish culturists believed natural spawning was inefficient. This belief was based on three assumptions. First, few eggs were actually fertilized during natural spawning. Second, most of the fertilized eggs were eaten by predators. And finally, they assumed that these two factors constituted the primary limits on the number of adult salmon. These assumptions logically led hatcherymen to believe they could increase salmon abundance if they fertilized the eggs and then incubated them in the protected environment of the hatchery. But once the eggs hatched and the juvenile salmon were released back into the river, the sack fry, which normally would have remained hidden in the gravel, found themselves in the stream, fully exposed to predators. By trial and error, some fish culturists did learn that releasing sack fry was counterproductive, but their findings failed to alter standard hatchery practices because fish culturists were generally slow to change in response to new information.[84] Instead, they debated for several decades about how long to hold young salmon in the hatchery. The arguments were often heated and extended beyond the small circle of culturists to include cannerymen, politicians, and the press.

Cannery owner R. D. Hume was one of the first to conclude that more of the salmon would survive after release from the hatchery if they were held in ponds, fed for several months, and released at a larger size. He tried unsuccessfully to convince the Oregon Fish Commission to accept his advice and change its program.

Hume had left his brothers on the Columbia River in 1876 and moved his cannery to the Rogue River on Oregon's southern coast, where he quickly gained control of the entire salmon fishery and became a strong advocate of hatcheries. With a monopoly on the Rogue fishery, he alone stood to benefit from an investment in artificial propagation. In 1879, Hume built a hatchery on the lower Rogue River with the help of Kirby Pratt, one of Livingston Stone's assistants, and in the fall of that year he released the first hatchery-produced juvenile chinook salmon into the Rogue River. Although Hume's investment in the Rogue hatchery was self-serving, his motivation was not that simple. Unlike most of his colleagues in the salmon-canning industry, R. D. Hume was genuinely interested in the conservation of salmon.

Hume was a keen observer of the salmon's natural history, although he did not always interpret his observations correctly. In 1904 he successfully experimented with feeding young salmon in the hatchery until they were five to six inches in length, as opposed to the normal practice of releasing fry immediately after hatching. With the enthusiasm of innovation, Hume pushed his new discovery aggressively. However, Henry Van Dusen, Oregon's master fish warden, took a dim view of Hume's ideas. Their disagreement about salmon rearing sparked an intense battle between the two feisty men. Relishing a fight that he knew he could win, Hume took his case to the press, convincing the *Oregonian* to run a series of articles supporting his position. Then he convinced Van Dusen's assistant, H. T. Webster, to publicly disagree with his boss in the newspaper. Then a string of poor fishing years on the Columbia River was attributed to failure of the state's hatcheries, increasing the public pressure on Van Dusen.

Hume thought he would prevail, but Van Dusen resisted change right up to the day he was fired in the spring of 1908. Hume victoriously recorded the event in his own local newspaper with the headline VAN DUSEN'S HEAD DROPS. DULL THUD.[85] But despite this colorful debate and Van Dusen's firing, the state hatcheries still refused to try holding their fry longer. Hume died soon after, frustrated to the end by the state hatchery program's resistance to change.

As catches on the Columbia River continued to decline in spite of the intensive artificial propagation program, public concern arose and eventually pressured state fish culturists to look for ways to improve their hatchery program.[86] With encouragement and financial assistance from several cannery operators, state hatcherymen finally began to hold and feed juvenile chinook salmon and release them at larger sizes. The first group of larger fish were released in 1912, and the first adults from this release were expected to return to the Columbia River two years later. The catch of chinook salmon did increase in 1914 and remained high for several successive years.

After only five years of improved cannery packs on the Columbia River, the Oregon Fish and Game Commission announced the success of these experiments: "This new method has now passed the experimental stage, and . . . the Columbia River as a salmon producer has 'come back.' By following the present system, and adding to the capacity of our hatcheries, thereby increasing the output of young fish, there is no reason to doubt but that the annual pack can in time be built up to greater numbers than ever before known in the history of the industry."[87]

Across the river, the Washington State Fish Commission was also claiming credit for the increased cannery pack, although it attributed the increase in chinook salmon to its own hatcheries.[88]

In reality, however, the release of only 1.4 million fingerlings could not have produced the large increase in cannery output in 1914. Although the increase in abundance was real, its true cause was more likely favorable climatic conditions.[89] The Umatilla River, which had no hatchery, also showed a large increase in the number of the chinook salmon in 1914.[90] Attributing this increase in the chinook salmon harvest to the hatchery program was just another example of how willing fish culturists were to place blind faith in technology and to draw favorable conclusions from inadequate data. As long as salmon remained somewhat abundant, and as long as claims of success were not based on careful evaluation, hatcheries could continue their operations, stripping millions of eggs from wild fish.

The continuing failure of fish culture to replenish rivers or forestall depletion prompted many biologists to call for a critical evaluation of hatchery programs, and by the 1930s it became clear that a scientific basis for fish management was needed.[91] For example, fisheries expert Paul Needham told managers that they should carefully evaluate the results of their hatchery operations. He drew an apt comparison with the banking industry by asking, "How long would a bank last that invested $500,000 without the slightest knowledge of what return it might expect annually from its investment? Just this sort of thing has been going on for years with regard to fish propagation."[92] Needham thought the lack of evaluation was an honest mistake, but if it was, it was an inevitable mistake with origins deep in the ideological roots and political maneuverings that underlay the beginnings of fish culture in the United States. And it was a mistake that would not be corrected in Pacific salmon hatcheries for several decades.

Scientific evaluations of hatchery programs were carried out in other regions, and they led some biologists to a loss of confidence in artificial propagation. Ralph Hile took a close look at the artificial propagation of the yellow-pike perch in the Great Lakes and concluded that the increases in the abundance of the fish were due not to hatchery programs but to natural fluctuations. He found absolutely no correlation between the number of fry planted and the resultant catch of adults in subsequent years. Hile concluded that in the Great Lakes, fish culturists had underestimated the value of natural production and overestimated the value of artificial production.[93]

Scientific skepticism about hatcheries had grown in Canada as well. In 1922, the British Columbia Fisheries Commission recommended that a scientific experiment be conducted over the course of several years to evaluate the success of hatcheries. After careful planning, biologist Russell Foerster started the experiment at a hatchery on Cultus Lake in the Fraser Basin in 1925. Foerster's study marked a major departure from all earlier evaluations of salmon hatcheries in that it sought to estimate the value of artificial propagation not by measuring the number of fish released but by comparing the efficiency of propagation to that of natural spawning. After more than fifty years of hatchery operations, Foerster would finally test the basic premise underlying their use—that artificial propagation was more efficient and successful than natural propagation.

The study lasted ten years, and in 1936 Foerster reported his findings: the efficiency of artificial propagation was not significantly different than that of natural propagation. In streams where the natural population had a reasonable chance of successful spawning, the Canadian government could not justify the additional expense of a hatchery for the negligible benefits it would produce. As a consequence of Foerster's study, the Canadian Fisheries Department immediately announced the closure of all its salmon hatcheries.[94] For the first time, a major decision regarding artificial propagation was based on a well-designed scientific experiment. This decision should have demonstrated to policymakers that science could be useful in the development of hatchery policy. It should have alerted the neighboring governments of Washington and Oregon to the flawed premise of their hatchery policies. But it did neither.

The ideological and political underpinnings of the American program prevented it from acknowledging the consequences of the Canadian study. In fact, Foerster's study was, by agreement, supposed to apply to both sides of the border. The United States had agreed that the Cultus Lake study would stand as the test for artificial propagation—that Americans would not duplicate the research. But when the results cast doubt on hatcheries, they were simply ignored.[95]

It's possible that Henry O'Malley, the U.S. fish commissioner, had a reasonable scientific justification for ignoring Foerster's study. The Canadians had studied a sockeye salmon hatchery, whereas chinook salmon were the species of prime importance on the Columbia River. Sockeye and chinook salmon have very different life histories, so their performance in hatcheries may not have been comparable. But that does not explain why O'Malley ignored an evaluation of chinook salmon hatcheries in the Columbia River made by one of his own employees, Willis Rich.

In the same year Foerster started his field study, Rich completed a statistical evaluation of the Columbia River hatcheries. He opened his report by reminding the reader that the Columbia River is commonly believed to "most conspicuously show the benefits of artificial propagation." Then he explained how his analysis showed "that the available statistics do not warrant the conclusion that the maintenance of the salmon pack is due to artificial propagation." Rich admitted that hatcheries did some good, but his research showed that their contribution to the runs was minuscule compared to the results of natural production. He was a careful scientist, so he emphasized the uncertainty inherent in his work. "It must be concluded, therefore, that there is no evidence obtainable from a study of the statistics of the pack and hatchery output that artificial propagation has been an effective agent in conserving the supply of salmon. The writer wishes again to emphasize the fact that the data here presented do not prove that artificial propagation may not be an efficient measure in salmon conservation. These data prove only that the popular conception, that the maintenance of the pack on the Columbia River is due to hatchery operations, is not justified by the available evidence."[96]

In a report published in 1930, John Cobb, dean of the School of Fisheries at the University of Washington, actually identified hatcheries as a major threat to salmon fisheries. Cobb worried about the region's "almost idolatrous faith" in artificial propagation, and he regarded the lack of critical evaluation as a serious threat to the Northwest salmon industry.[97]

Despite its roots in fish culture, even the American Fisheries Society, at its seventy-first annual meeting in 1941, strongly criticized hatcheries. During the business session, when John Gottschalk delivered the report for the society's Division of Fish Culture, he concluded that "the most significant trend in fish culture is the increasing doubt in the ability of hatcheries to perform the task assigned to them." According to Gottschalk, "Should an outsider decide to investigate the reference literature on fish culture he would wonder what the word 'progress' means, when he can find statements of 30 years ago which coincide, almost word for word, with statements made during the past 5 years."[98]

Gottschalk told his colleagues that in spite of the overly optimistic promises made by the fish culturists of 1941, which were the same as the claims made thirty years earlier, the public recognized the failure of hatcheries. It was high time, he said, that the profession step in to correct the situation before the public had to. If fish culturists did not admit to past mistakes and give a fair evaluation of the goals hatcheries could achieve, he warned, they would be out of a job.

The mounting scientific criticism of hatcheries was on the brink of making a difference, but at the same time, biologists, managers, and culturists were confronting major new threats to the Pacific salmon. The region was on the verge of rapid development of its hydropower resources, which would stimulate a population explosion, a boost in timber harvest, an increase in water withdrawals from rivers, and a rise in fishing pressure from a growing sport fishery. The salmon managers of the Pacific Northwest faced these changes with a toolbox containing essentially just one tool, the hatchery—a tool whose real performance for seventy years had been hidden behind a shield of ideology, blind optimism, political maneuvering, and, with few exceptions, self-serving science.

CHAPTER

The Winds of Change

The morning when I began work on this chapter, I was in a Portland hotel room trying to squeeze in a few hours of writing before attending a meeting about the Pacific salmon. The *Oregonian* lay on the floor next to the desk. Its boldfaced headline stared up at me, pulling my attention away from the stack of notes and the blank piece of paper: "[Governor] Kitzhaber Convenes a Summit on Salmon," the headline said. "Four Northwest governors and leaders of nine Indian tribes sit down to study the problem."[1] As I read the article, I couldn't help but think about Santayana's warning that those who fail to understand history will be condemned to repeat it.

For several days after that headline caught my attention, various pundits described the meeting as "unprecedented and historic." Obviously, they knew little about the history of salmon restoration. The meeting may have been historic and important, but it certainly was not unprecedented. When the summit participants stated that current efforts to restore the Columbia Basin salmon were not working, the *Oregonian* reported that conclusion as news. But concern about the failure of salmon restoration was news seventy-two years ago; today such reporting merely proves Santayana's admonition.

Nowadays, hundreds of meetings are convened each week to discuss salmon, their management, their status, their cultural importance, harvest rates, endangered species implications, agency budgets, politics, and so on. Regional and international meetings have become so commonplace that salmon managers consider them part of the basic fabric of their work. Given this sea of meetings, rarely if ever can any one of them be accurately described as historic or unprecedented. In 1925, however, the situation was different: a historic, newsworthy, and truly unprecedented meeting about salmon did take place.

Groping in the Dark

By the mid-1920s, even the most die-hard optimists recognized that the salmon industry needed help. Throughout the Pacific Northwest, salmon harvests were clearly declining. For the first time, salmon managers began to question the effectiveness of hatcheries. After sizing up the situation, Ed Blake of the Washington State Fisheries Board stated that by 1925, "the decline of the runs had been constant, and had approached the danger point." Blake concluded that "either artificial propagation was ineffective as then conducted by the state, or that the magnitude of the operations was inadequate to meet the drain on the annual runs."[2]

Against this backdrop of declining salmon abundance and eroding confidence in contemporary management practices, Miller Freeman, publisher of *Pacific Fisherman* magazine, organized a meeting of fisheries executives and scientists from all the salmon-producing states (except Idaho) and British Columbia. Twenty-six men from Alaska, British Columbia, Washington, Oregon, California, and Washington, D.C., met for two days at Seattle's Olympic Hotel to discuss their mutual salmon management and conservation problems. When Freeman opened the first session, he called the meeting unprecedented and historic. It was the first time that representatives from all the institutions concerned with salmon had come together in a single place.

The biologists and executives willingly came to Seattle because, like the people attending the meeting that opened this chapter, they recognized that their attempts to maintain salmon abundance were failing. Salmon harvest in most major subregions of the Pacific Northwest had already peaked (Table 5.2, Chapter 5). In fact, the salmon were beginning a century-long slide toward depletion, regional crisis, and extinction, although that was all in the future; in 1925, the meeting participants did not foresee the general collapse of the salmon fisheries. The executives and biologists realized they had a problem, but the transcript of the meeting's discussions does not show the sense of urgency they would have felt had they known the real dimensions of the crisis they were facing.

Through their often freewheeling discussions, meeting participants identified four primary problems they needed to solve to stop the salmon's decline: facilitating the transfer of salmon eggs from one region to another, eliminating the enemies of salmon (predators), controlling the ocean troll fishery, and securing salmon passage at high dams.

When the participants discussed ways to improve the transfer of salmon eggs from areas of high abundance to areas with shortages, a fundamental split in the thinking about the use of artificial propagation became apparent. J. R. Heckman of the Alaska Territorial Fish Commission showed a particular largesse with his state's salmon eggs, explaining that he would "always feel like helping the State of Oregon, or the State of Washington by letting them have eggs."[3] But an exchange between E. A. Sims of the Washington State Fisheries Board and Norman Scofield, a biologist in charge of California's commercial fisheries, revealed that different attitudes toward artificial propagation were emerging. Here is the heart of their exchange from the transcript:

> *Sims:* Mr. Scofield, what do you think about the interchange of eggs?
>
> *Scofield:* I do not know that we are especially interested in California.
>
> *Sims:* Well, maybe I have not just stated it right, maybe it might not be an interchange. For instance, from the statement made by Mr. O'Malley this morning, I take it that the Sacramento River is in need of seeding now, unless you have an ample supply from other sources, you are going to require seeding from some source for that river, are you not?
>
> *Scofield:* I do not know. I have not looked at it in that way. My idea was that by stopping the outside trolling we would get enough salmon into the river to take care of it.[4]

Heckman and Sims represented the long-held view that artificial propagation could overcome all problems. To them, a salmon was a salmon. Alaskan salmon were no different than California salmon, and maintaining abundance was a simple logistical issue—the transfer of eggs from areas of abundance to areas with shortages.

Scofield, a biologist trained by pioneering salmon researcher Charles Gilbert, reflected a different view. From his perspective, ensuring an abundant supply of salmon was not a mere matter of logistics but an enterpise that required stewardship and effective regulation of human activities such as fishing. Scofield's attitude reflected the growing skepticism toward hatcheries among the small group of trained fisheries biologists working in the Pacific Northwest.

In 1925, salmon managers did not view rivers as ecosystems composed of predators and prey interacting in natural, healthy ways. To them, salmon rivers contained friends and enemies. Friends should be cultivated and protected, their thinking went, and enemies should be killed. The salmon were friends, but many species native to Northwest ecosystems made the enemies list, including sea lions, seals, killer whales, eagles, terns, gulls, Dolly Varden trout, beluga whales, beavers, kingfishers, and even water ouzels. When conference participants talked about predators—enemies of the salmon—their discussion generally reflected contemporary attitudes in natural resource management. Such attitudes are best illustrated by William Hornaday's 1935 book, *Hornaday's American Natural History,* which put most birds of prey on the enemies list. For peregrine falcons, Hornaday recommended, "First shoot both male and female birds, then collect the nest"; for the sharp-shinned hawk, he wrote, "There can be no question regarding the necessity for the destruction of this bird, wherever it is found." Hornaday called the Cooper's hawk "a companion in crime to the proceeding species [sharp-shinned hawk], equally deserving of an early and violent death."[5]

This attitude made the killing of predators an essential part of fish and wildlife management. The number of "enemies" killed in the name of salmon conservation was staggering. In 1923, Oregon destroyed about 7,000 seals or sea lions; Alaska killed 3,100 seals in the Copper River Delta with high-powered rifles and dynamite, and shot 20,000 eagles as well; and British Columbia machine-gunned 2,700 hair seals and sea lions.[6]

On the issue of predators, Norman Scofield's attitude again differed from that of most of the others at the meeting. Scofield was interested in facts, not myth or speculation, and he was not convinced there was enough evidence that marine mammals were actually reducing the abundance of salmon: "The whole matter, as it strikes me, is a good subject for investigation, to get the definite proof." And when questioned about predation by birds, Scofield responded, "Oh, I do not think we have any good evidence against the birds."[7]

Most of the meeting participants found common ground on ocean trolling, however. Ed Blake, from the Washington State Fisheries Board, considered the offshore trollers who fished for immature salmon to be another of the enemies of salmon. Blake did not propose to treat the trollers like eagles or seals, but he pressed to eliminate them through regulation of offshore fishing. But while the participants recognized that the troll fishery was a developing problem, they felt their hands were tied. Because the trollers were fishing beyond the three-mile limit, there was little the individual states could do by themselves, short of an interstate compact and an international treaty.

Finally, on the fourth issue, John Cobb, dean of the College of Fisheries at the University of Washington, discussed his research on the difficulty of passing salmon over high dams—a problem of increasing urgency. Then he gave the group his pessimistic conclusion. "There is only one sure way of getting fish up [the rivers], and that is to have an open river," he told the meeting participants, "but as we apparently cannot get away from the development of the power and irrigation projects . . . , I feel we have got to go as far as we possibly can in this work of investigation."[8]

The discussion about dams and fish ladders produced a remarkable exchange between John Babcock of British Columbia and Captain Harry Ramwell of the Washington State Fisheries Board. Their words accurately predicted the future of the two great salmon rivers of the Northwest, the Fraser and the Columbia, and revealed the different attitudes toward development held by Canadian and American participants at the meeting. When Babcock asserted that the construction of dams for power production on the Fraser River would be limited to sites in the headwaters beyond the distribution of salmon, Ramwell expressed disbelief. Reflecting the American attitude that development was inevitable, Ramwell thought that every possible site for power production on the Fraser would be developed. He pressed Babcock, "What is the matter with Hells Gate? Suppose they dammed Hells Gate?" Babcock responded, "If a dam were put in Hells Gate, we would take it out as quick as we could get it out." Not satisfied, Ramwell pushed Babcock harder: "But suppose someone came along and said they wanted to dam Hells Gate just like they did Priest Rapids?" Babcock held firm: "They will never get it."[9]

John Cobb then shifted the conversation in a new direction, asking what dams such as the one slated for Priest Rapids would mean to the salmon. Carl Shoemaker of the Oregon Fish Commission gave him an answer: "It simply means you are through with your salmon, that is all."[10]

After the discussions of egg transfers, predators, ocean trolling, and fish passage, John Babcock and Charles Gilbert of Stanford University introduced the most important subject discussed at the 1925 meeting—a subject that underlay all the others. There was a fundamental need for more knowledge about the salmon. Babcock stressed, "We have got to have the facts . . . scientifically ascertained." Gilbert reinforced Babcock's concerns, telling the group that in spite of some progress, "We cannot any of us point with pride to the advance that has been made."[11] He then described ongoing efforts to manage salmon without basic facts as merely groping in the dark. Comparing salmon management to agriculture, he explained that a farmer would quickly fail if he tried to practice his profession with the same ignorance of basic facts that plagued salmon managers.

Gilbert was obviously frustrated by the low priority that state and federal agencies had given to research. The Seattle meeting would bring about some changes, but Charles Gilbert did not live to see them fully implemented. He died at age 68, three years after the Seattle meeting.

In the Pacific Northwest, March is often a cold and blustery season. While brisk March winds blew off the waters of Puget Sound, inside the Olympic Hotel a different kind of wind began blowing—a wind of change for salmon management. In spite of their persistent attachment to hatcheries, the practice of killing predators, and the massive transfer of eggs between regions, biologists and executives at the Seattle meeting did open the door, however slightly, to a different future for the Pacific salmon. Scofield's views on egg transfers and killing predators, together with Babcock and Gilbert's plea for better scientific knowledge, signaled the potential for real changes in state and federal salmon management programs.

Science and Salmon Management

Less than two years after attending the 1925 Seattle meeting, Henry O'Malley, the U.S. fish commissioner, assembled his scientific staff to discuss the need for a new approach to fisheries management. It was the first time the Division of Scientific Inquiry had come together in the bureau's forty-five-year history.

O'Malley opened the meeting with what amounted to a confession of failure. In a typical bureaucratic softening of the obvious, he told his scientists, "I should not care to say that the Bureau's work had amounted to nothing, but we must conclude that our efforts have not been sufficient to maintain the fisheries in their former state of productiveness."[12]

O'Malley went on to explain that the declining fisheries had brought about a broad change of attitude. Congress, which in the past had wanted to fund only hatcheries, was now willing to fund biological investigations to gain the information needed to reverse the slide. With apparent enthusiasm, O'Malley proclaimed that the time when fisheries investigators were looked upon as odd "bug hunters" was now over.

As the leader of the Division of Scientific Inquiry, Elmer Higgins continued with O'Malley's theme. Fifty years ago, he told the scientists, North America was young, with virgin natural resources. Those who had worked to develop the nation's seemingly unlimited resources simply had not considered the possibility of depleted fisheries. Then Higgins went straight to the heart of the matter, explaining that "the great promise and popular success of fish culture induced a complacent confidence, as it was believed that the control of the fish supply was in easy grasp."[13] Higgins reminded the scientists of the outlandish claims of success made by George Brown Goode fifty years earlier. "Here the fish culturist comes in with the proposition that it is cheaper to make fish so plentiful by artificial means, that every fisherman may take all he can catch, than to enforce a code of protection laws. The salmon rivers of the Pacific slope . . . and the shad rivers of the East and the whitefish fisheries of the lakes are now so thoroughly under control by fish culturists that it is doubtful if anyone will venture to contradict his assertion."[14] Higgins proceeded to explain what had actually happened:

> How well founded was this faith in the all-effectiveness of fish culture in maintaining or restoring the fisheries in the face of all possible destructive influences may be seen by the fate of the three great fisheries that he [Goode] chose as illustrations. Despite fifty-five years of artificial propagation the Pacific salmon fishery has declined alarmingly, the pack in Puget Sound in 1924 amounting to but 12% of the peak production in 1913. The shad fisheries of the East Coast declined, on the whole, 74% from 1896 to 1923, and now are totally destroyed in many of our rivers. The whitefish fisheries of the lakes, despite an annual distribution of 409,000,000 eggs and fry per year, have declined in yield from first place in 1880 to fourth place in 1922, with a yield only slightly greater than that of suckers.[15]

Yet even with these failures, fish culture was too strongly entrenched in the state and federal programs for Higgins to be too critical of it. He placed part of the blame for depleted fish stocks on fisheries biologists for neglecting fish culture in their research and on fisheries administrators for accepting the promises of fish culture uncritically.

During the meeting, the need for better scientific knowledge in order to wisely manage fisheries emerged as a painfully obvious problem. At the same time, a similar awareness grew throughout the region as other salmon management institutions, academics, and research organizations confronted the salmon's persistent decline.

Higgins and O'Malley were probably sincere in their call for more research. They had little choice: faced with a looming problem, they could not simply propose the status quo as a solution. They needed new ideas, and to generate them they needed more and better information. What new information they did receive, however, was selectively used. If it did not support the continued use of fish culture, it was ignored. For example, O'Malley took no action on the findings of two studies that called into question the efficacy of hatcheries.

While the winds of change were not blowing hard enough to shake the ideological roots of fish culture, stirrings of change did gradually spread from the federal fisheries agency to the entire Northwest, including Canada. Biologist William Ricker, who compiled an account of the Fisheries Research Board of Canada's first seventy-five years of operation, considered 1925 to be a transitional year for the board. At that time, its function changed from providing temporary facilities for academics researching aquatic biota to employing a full-time scientific staff to conduct research aimed at solving problems in fisheries management.[16]

Change also came to the University of Washington, one of the first schools in the Northwest to train fisheries professionals. After the death of John Cobb in 1929, the new dean of the School of Fisheries, W. F. Thompson, strengthened the students' training in science. He canceled classes in fishing methods and canning technology and instead adopted a curriculum emphasizing fisheries biology and conservation with close ties to mathematics and chemistry.[17]

State fish commissions also began to face the failures of their salmon programs during this period. But in those agencies, competition among different points of view made the direction of any change uncertain. Some state fisheries managers favored natural production and called for the closure of hatcheries; they proposed to spend the savings on stream and lake restoration. But proponents of artificial propagation, such as O'Malley, pushed for more research on hatchery operations and an increase in the capacity to restock waters.[18] Still other managers called for a third approach that relied on basic business principles and reduced salmon management to issues of supply and demand.

The earliest attempts to introduce ecological thinking into institutions dominated by fish culturists ran into stiff resistance, and sometimes the conflict between the fish culturists' agricultural approach to management and the biologists' ecological approach erupted into the public forum. In 1917, the Oregon Game Commission established its first permanent position for a biologist and hired William Finley, a nationally known naturalist and wildlife photographer, to fill it. Finley and the commission did not see eye-to-eye, however; he did not blindly accept the success of hatchery operations.[19] His view of fish and wildlife management encompassed whole ecosystems, a view that was too broad for the game commissioners. Consistent with his ecological view, Finley considered nongame species, such as eagles, hawks, and chickadees, as worthy of study. Moreover, he believed their protection was part of the commission's mission. But the commissioners thought management efforts should focus exclusively on providing quarry for fishermen and hunters.[20] Friction between Finley and the commission grew, and on December 11, 1919, he was fired.

Finley's dismissal caused a huge public uproar. Oregon grade school students signed petitions to reinstate the fired biologist, and the governor called for an investigation.[21] But the firing stuck. Nevertheless, an editorial in Portland's *Oregon Journal* indicated that Finley's approach to fish and game management was shared by many Oregonians:

> Members of the Fish and Game Commission seem to miss the point in the popular demand for retention of Mr. Finley.
>
> The protest is not merely from sportsmen. It is not a mere question of protecting wild game.
>
> It is the much higher and more important thought of bringing all people into touch with the great outdoors and nature and true living. . . .
>
> When you see the outdoors as Finley's peculiar genius reveals it, . . . when you are brought in closer contact with the hills and valleys and the eternal things out in the wilds, a new world is opened with new charms and new prospects and new hopes that you had never visioned or imagined.
>
> It is life made more beautiful and the world painted in lovelier color. It awakens a new pride of country, a new realization of the abounding things to please and delight us out there in rural Oregon. . . .
>
> All these things in the wild life and mountain habitations were created as a part of the great scheme of the world for the happiness and benefit of man. . . .
>
> They are quite as important as salmon and salmon canning. They are more so because they can touch for the better the lives of all the people, not a mere few.[22]

Finley's approach to fish and wildlife management may have been ahead of its time, but the ideas expressed by the *Oregon Journal* suggest that his values were shared by other Oregonians. At least some people were beginning to realize that nature was worth more than the sum of its commodities. But the Oregon Game Commission failed to recognize these new values and held firmly to the status quo.

Eighty years later, the winds of change still have not shaken the foundations of the status quo. In Oregon, with coho salmon in coastal streams in danger of extinction, an overwhelming majority of Oregonians support salmon restoration based on ecological values. But the state's Fish and Wildlife Commission is once again out of step with the public, regularly devising excuses to permit commercial and sport harvest of the few remaining endangered fish.[23]

In June 1998, the Washington Fish and Wildlife Commission fired Bern Shanks, the agency's director. While budget problems were given as the reason, many believed Shanks was fired because he was "the first director in twenty years to look out for the fish. The budget deficit is a red herring."[24] Shanks was an advocate for wild salmon. He proposed cutting back commercial fishing, more protection for streams, and changes in hatcheries. Those proposals to change the status quo created too many opponents. Shanks's ideas, like Finley's, were ahead of their time.

In 1931, Oregon adopted a business approach to the management of fish and game, launching an ambitious ten-year program whose goal was to saturate Oregon's habitat with artificially propagated fish and game. This abundance would attract tourists, which in turn would provide economic benefits that would more than pay for the program. Although the state fish and game commission ostensibly recognized the need for research, its policies were guided foremost by business principles. In real terms, science was given a low priority. At that time, the amount of money allocated to scientific investigations (2 percent of the budget) was less than the amount allotted to the business section for accounting purposes. Artificial propagation of fish and game received 84 percent of the budget.[25] Although adopting a business approach to salmon sounded like a new development, the state's fundamental management practices stayed the same. Maintaining the supply of commodities for harvest was still the highest priority, and artificial propagation remained the way to achieve that goal.

One of the greatest impediments to developing a science-based approach to salmon management was the absence of trained biologists in state management agencies. Universities were just beginning to graduate qualified fisheries biologists, but management agencies, which were dominated by fish culturists, were reluctant to hire them. The culturists had grown used to not verifying their claims of success, and they viewed biologists' skepticism with alarm.[26]

In the 1930s, Oregon's fish and wildlife management was a centralized and political operation. It emphasized two programs: enforcement and artificial propagation. There were no inventories of fish and wildlife and very little research.[27] As late as 1936, out of 271 positions in the Oregon Fish Commission, only one was held by a fisheries biologist.[28]

The "management" of ecosystems for desired species only, which led to the killing of predators and to hatchery stocking, was so deeply and thoroughly entrenched that it was heresy to question it. The old unfounded beliefs, especially those regarding artificial propagation, were held so strongly that they effectively undermined most attempts to develop a scientific basis for salmon management in the 1920s and 1930s. Wildlife biologist Durward Allen described the situation within fish and game departments this way:

> Their old-line game keepers and fish culturists were skilled artisans who had learned through long apprenticeship, the trade of raising game birds and fish. In public utterances, they did not hesitate to interpret their experience (frequently excellent) in terms of on-the-land management, game in the bag, fish in the creel. They got away with it because we, the sportsmen, did not know where their livestock rearing knowledge ended and their ecological ignorance began. They didn't either; and they reacted instinctively against anything that appeared to call their art to question. . . . The effect of it all was to put the biologist in a difficult position when he arrived on the scene and found things that needed to be changed.[29]

This was unfortunate because the Pacific Northwest was on the verge of massive hydroelectric development and broad expansion of irrigated agriculture. These changes would radically alter salmon habitat in the Columbia and other major salmon-producing rivers. So at a time when scientific expertise was needed most, state management agencies were only just beginning, and then reluctantly, to hire appropriately trained personnel.

Old Myths Dispelled

Although state agencies were slow to reap its benefits, the science of fisheries biology had been slowly growing in universities for some years. The center of this new science was Charles Gilbert's program at Stanford University. Of the men seated around the meeting table at the Olympic Hotel in Seattle for those two days in March 1925, Gilbert understood, more than anyone, that salmon management suffered from the lack of basic knowledge about the salmon's biology and natural history. For more than thirty years before the Seattle meeting, he had led a group of scientists researching the biology of the Pacific salmon. The professional field of fisheries biology in the United States was launched by Gilbert and his students; according to some, Charles Henry Gilbert was the first practicing fisheries biologist in this country.[30]

Gilbert received the first doctorate awarded by Indiana University, where he specialized in ichthyology and the systematics of fishes. When a colleague at Indiana University, David Starr Jordan, became the founding president of Stanford University, Gilbert went west with him. In 1891 he was appointed chairman of Stanford's zoology department.

Shortly after arriving on the West Coast, Gilbert and his students began their studies of the Pacific salmon. Many of those students went on to make major contributions to fisheries science, including Willis Rich, Cludsley Rutter, Norman Scofield, Carl Hubbs, Harlan Holmes, Frank Weymouth, Francis Clarke, and W. F. Thompson. One of Gilbert's students remained interested in fish and became an avid fisherman, but took a different and particularly notable career path. That student was Herbert Hoover, the thirty-first president of the United States.

When Gilbert began his studies, knowledge about the biology of the Pacific salmon was rudimentary. The nature of the salmon's life cycle contributed to the paucity of information. The salmon were easily observed at only a few points during their lives—when they entered the rivers and filled the waiting nets and traps, and when the big fish spawned in the small tributary streams. As a result, upstream migration and spawning became the focal points of knowledge about the fish. The rest of the salmon's life cycle, the part between spawning and adult migration, remained mysterious and hidden from view. Given the economic importance of the salmon-canning industry, however, it is not surprising that people used self-serving theories and unfounded myths to fill the gaps in the available information. For example, when cannery operators theorized about the salmon, their speculations were often contrived to rationalize intensive fishing and to support artificial propagation as an alternative to stewardship. In some cases, myths continued to influence salmon management long after science showed they were false.

One key issue, hotly debated as late as the 1930s, was the homing ability of the Pacific salmon. Even respected scientists such as David Starr Jordan did not believe that the salmon had any special ability to find their home stream. He thought that once the salmon entered the ocean, they migrated only twenty to forty miles from the river they had left as juveniles; when they matured and began their spawning migration, they simply swam up the first river they came to, and because it was close by, it was usually their home stream.[31] If the salmon randomly selected their spawning streams, as Jordan believed, that meant the different species of salmon were genetically uniform over large parts of their range.[32]

The beliefs that the salmon could not home to their natal stream and that each species was genetically uniform went hand in hand. By accepting these speculations as truth, salmon managers in effect rationalized the widely used but detrimental hatchery practice of transferring salmon eggs between hatcheries and rivers. Because fish culturists sought to keep hatchery ponds fully stocked (full ponds and the number of juveniles released, not returning adult fish, were the standard measure of success), they moved eggs and fry from hatcheries with surpluses to those with shortages. Belief in genetic uniformity precluded biological arguments that might call this widespread practice into question.

The structure of salmon populations within a species was also the subject of debate through the 1920s. For example, some biologists, including Hugh Smith of the U.S. Bureau of Fisheries, argued that there were no genetic differences among the races of chinook salmon. The chinook were divided into four races according to the time of their spawning migration: spring, summer, fall, and winter. The spring and fall races were most common, although many rivers supported a summer chinook run as well. The winter race was restricted to the Sacramento River. According to Smith, the return of a chinook salmon in either the spring or fall was a random event; the salmon's race was not an inherited trait.[33]

This theory had important implications for the commercial fishery. The fishermen targeted spring-run fish because their higher fat content made the best canned product, and the fall chinook with their lower fat content were considered inferior. If all chinook were part of the same genetically uniform pool of fish, and random choice determined the time of their return to the river, then cannerymen could fully harvest the spring run with little or no regulation, as long as enough of the inferior fall-run fish were allowed to spawn and replenish the pool. Hatcheries had only to propagate the fall fish for the same reason.[34] Because these unfounded assumptions about salmon biology perpetuated the status quo, that is, little or no harvest regulation, they remained widely held even after the pioneering studies of Charles Gilbert and his students proved them wrong.

As early as 1913, Gilbert had used an ingenious method to work out the basic life histories of the Pacific salmon. Life history simply means what the salmon do, where they do it, when they do it, and how they do it. Gilbert's methodology was similar to the use of tree rings to analyze the growth history of individual trees. The salmon's scales grow roughly in proportion to the growth of the fish, and as the scales grow, circular ridges are deposited on their surface. These rings, called circuli, are similar to the rings in a cross-sectional cut of tree. Narrow spacing between tree rings means slow growth, whereas wide spacing indicates rapid growth. The same is true for the pattern of circuli spacing on the scales of adult salmon. Thus each fish carries on its scales a capsule history that includes information about its past growth; the age at which it migrated to sea; the relative time it spent rearing in fresh water, estuary, and ocean; and whether its parents spawned naturally or in a hatchery. It's as though each salmon has kept a diary that can be conveniently retrieved by biologists.

Much of what Gilbert learned through his analysis of scales still forms the basis of our knowledge of the salmon's life history. For example, Gilbert divided juvenile salmon into two general categories, stream and ocean, based on their life histories. Stream-type fish remained in fresh water for a year and migrated to sea in the spring of their second year of life. Ocean-type fish migrated to sea shortly after emerging from the gravel during their first year of life.[35]

Gilbert published one of the earliest reports supporting the home stream theory based on the information he derived from the analysis of scales. In his 1914 study of sockeye salmon in British Columbia, he showed that salmon from different rivers laid down unique patterns of circuli spacing on their scales, that is, they had different life histories. According to Gilbert, a biologist examining those patterns could identify which river a sockeye salmon came from. This led him to conclude that the sockeye salmon from Rivers Inlet and the Skeena and the Nass Rivers were each from a distinct biological race—a finding that disposed "effectively of the general question concerning the return of our salmon at maturity to the river basins from which they were hatched." Gilbert wrote, "It can now be affirmed with entire confidence that they do so return, that they are effectively isolated, and that they interbreed thus within the limits of their colony."[36]

Although Gilbert was unequivocal in his conclusion, the debate about whether the Pacific salmon returned to their home stream to spawn continued for another twenty-four years.[37] It persisted because a key piece of information was missing: How far did the salmon migrate once they entered the ocean? If the salmon undertook a long ocean migration, then their return to the stream of their birth could not be a matter of chance, as Jordan hypothesized; it would require a special homing ability. But in 1914, the salmon's ocean-migration process was still a mystery.

Aside from his work on salmon homing, Gilbert's studies led to practical recommendations for management. It was the common practice of hatcherymen to release juvenile salmon to the rivers as sack fry shortly after hatching. However, Gilbert's life-history studies showed that most of the wild sockeye salmon surviving to the adult stage had spent an entire year in fresh water before migrating to sea. Gilbert reasoned that artificially propagated salmon, too, might survive at a higher rate if their release from the hatchery mimicked the natural pattern of spending a year in fresh water.[38] But although hatchery managers made a few experimental releases of older juvenile salmon, extended rearing in the hatchery did not become a common practice for another quarter-century.

To accept Gilbert's findings and release juvenile salmon at the time of their natural migration would have required a significant shift in the way salmon managers and fish culturists thought about nature. Hatcheries were based on the assumption that human control over production was more efficient and productive than natural ecological processes. To them, the natural river was a dangerous place full of enemies, so their primary objective was to circumvent and improve on nature, not to copy it.[39] The idea of mimicking the natural relationship between the river and the fish was fundamentally at odds with the very premise of hatchery work. Because salmon managers and fish culturists were not prepared to deal with this conflict between science and their own worldview, they ignored Gilbert's findings.

While Gilbert contributed to our understanding of broad issues such as the homing of salmon, he also examined more specific questions. He learned that salmon populations were composed of individual fish that exploited their environment in different ways. He speculated that this life-history diversity had an ecological basis, and that it was an integral part of the relationship between the fish and its environment. Gilbert was beginning to discover the secrets of the salmon's evolutionary history and map the life-history pathways the fish used to survive their trip through freshwater habitats. Recently, biologists have rediscovered the Pacific salmon's rich biodiversity and are beginning to understand that it is a critical part of the salmon's evolutionary history, necessary for the long-term survival of individual populations.[40]

Gilbert died before all the pieces of the salmon's life-history puzzle were put together, but many of his students continued to make major contributions to fisheries biology and management. Willis Horton Rich stands out for his contribution to our understanding of salmon biology and the role he played during the era of large-scale hydropower development in the Northwest. Though he remained affiliated with Stanford University throughout his career, Rich often held additional positions with the U.S. Bureau of Fisheries, U.S. Fish and Wildlife Service, and Oregon Fish Commission. Many of the early salmon managers had little education in the biological sciences, and their approach had more in common with agriculture than with ecology or zoology. Rich, however, was a Stanford-trained biologist whose background gave him an entirely different perspective on the salmon and their management.

Rich believed life-history studies were the key to understanding the salmon and developing effective management programs. He continued and expanded his mentor's earlier work. As a result of studies he conducted between 1914 and 1929, Rich learned that juvenile chinook salmon migrated through the mainstem of the Columbia River all during the year. He described this continuous migration as a well-orchestrated event composed of many independent populations. Each population of salmon left its tributary spawning and nursery areas and began its migration to sea according to unique environmental cues.[41] The salmon, as Rich began to understand them, were richly diverse in their life histories and in how they responded to the complex mix of habitats in the Northwest.

Rich then teamed up with Harlan Holmes on a study that debunked the myth that the spring and fall runs of chinook salmon were random events rather than inherited traits. They conducted a series of experiments to determine if it was "necessary to breed from fish of the spring run in order to produce spring-run fish," or if it was possible, "by proper handling of the progeny of the fall run, to produce fish that will return as adults to fresh water early in the spring?"[42] They proved that the races of chinook salmon were determined by heredity: adult spring-run salmon produced offspring that returned to the river in the spring, and adult fall-run salmon produced offspring that returned in the fall.

While Rich was examining the salmon's freshwater life, Canadian biologists were tracking their ocean migration. When they found chinook salmon from the Columbia River off the west coast of Vancouver Island, they confirmed that salmon did indeed make extensive ocean migrations. The discovery of Columbia River salmon so far from their home stream was the key piece of evidence needed to prove the remarkable homing ability of the Pacific salmon.[43]

By 1939, Willis Rich had amassed enough information to present his findings to the American Association for the Advancement of Science. He concluded that the Pacific salmon returned to their natal river to spawn and that each species of Pacific salmon was composed of many local, self-perpetuating populations. He then constructed a new and revolutionary vision of salmon management:

> In the conservation of any natural, biological resource it may, I believe, be considered self-evident that the population must be the unit to be treated. By population I mean an effectively isolated, self-perpetuating group of organisms of the same species regardless of whether they may or may not display distinguishing characters and regardless of whether these distinguishing characters, if present, be genetic or environmental in origin. Given a species that is broken up into a number of such isolated groups or populations, it is obvious that the conservation of the species as a whole resolves into the conservation of every one of the component groups; that the success of efforts to conserve the species will depend, not only on the results obtained with any one population, but with the total of individuals in the species that is contained within the populations affected by the conservation measures.[44]

In this single paragraph Rich proposed a complete overhaul of salmon management. Instead of treating all salmon of the same species as though they were part of one homogenous pool, Rich proposed an approach that focused on the individual population and its home habitat. According to Rich, the watershed and its unique salmon population should become the basic unit of management. Hatchery operations could no longer treat salmon as so many uniform widgets. It was no longer acceptable to devise myths about the salmon's biology and ecology in order to perpetuate preexisting harvest and hatchery strategies.

These implications were immediately evident to some salmon managers. In the same year Rich published his paper, Hugh C. Mitchell, director of fish culture for the Oregon Fish Commission, recognized that "the older system of transferring by truck fish raised at a station on one stream to another stream for liberation, is now considered undesirable" and that changes would have to be made. "With this in mind," he wrote, "the policy has been adopted, insofar as the available funds will permit, to establish and operate small stations on such streams of the state as are suitable for salmon runs."[45]

Old practices and myths are difficult to change, however. In spite of Mitchell's new policy, the transfer of salmon between rivers and hatcheries continued. For example, the Oregon Department of Fish and Wildlife continued to stock winter steelhead from the Alsea River hatchery in nearly all its coastal streams, and the Washington Department of Fish and Wildlife distributed Chambers Creek fish throughout its western streams.

In 1995, following a study of the effects of hatcheries on native coho salmon in the lower Columbia River, biologists from the National Marine Fisheries Service concluded that one of the factors contributing to the loss of the native coho was the transfer of hatchery fish between tributaries in the lower river. They advised that such transfers be stopped.[46] Their recommendation came sixty-nine years after salmon managers first recognized that the practice was detrimental. The obvious question is why did it take so long?

By the 1940s, scientists had made significant progress toward unraveling the mystery of the salmon's biology. Many of the old myths had been dispelled by facts. The homing of salmon to their natal streams, the hereditary basis of the races, the subdivision of the species into uniquely adapted populations, and the diversity of life histories within a population were giving biologists a solid picture of the salmon's biology and ecology.

Nonetheless, the myths—composed as they were of self-serving speculation bent to conform to the ideology of unconstrained development, the economics of the canning industry, or the wishful thinking of fish culturists—did not give way entirely to the new, scientific understanding of the salmon. The winds of change, while substantial, were not strong enough to blow away the false beliefs deeply embedded in the existing management programs. It was easier to believe that the salmon could be transferred from river to river to keep hatchery ponds full and to ignore the uniqueness of the individual populations. It was easier to believe that human control over reproduction was more efficient and productive, even though evidence to the contrary was accumulating. It was easier to continue groping in the dark.

In the 1930s and 1940s, several forces converged on the Northwest that made that period a crossroads for the Pacific salmon. First, science brought a greater understanding of the salmon's biology and pointed toward better management practices. At the same time, persistent drought, excessive harvest, and habitat degradation sent salmon stocks into significant decline. This should have caused salmon managers to reassess their existing programs and use the new scientific information to improve their programs. But the same drought that depressed salmon stocks created a dust bowl out of farmlands of the Great Plains, while the Great Depression idled America's factories. Millions of farmers and factory workers became homeless.

The government saw the Pacific Northwest and the hydroelectric potential of its rivers as a part of the solution to the nation's social and economic problems. Abundant electricity and expanded irrigation would enable the displaced farmers and workers to build new lives in the deserts of Washington, Oregon, and Idaho.

With this vision, the federal government authorized the construction of Bonneville and Grand Coulee Dams. Those two projects were only the first in an orgy of dam building. In the end, the John Day and Salmon Rivers would be the only large salmon-producing rivers of the Pacific Northwest to remain free of major dams.[47]

Although the work of Gilbert and Rich laid a foundation for understanding the salmon's biology, the implications of their work would not be recognized before the rivers of the Northwest went through another disastrous era of development. The emerging understanding of salmon biology was not compatible with the ideology of development and the vision of rivers brought fully under control of humans. So once again, science was bent to accommodate industrial economic goals.

Fisheries managers entered the crossroads of the 1930s and emerged on the other side still on the same path: hatcheries would make salmon compatible with fully developed rivers and degraded ecosystems. The opportunity for change was lost. Some would argue that there was no real crossroads, that there were no real choices, that full development of the hydroelectric and irrigation potential of rivers like the Columbia was inevitable, and that the status quo—continued reliance on hatcheries—was the only way to accommodate the difficult hand dealt to the Northwest's rivers. They are wrong. The best way to understand that there were other choices is to compare what happened on the region's two greatest rivers—the Columbia and Fraser. The stories of those two great rivers are taken up in the next chapter.

Chapter 8 A Story of Two Rivers

The Fraser and the Columbia Rivers are the two largest watersheds in the Pacific Northwest. Together they drain about 343,000 square miles—an area equivalent in size to the combined land masses of France and Germany. Each of these giants was the world's best at producing a species of Pacific salmon—sockeye for the Fraser and chinook for the Columbia.

The Fraser and the Columbia Rivers are very different ecosystems, characterized by different habitats. For example, the Fraser watershed contains an extensive system of large lakes, which are critical rearing areas for juvenile sockeye salmon. Their ecological differences notwithstanding, the story of their development and salmon management is instructive. Both rivers supported major salmon fisheries, and both showed signs of deterioration by the turn of the century and then rapid decline by the 1920s. The stories of these two great salmon rivers, the depletion of their salmon runs, and the subsequent attempts to restore them demonstrate that the current crisis has at least part of its roots in the failure of science to overcome nineteenth-century myths.

Key to this story are the major restoration programs initiated in each river within ten years of each other: the Fraser in 1937, the Columbia in 1947. Each effort took a very different approach to restoration, and over the course of fifty years—about twelve salmon generations—each has yielded very different results.

The Fraser River

The cornerstone of the Fraser River restoration program was the ratification of a treaty creating the International Pacific Salmon Fisheries Commission.

Before discussing what happened after the commission came into existence, some historical background on what led to the treaty is in order.

The Fraser is the largest river in British Columbia. With a catchment of 84,000 square miles, the Fraser (like the Columbia) penetrates the coastal mountains and drains a large part of British Columbia's intermountain interior. The Fraser starts its 850-mile journey in the Canadian Rockies and then flows northwest through the interior of British Columbia before taking a sharp turn to the south. As the river approaches the United States–Canada border, it takes a sharp turn to the west, eventually reaching salt water in the Strait of Georgia (Figure 8.1).

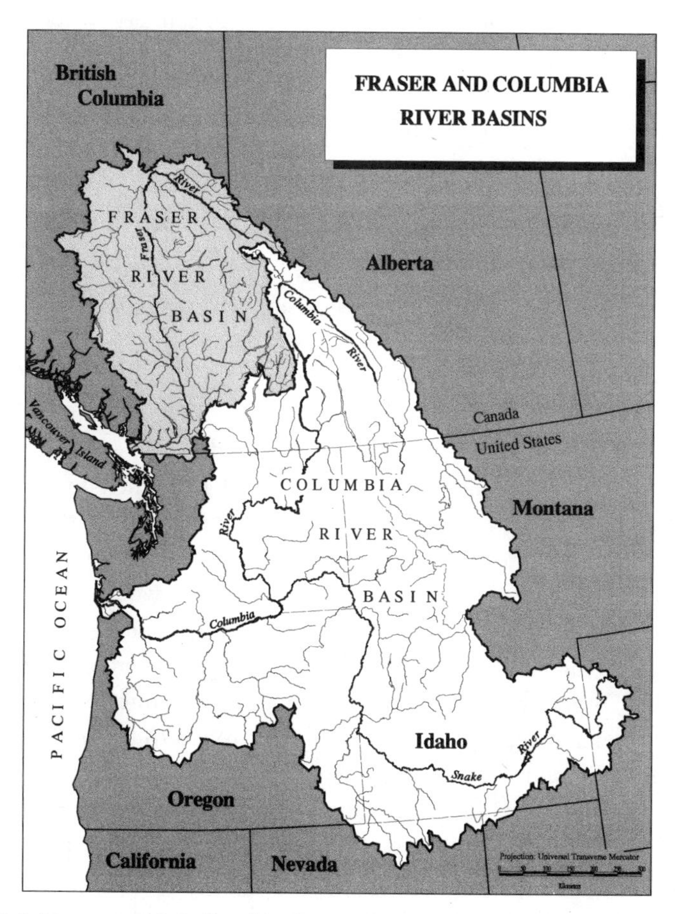

Figure 8.1. Fraser and Columbia River Basins. (Source: Dorie Brownell, Interrain Pacific)

Embedded in the Fraser's tributaries—including the Quesnel, Chilcotin, Thompson, and Harrison—are several large lakes that provide critical rearing habitat for juvenile sockeye salmon. Like the other Pacific salmon, adult sockeye deposit their eggs in the gravel of rivers, but they can also spawn on lake beaches where groundwater wells up through the gravel. No matter where the eggs develop, the juvenile sockeye nearly always spend their freshwater residence in a lake.[1] Because the Fraser River's extensive lake system makes ideal rearing habitat for young sockeye, the watershed is one of the two largest producers of sockeye salmon in the North Pacific (the other is the Bristol Bay watershed of southwestern Alaska).[2]

Before 1913, the Fraser River runs of sockeye salmon gave fishermen and canners every indication that they were truly inexhaustible. So as spring eased into the summer of 1913, the fishermen and cannery workers in north Puget Sound, the Strait of Georgia, the Strait of Juan de Fuca, and the lower Fraser River estuary waited with great expectations. Although no one knew exactly how many salmon would return that summer, they all expected the run to be particularly huge because every four years, like clockwork, an especially large run of sockeye came up the Fraser, and 1913 was the next big year in the four-year cycle.

The difference between the dominant, large-run years and the intervening three years was often striking. In the sockeye run that went up the Fraser to the Adams River, there may have been as much as a thousandfold difference between the largest dominant run and the smallest off-year run.[3] In the big year of 1901, fishermen caught 25 million sockeye salmon; in 1905 and again in 1909, 20 million sockeye were harvested.[4]

Thus, in the early summer of 1913, 4,000 fishermen in 2,500 boats waited for the next big run to begin. Down on Puget Sound, the Americans were also waiting. They had built huge traps across the known migration routes of the sockeye. The salmon's natural instincts would lead them into those meshed labyrinths from which there was no escape.

The size of the run four years earlier had caught the Fraser River canners unprepared. When the flood of sockeye came pouring into the canneries, the processors were so overwhelmed that excess fish had to be shipped south to be canned by Columbia River packers. This time the Fraser canners had built a large inventory of sanitary cans. Canneries that had been idle during the off-years between 1909 and 1913 were reopened, and their machines were oiled, tested, and put into good running order. The cannery workers rested, knowing that when the fish arrived, there would be little sleep until the last pound of pink flesh was sealed in tin.

Along the piers and wharves of the lower Fraser River, gillnetters mended nets and arranged gear in their small wooden boats. They looked out toward the salt water, beyond where they knew the salmon were massing for their run upriver. The fishermen waited. The seagulls, bears, and eagles waited. The fishermen, cannery owners, trapmen, butchers, can packers, warehouse workers, and cooks—all of them knew this was going to be a big year.

The four-year cycle of big runs held true. The catch exceeded all expectations, and despite the preparations, the canneries were again swamped with sockeye coming in faster than they could be canned. Fishermen tried to sell whole salmon on the docks at three for a quarter, but there were not enough outlets for the fish. Even a short-lived strike by some fishermen failed to slow the flood of fish into the canneries. As the catch overwhelmed the canneries' capacity, heaps of rotting salmon piled up everywhere.

When the season was over, the catch had broken all records. Thirty-two million sockeye were canned—22 million from the American traps and 10 million from the Canadian gillnetters. Adding an estimate of the wasted fish to those canned could push the total catch to about 50 million salmon. Despite that intensive harvest, about 6 million sockeye escaped the nets and traps and headed for the spawning grounds. The Pacific Salmon Commission's official statistics, which do not include waste, set the total 1913 run at 38 million sockeye.[5] In a recent report, however, Canadian biologist William Ricker reanalyzed the catch and escapement figures for the early years of the Fraser River sockeye fishery. Using a different but plausible set of assumptions, Ricker concluded that the size of the dominant runs in 1901, 1905, 1909, and 1913 might have reached 100 million fish.[6]

The frenzy of fishing and canning in 1913 reaffirmed the Fraser's status as the world's largest salmon producer. However, an incident in 1912, which went largely unnoticed at the time, would soon illustrate the vulnerability of the river and its salmon to the irresponsible use of technology. The sockeye run in 1913 would be the last large run for many decades.

From June 1911 until the end of 1912, the Canadian Northern Railroad laid tracks along the east bank of the Fraser River canyon. At a place known as Hells Gate, where water draining thousands of square miles funneled through a canyon only 110 feet wide, the construction crew blasted thousands of tons of rock into the river below the railroad grade. This constricted the channel even more, creating extremely turbulent conditions. Sockeye salmon that had escaped the gauntlet of traps and nets now found themselves blocked by the high-velocity currents at Hells Gate. A few fish got past Hells Gate early in the run, when the flows were more favorable, but most of the 6 million fish that escaped the fisheries could not surmount the obstruction.

Millions upon millions of sockeye salmon beat themselves to death in a futile effort to swim against the strong currents at Hells Gate. As the season advanced, the sockeye turned on their bright red spawning colors, which made the blockage and their futile struggle more evident. For ten miles below Hells Gate, millions of dead sockeye salmon filled every eddy and lined the banks of the river. Once people recognized the full extent of the disaster, action was taken to reduce the velocity of the river rushing through the narrow canyon. Work started in September 1913, and by the end of the year, workers had removed 2,000 cubic yards of rock from the river. Seeking to put a positive spin on the disaster, government officials declared that the blockage had been corrected, failing to note that it was already too late for the sockeye to reach their spawning grounds that year.

To make matters worse, the next year railroad workers triggered a massive landslide just upstream, squeezing the river into a seventy-five-foot-wide channel. This time the Canadian fisheries officials did not need millions of dying sockeye to spur them into action. Within a few weeks, they directed workers to remove the new obstacle. The men trying to remove the landslide had to contend with extreme conditions. Wet winter weather made work in the icy river difficult, and blasting from continued railroad construction sent rocks down the slope, injuring some of the men trying to save the salmon from another disaster. A year later, after workers had removed and redistributed about 60,000 cubic yards of rock, fisheries officials believed that the impassable conditions had been corrected, so work was suspended.[7]

Not everyone agreed that the situation at Hells Gate had been corrected. But believers and nonbelievers alike held their breath and wondered if the Fraser River would recover its tremendous fertility—if the huge sockeye runs would return, bringing back the enormous profits of previous years.

The full impact of the disruptions at Hells Gate was not recognized until 1917, when the adults produced from the 1913 run returned to the river. It was a disaster. Only 8 million fish returned to the Fraser in 1917, compared to 38 million in 1913. Of those, only 615,000 escaped the fishery to spawn. In the same cycle, four years later (1921) the total run dropped to only 2 million fish. Through the 1920s and 1930s, the catch in the "big" years plummeted to less than 10 percent of the harvests before the Hells Gate disaster. Hardest hit by the depleted runs were the Indians living above the blockage, who depended on the Fraser's salmon for their sustenance throughout the year.

Salmon depletion is nearly always the result of cumulative stresses on the fish's life cycle in the river, estuary, and ocean. In healthy, resilient ecosystems, stresses are absorbed with little discernible change in gross measures of production. As stresses accumulate, however, the resiliency of the ecosystem is slowly and invisibly lost. At some point, one more stress added to all the other stresses on an ecosystem triggers a major collapse. Debate may then ensue about which specific stress caused the catastrophe. For example, William Ricker believed that overharvest provoked the collapse of the Fraser River sockeye, while W. F. Thompson considered the blockage at Hells Gate to be the culprit.[8] In truth, the careless action of railroad construction crews triggered an avalanche of problems, only some of which were the rocks that fell into the river.

Although the blockage at Hells Gate posed a significant threat to the Fraser's sockeye, other types of habitat degradation and overfishing had already put increasing strains on the ecosystem. The runs of sockeye in the off-years had started to show a steady decline as early as the turn of the century.[9]

As in many rivers farther south in the United States, human activities had already degraded habitat and severely reduced the abundance of some salmon stocks in the Fraser River. For example, two of the most important sockeye runs in the Fraser system, the Quesnel-Horsefly River and Adams River populations, were devastated by environmental problems years before the Hells Gate disaster.

In the big years of the four-year cycle, the Quesnel-Horsefly system may have produced 10 million adult salmon.[10] Then, in 1896, a mining company built a dam on the Quesnel River, blocking access to the spawning areas in the Horsefly River and the rearing areas in Quesnel Lake. A few fish managed to pass the dam using an inadequate fishway, but the dam blocked most of the run. After building the dam, the mining company found it to be unsuitable and never used it; however, incredibly, it remained in place until 1921, even though the mine was closed in 1905.[11]

Similarly, a logging company built a dam across the Adams River in 1908. This dam also had an inadequate fishway that blocked salmon migration into the upper river, rendering useless thirty miles of the most valuable spawning gravel in the world.[12] The water and logs flushed downstream from the large splash dam gouged salmon eggs from the gravel and killed them. This occurred once a day, six days a week. In between each "splash" of water and logs, the flow was cut off, leaving the river below the dam nearly dry. The rich sockeye run below the dam was nearly exterminated, while the run above the dam was driven to extinction. This dam was not removed until 1945.[13]

In addition to habitat destruction, the intensive fishery also put tremendous stress on the Fraser River sockeye population. The amount and array of fishing gear used to harvest the big runs every four years overwhelmed the smaller runs in the off-years. Reconstructed harvest rates for the years before 1935 suggest that the fishermen may have taken 90 to 94 percent of the sockeye salmon in the off-years.[14] Harvest at that rate was not sustainable—it left too few fish to spawn and produce the next generation.

Traps set in U.S. waters were particularly effective in catching the salmon. The sockeye bound for the Fraser River generally followed predictable migration routes through the Strait of Juan de Fuca and the San Juan Islands to their home river. The Lummi Indians knew these migration routes well and had long used their knowledge to harvest salmon with reef nets.[15] When commercial fishermen in Washington State began harvesting the Fraser's sockeye, they used the Lummi's knowledge. But instead of reef nets, the white men built large traps—often placed directly in front of Indian reef net sites, effectively blocking the Lummi's access to the fish.[16] The Americans were soon catching more of the Fraser's salmon than were Canadian fishermen. In 1901, Washington State issued 149 licenses for traps to intercept sockeye salmon bound for the Fraser River, and the Americans canned 1,105,096 cases of Fraser River sockeye compared to the 928,669 cases canned by Canadians.[17]

The traps had quarter-mile-long leads, or fences, that crossed the migration route of the salmon. Once the salmon hit a lead, they moved along it, trying to find their way around the obstacle. But the lead guided the fish into large, corral-like enclosures that caught everything; one writer called them the fisheries version of clearcut logging.[18] A single trap operated by six men could catch as many fish as 150 gillnet boats operated by 300 men.[19] The Americans' traps caught nearly 60 percent of the Fraser River's sockeye salmon.[20]

When Washington's commercial fishermen entered the fishery with their massive traps, questions of fairness were quickly raised: Why should Americans harvest the largest share of Canada's premier fisheries resource?[21] This disparity led to discussions between the state of Washington and the Dominion of Canada and then between the two federal governments; the final result was the Bryce-Root Treaty of 1908. The next year, professors Edward Prince (for Canada) and David Starr Jordan (for the United States) drew up regulations to manage the Fraser River sockeye harvest so that each country would receive a fair share of the fish. The Canadians quickly passed enabling legislation and gave the regulations the force of law, but there was a snag in the U.S. Congress. American lawmakers never approved the enabling legislation, and in 1914, tired of waiting, Canada abandoned the treaty.[22]

The treaty failed in part because of the difference in the way Canada and the United States manage their fisheries. Management of Canada's fisheries is largely centralized in the federal government, whereas in the United States, individual states control the fisheries in their waters.[23] So with Bryce-Root, the U.S. government essentially negotiated a treaty governing a fishery over which it had no jurisdiction. The state of Washington was keenly aware of its "states' rights" and successfully lobbied Congress against the treaty.[24] Because of Washington State's intransigence, it would take another thirty years to ratify a treaty and bring the harvest of Fraser River sockeye under sensible, unified management.

The decimation of the sockeye fishery in 1917 pushed the United States and Canada into new treaty talks. In the space of just a few years, a canning industry worth $30 million per year had been reduced tenfold, and people feared that it was on the verge of total collapse. In 1918, the Hazen-Redfield Commission, whose members included representatives of both Canada and the United States, drafted a report that called for the establishment of a new international commission to regulate the fishery and to conduct studies of the salmon's life history, hatcheries, and spawning-ground conditions. If approved, the new commission would include two members from each country. The Canadian House of Commons quickly approved the treaty, but once again, the state of Washington intervened and, through strong lobbying, convinced Congress not to ratify the agreement.[25]

For Washington State, more was at stake than the principle of "states' rights" over fisheries resources. When the Hazen-Redfield Commission drafted its treaty, salmon fishing in Washington State was regulated directly by the state legislature, which had apparently made little use of science in carrying out its management responsibilities.[26] As a result, the legislature was easily swayed by "selfishly interested" parties to defeat needed conservation measures.[27] For example, in 1913, there were 168 traps in Washington waters catching as much as 60 percent of the Fraser River sockeye salmon. Washington's battle against a rational management program for the Fraser probably had more to do with protecting special financial interests than with the principle of states' rights. The primary victim of Washington's intransigence was the salmon; but in the long term, the state's fishermen also paid a high price.

The sockeye stocks, badly damaged by the Hells Gate blockage, could not recover as long as overfishing prevented the salmon from reaching their spawning grounds in adequate numbers. Prior to the Hells Gate fiasco, the big-run years had sent as many as 12 million sockeye (in 1901) to the spawning grounds. The smallest number of spawners to escape the fishery was 4.7 million in 1909. But starting in 1917 and continuing through the 1920s, the distinctively large runs in the four-year cycle disappeared. In 1918, only 85,000 fish escaped the fishery and reached spawning areas, and from 1912 to 1929 an average of only 320,000 made it past fishermen to spawn. The low escapements created two problems. First, the depleted runs could not fully seed the Fraser's extensive lake system with juveniles. Second, the lack of spawners altered the ecological process, which determines the productivity of the Fraser's lake systems. Spawning salmon not only deposit eggs into the gravel but also fertilize the aquatic ecosystem with nutrients released by their decaying bodies. By dying after spawning, the adults nourish their offspring through the freshwater stage of their life cycle. Although the loss of nutrients from millions of decaying carcasses was invisible, it likely had a detrimental effect on the rearing juveniles and was probably a significant form of habitat degradation.[28]

Treaty negotiations continued unsuccessfully through the 1920s. Out of frustration, the Canadians even considered physically rearranging the mouth of the Fraser River—by cutting off the southern outlets—with the hope of redirecting the migration route of the fish away from American waters.[29]

When treaty negotiations resumed in 1930, a new development in the state of Washington gave politicians there even more incentive to oppose rational management. American purse seiners had begun fishing in international waters outside the Strait of Juan de Fuca, and within a few years the seiners had learned how to use their gear to hunt and catch sockeye salmon heading for the Fraser River. In 1930, the year a new treaty was drafted, the American purse seine fishery made its largest catch of sockeye. Predictably, the purse seiners disliked two provisions of the new treaty: the establishment of an independent commission that would be given jurisdiction over fisheries in international waters, and the stipulation of a fifty-fifty split in the harvest between the United States and Canada.[30] So to defeat the treaty, selfish economic interests employed power politics.

The point man for the political maneuvering against the treaty was Washington governor Roland Hartley. In a speech broadcast over radio to the citizens of Washington, he flatly refused to admit that the Fraser River salmon were in trouble. Hartley assured people that what they had heard was "untruthful propaganda," and he considered it his "duty to truthfully set forth the situation." Then he proceeded to manipulate statistical information to hide the collapse of the Fraser River sockeye stocks behind a thick layer of political fog. Although he did admit that the Fraser River salmon needed some help owing to the Hells Gate incident, he condemned the treaty as "sugar-coated for our public with that hackneyed word 'conservation,' behind which so many of the people's rights are being taken from them." Instead, Hartley promised that the state of Washington would take action by funding more hatcheries in the headwaters of the Fraser River.[31] Miller Freeman, publisher of *Pacific Fisherman* and a longtime advocate for the treaty, rebutted Hartley's misleading statements a few days later in a speech to the Seattle Bar Association. Hartley's obfuscation achieved its goal, however. The treaty sat in limbo for seven more years.

Then two unusual events—a political one in 1934 and a natural one in 1936—combined to rescue the treaty. In the fall of 1934, Washington voters approved Initiative 77, which prohibited the use of traps or any fixed-fishing gear in the state's waters and placed severe restrictions on commercial fishing in parts of Puget Sound. (Initiative 77 was not entirely motivated by a desire to conserve salmon; it was another battle in the "gear wars" that characterized the salmon fishery, which were described in Chapter 5.) After an unsuccessful challenge in the Supreme Court, the initiative became law, and the effect on the Fraser River sockeye fishery was immediate. Prior to Initiative 77, Washington's share of the harvest had risen to about 70 percent of the total catch. In 1935, the year after the traps were banned, Canada caught 57 percent of the Fraser-bound sockeye. Because Initiative 77 unwittingly shifted the catch in favor of Canada, the fifty-fifty sharing provision of the proposed 1930 treaty became more appealing to Washington fishermen. In 1936, when the Canadians caught a whopping 82 percent of the sockeye, American fishermen and cannery owners began to seriously reconsider the treaty.[32] Finally, Washington governor Clarence Martin indicated that his state was ready to cooperate.[33]

The big jump in Canada's 1936 catch probably resulted not from Initiative 77 but from a natural variation in the migration route of sockeye salmon. Sometimes the sockeye migrate down Johnstone Strait and approach the Fraser River from the north instead of using their normal southerly route through the Strait of Juan de Fuca. The sockeye appear to have taken the northern route in 1936, giving the Canadian fishermen a great advantage.

Happily, the fortuitous shift in migration and Initiative 77 were enough to push the state of Washington into supporting the salmon treaty, and it was finally ratified by both countries on August 4, 1937. With the adoption of the treaty, the joint Canada–U.S. program to restore the sockeye salmon in the Fraser River was under way, and the International Pacific Salmon Fisheries Commission would become the primary management agency for the Fraser's sockeye.

The Columbia River

The Columbia River drains an area about the size of France in its 1,200-mile journey to the sea. Historically, the Columbia produced more chinook and coho salmon and steelhead trout than any other river in the world.[34] Prior to the arrival of Euro-Americans, 10 to 16 million adult salmon of all species entered the river each year.[35] Though all five species of salmon spawned in the Columbia River, the royal chinook was by far the most abundant (Figure 5.1, Chapter 5). About 8 to 10 million chinook entered the river each year. Many of them, especially the "June hogs" of the summer run, weighed in at fifty to sixty pounds each.

In 1883, the Columbia River was full of the large spring and summer chinook; by the end of the summer, the canneries had processed 43 million pounds of the salmon. It was the largest harvest of Columbia River chinook ever recorded. Following that bonanza year, the catch declined to roughly 25 million pounds annually and then appeared to stabilize for about thirty years. After 1920, the harvest of chinook salmon went into a decline that continues today (Figure 5.1, Chapter 5).

The stable harvests over the thirty years from 1889 to 1920 created a numerical illusion that apparently fooled even experienced biologists such as Willis Rich. In a 1940 publication he remarked that "the chinook salmon catch has held up remarkably well" in spite of an intense fishery, "but the record since 1920 is one of constantly decreasing catches."[36] The raw harvest numbers hid important qualitative shifts in the annual runs of chinook salmon—shifts that signaled early depletion of the prime spring and summer chinook. The fishermen actually maintained an average harvest of about 25 million pounds between 1889 and 1920 by shifting from the spring and summer runs, once they became depleted, to the less desirable fall chinook. Figure 8.2 shows that in 1878 most of the harvest of chinook was taken from the spring and summer races, May through July. By 1919, most of the harvest occurred in August, when the fall race of chinook salmon was migrating through the river.[37]

If the Columbia River had a single-year turning point, an equivalent to

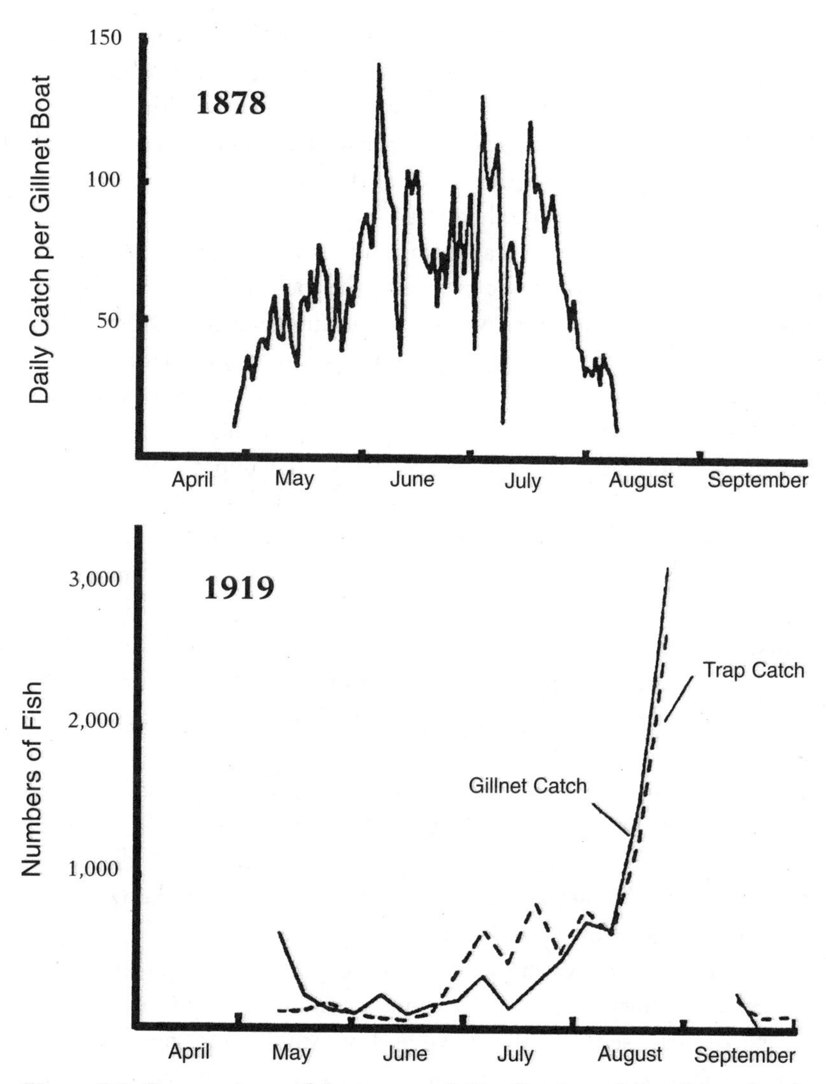

Figure 8.2. Comparison of the seasonal distribution of the chinook salmon harvest on the Columbia River, 1878 and 1919. (Source: Thompson 1951)

the infamous 1913 on the Fraser, it would be 1920. By that time, habitat degradation and overharvest had been reducing the productivity of the Columbia River's chinook salmon for over half a century, and the cumulative effects of poor stewardship and the gradual impoverishment of the watershed could no longer be hidden in the aggregated harvest statistics and shifting fisheries. Even the most die-hard optimists had to concur with a comment made twenty-six years earlier by Hollister McGuire, Oregon's first State Fish and Game Protector. "Oregon has drawn wealth from her streams," he said, "but now, by reason of her wastefulness and lack of intelligent provision for the future, the source of that wealth is disappearing and is threatened with complete annihilation."[38]

The depletion of the Columbia's spring and summer chinook salmon was not triggered by a single dramatic event like the rock slide at Hells Gate on the Fraser. The decline followed years of continuous attrition from several sources: excessive harvest, mining, logging, irrigation ditches that killed millions of juvenile salmon, and hatcheries that mined the eggs from wild populations. The spring and summer chinook were particularly vulnerable to habitat degradation because they migrated into the upper reaches of the Columbia and Snake Rivers and into the headwaters of the tributaries where habitat degradation was most intensive. By 1930, 50 percent of the best habitat for spring and summer chinook in the Columbia's tributaries had been blocked or severely degraded. The fishery was supported by fall chinook that spawned in the mainstem of the Columbia and Snake Rivers and in the lower mainstems of the larger tributary streams. Their mainstem habitat remained largely intact until the major dam-building era of the 1930s to the 1970s.

The problem created by overfishing and habitat degradation in the tributaries would soon be overshadowed by another major crisis. In the early 1930s, America was reeling from major economic and social disruption caused by the Great Depression and the environmental tragedy known as the Dust Bowl. With his New Deal, President Franklin Roosevelt set out to remedy the situation by relocating people from depressed areas to the Pacific Northwest. The Northwest would become the nation's escape valve, a place where the dispossessed and the down-and-out could start new lives and engage in productive work. Accomplishing this vision would require the full development of the hydroelectric power and irrigation potential of the Columbia River. As Roosevelt told the people of the region, "You have acreage capable of supporting a much larger population. . . . By proceeding with these great projects [Bonneville and Grand Coulee Dams], it will . . . give an opportunity to many individuals and many families back in the older, settled parts of the nation to come out here and distribute some of the burdens which fall on them more heavily than fall on the West."[39]

Massive development of the Columbia's hydroelectric and irrigation potential would eliminate the river's mainstem salmon habitat, the least changed part of the watershed. Rock Island Dam was the first to span the mainstem of the upper Columbia River in 1933. That same year Congress authorized construction of Bonneville Dam, to be followed two years later by Grand Coulee Dam. These three dams, and the prospect of many others, compelled some state politicians and state and federal fisheries agencies to work together on the problems the mainstem dams would create for the Columbia River salmon.

With the authorization of Bonneville Dam in the lower river, state fish and wildlife agencies recognized that the design of adequate fish-passage facilities was critical. As one official explained, such work was "too great a task, and the ultimate results to be achieved of too vast importance, to be carried out by any one state department or individual."[40] Therefore, a committee of eleven was created with representatives from the state fish and wildlife agencies of Oregon, Washington, and Idaho; the U.S. Bureau of Fisheries; and the packers and fishermen from Oregon and Washington. The purpose of the committee, which called itself the Interstate Fish Conservation Committee, was to "consider the Bonneville project and to decide upon the general types of passageways and other safeguards for the protection, preservation and continued perpetuation of the Columbia River salmon."[41] This was the first, tentative step in organizing a comprehensive interstate salmon protection and restoration program for the Columbia. However, it was also the beginning of what would eventually become a complex maze of committees and subcommittees addressing the problems related to salmon and dams on the Columbia River.[42]

Rock Island, Bonneville, and Grand Coulee Dams were only the first dams to block the Columbia. Another twenty-four dams were slated for construction on the mainstems of the Columbia and Snake Rivers over the next forty-five years. By 1933, engineers had already designed many of the dams, and these plans were a source of grave concern for biologists as well as the fishing industry. At the time, engineers and biologists had not yet demonstrated that salmon could be moved past dams as high as those planned or already under construction. The threats of massive dams, together with the still-declining harvests, prompted political leaders in Oregon and Washington to take an interest in the fate of the salmon.

In Oregon, Governor Charles Martin asked the Oregon Planning Board to investigate the salmon fisheries of the Columbia River. In its 1938 report, the board observed that the responsibilities of salmon management were much broader in scope than one state could resolve alone, and it recognized that management fragmented among too many state and federal agencies was ineffective. In effect, the board rediscovered the problem highlighted in Theodore Roosevelt's 1908 State of the Union speech. It recommended that the Oregon, Idaho, Washington, and California legislatures enact an interstate compact to establish a joint Columbia River Fisheries Commission. The commission would be given the responsibility and authority to regulate the total catch, set the seasons, and proscribe types of fishing gear. The planning board also recommended giving the new commission the job of directing needed scientific research, including studies of the effects of pollution on salmon production, potential improvements in fish culture, the effects of harvesting sardines and other food fishes of the salmon, and the need to set aside tributaries to the Columbia as salmon sanctuaries.[43] The Oregon Planning Board's recommendation for an independent commission, which was modeled after the commission that had been created to manage the Fraser River salmon, was never implemented.

Oregon was not alone in its concern for the future of the Columbia River salmon-canning industry. In 1941, the Washington State Senate created a committee to investigate the salmon problems on the Columbia River—the Columbia River Fisheries Interim Investigative Committee (CRFIIC). The committee was to work with similar groups in Oregon and Idaho and report back with recommendations.

In its 1943 report, the CRFIIC viewed the condition of the Columbia River salmon with alarm. Just sixty years earlier, in 1883, the fishery had harvested 43 million pounds of fish, all from the prime spring run of the chinook salmon. But by the early 1940s, the average annual harvest had dropped to 16 million pounds, and most of that was from the less desirable fall run of chinook. The committee speculated that with adequate regulation, the salmon harvest could have approached 75 million pounds a year, thus calculating the annual loss at about 59 million pounds. From a wholly economic perspective, the CRFIIC explained, "by our negligence and refusal to meet the issues involved, we have sacrificed millions of dollars of revenue to the State of Oregon and the State of Washington and, in turn, to the United States."[44]

According to the committee, three fundamental factors had caused the depletion and prevented the restoration of the Columbia River chinook salmon. The first was overfishing. The CRFIIC estimated that fishermen harvested 90 percent of the summer-run chinook salmon. Incredibly, the committee laid the primary blame for overharvest on the Indian fishermen, who provided a convenient scapegoat for the deleterious effects of the industrial economy on the salmon and their habitats.[45]

The second factor was habitat degradation. After touring the Columbia Basin, the CRFIIC concluded that the Salmon River in Idaho was the only tributary stream not suffering from severe habitat problems. Even there, the sockeye run into Redfish Lake had been blocked by a mining dam, and mining pollution and irrigation had degraded several of the upper tributaries and headwaters. The gravest threats were to the mainstems of the Snake and Columbia Rivers, however—important spawning habitat for the late summer and early fall runs of chinook salmon, which supported the commercial fishery. The Army Corps of Engineers and the Bureau of Reclamation were already beginning construction of an extensive series of dams that would flood all the major spawning areas in the mainstems.

Finally, the committee warned that responsibility and authority for salmon management was fragmented among too many uncoordinated agencies and institutions. Like the Oregon Planning Board, the CRFIIC considered this a serious threat to the salmon industry because "our [salmon management] activities are too involved in inter-state problems and lack the cooperation and policies to permit an improved situation." The committee concluded, "We are hopelessly defeated in obtaining any solution to the Columbia River fisheries unless we simplify our administration over this resource."[46] Over the next fifty years, politicians and salmon managers would make hundreds of predictions regarding restoration of the Columbia River chinook salmon. Unfortunately, the CRFIIC's prediction was one of the few that has been realized.

Echoing the recommendations of the Oregon Planning Board, the CRFIIC also proposed an interstate compact that would establish a new commission with complete control of anadromous fish in the Columbia River watershed. The commission would have full power to regulate harvest, to carry out research, to build, to operate or close hatcheries regardless of ownership, and to require fish protective devices at all dams and diversions. The independent commission called for by Washington, like that recommended by the Oregon board, was modeled after the International Pacific Salmon Fisheries Commission on the Fraser.

In response to concern about fragmented management and calls for management by an independent commission, B. M. Brennan, director of the Washington Department of Fisheries, did something that few bureaucrats ever do. He volunteered to make the "considerable sacrifice" of relinquishing a major part of his agency's authority over the Columbia River fisheries to an independent commission because he could "see no other method of reaching the desired end"—the restoration and protection of the Columbia River salmon. Seemingly free of the suspicion many westerners had about federal involvement in their affairs, Brennan believed that Congress could set up a commission to create an "almost utopian situation wherein the salmon and steelhead runs of the Columbia River will be properly protected and built up towards their former magnitude."[47]

Regrettably, no such commission was ever established; despite their remarkable consensus, the recommendations of the Oregon State Planning Board, the Columbia River Fisheries Interim Investigative Committee, and Brennan were uniformly ignored. As a consequence, fragmented management has continued to hamper salmon protection and restoration efforts in the basin up to the present day.

The threat of mainstem dams on the Snake and Columbia Rivers not only pushed politicians to think more about the salmon but also prompted the few biologists working in the basin to synthesize the available scientific information concerning the salmon's biology and use it to predict the effects of the dams.[48] The results of their analyses led to a great deal of apprehension and pessimism. The proposed dams were much larger than any the biologists had experience with. No one could predict exactly what would happen when these massive concrete structures blocked the free flow of the Columbia and Snake Rivers, but one thing was certain, according to biologist Lawrence Griffin: "Dams are not going to do the fisheries any good."[49]

Although Grand Coulee, Bonneville, and Rock Island were just the first of many dams planned for the Columbia and Snake Rivers, the salmon were spared further construction and river development for several years while America directed all of its energy to fighting World War II. Still, salmon biologists knew that when the nation turned its attention to domestic matters once again, development of hydropower on the Columbia River would be a high federal priority. Before the end of the war, in 1944 the U.S. Army Corps of Engineers described its construction plans for the Columbia River at a series of meetings in Portland, Oregon. Four biologists from the U.S. Fish and Wildlife Service—Willis Rich, Federic Fish, Mitchell Hanavan, and Harlan Holmes—were invited to those meetings. For the first time, the engineers gave biologists advance warning of their plans. Rich and his colleagues believed that it was critical to take full advantage of the warning. They realized the Fish and Wildlife Service needed a policy or program to deal with the fish conservation problems that the new dams would create. So the four biologists wrote a lengthy letter to Elmer Higgins, their supervisor in Washington, D.C., in which they gave their professional judgment but also expressed their personal feelings about the fate of Columbia River salmon.[50]

The biologists were uncertain about what could be done to save the salmon, and they felt frustrated by the situation. For example, they questioned the ability of technology such as fish ladders to overcome the cumulative change in the ecosystem that would occur with each new dam. Eventually, they feared, the ecological change in the river would be too severe and the salmon would disappear. In the end, they believed the fish ladders and other engineered solutions would be only "monuments to a departed race."[51] Doubting that anything could actually be done to save the salmon, they wrote, "No competent fishery biologist is willing to assert that the salmon runs can be preserved if the full program of construction does go through." Yet they were not willing to give up. As they explained to Higgins, "It goes without saying that *some effort must be made to save the salmon runs.* We cannot take the stand that the situation is hopeless and that it is useless to put time and money into the effort" (emphasis in original). They suggested an all-out "frontal attack" on the construction program. But having proposed such a radical approach, they quickly retreated into the safe haven of the bureaucracy, explaining that it was "not a job for fishery biologists—it is something that must be handled higher up."[52]

In the end, the biologists decided to accept the dams as inevitable but to offer their best ideas for a program to save the salmon. Rich and his colleagues proposed to abandon the upper river to development and move the most important salmon runs from the upper basin to the tributaries of the lower river, below the proposed Umatilla (later named McNary) Dam. They recognized that the success of this massive transplantation of upriver stocks would wholly depend on artificial propagation, a technique that, they admitted, had not yet been proven effective.[53]

Though hatcheries had not yet proved that they could add to natural production or replace production lost due to habitat degradation or overharvest, the biologists believed that full hydroelectric development of the basin was inevitable and saw little choice but to recommend the use of artificial propagation to compensate for the effects of the dams. Rich hoped that hatchery research would eventually find ways to produce satisfactory salmon returns.[54]

A little over a year later, the same biologists helped to prepare the official proposal for a fisheries program to deal with the planned construction of forty dams, many on the mainstems of the Columbia and Snake Rivers. As their proposal passed through bureaucratic channels, the trappings of officialdom were added, and the passion of the earlier letter was watered down. For example, contrast the earlier declaration, "No competent fishery biologist is willing to assert that the salmon runs can be preserved if the full program of construction does go through," with the version in the official final proposal: "It appears that the losses incurred . . . might be serious enough to make continued propagation in the headwater tributaries impracticable." Moreover, the final report referred to the plan to transfer salmon stocks from the upper river to the lower tributaries as an insurance policy to protect the salmon industry in case "at some unpredictable time, the fish protective devices [fish ladders over dams] fail in their purpose."[55] Labeling the plan as insurance against failure of the fish ladders clearly implied more confidence in artificial propagation, stock transfers, and fish ladders than the original letter. In the official proposal, the discussion of the consequences of river development took on a much softer tone.

How did the biologists reconcile the mass relocation of salmon from the upper to the lower river, since they also understood that salmon returned to the same gravel where they were born and that they were adapted to the environment of their home stream? How could they rely on hatcheries to make the transfer from upper to lower river, when they acknowledged that hatcheries had yet to demonstrate their effectiveness? Were they just fatalistic? Did they believe that development was inevitable and that applying the new scientific understanding was useless? Or did this capitulation to the status quo demonstrate the power of the development ideology to bend science to its purpose? The answer to those questions cannot be found in the documents. As we will see later, however, there *were* other choices, such as the options followed on Canada's Fraser River. But before American biologists could even conceive of alternatives to artificial propagation, they would have to embrace a different worldview—one capable of challenging the conventional wisdom that humans had to manipulate and control ecosystems and that technology such as hatcheries could replace natural ecological processes. Even the threat of massive dams, it seems, could not change their deep-seated assumptions about man's relationship with nature.

In spite of their knowledge of the consequences and their personal opposition to the full development of the Columbia River, by proposing hatcheries the biologists handed politicians and federal construction agencies the leverage they needed to dam the Columbia River without making the hard choices between development and salmon. But they avoided those choices only temporarily. The listing of salmon under the federal Endangered Species Act in the 1990s has finally forced the region to confront the choices it avoided in the 1940s.

The plan to relocate upriver stocks to the lower river using artificial propagation was a straw that politicians readily grasped to promote the belief that power and salmon were compatible. The biologists' skepticism about hatcheries was pushed into the background, allowing optimism about the future to move to center stage.[56] For example, to win approval for its dams on the Cowlitz River, a tributary of the lower Columbia, the city of Tacoma, Washington, distributed an information packet whose cover showed an electric transmission tower, a salmon, and a cheery message in bold print: "power and fish / you *can* have *both*.[57] Having won World War II, Americans had supreme confidence that they could engineer a better environment. As historian William Robbins pointed out in his environmental history of Oregon, in the early decades of the twentieth century, some Americans had an almost "transcendental belief in the efficacy of the unlimited manipulations of the natural world."[58] There was no reason to doubt that technology could and would make salmon and dams compatible.

Despite the biologists' somewhat accommodating proposal to compensate for the effects of development, the fisheries agencies did not give up their fight against the dams. In fact, in 1945 they did indeed launch a "frontal attack" challenging the economic justification for the additional power, irrigation, and transportation the dams were supposed to produce, in a report titled "The Wealth of the River."[59] The agencies were joined by legislative committees from Oregon and Washington, as well as tribal organizations and fishermen, in opposing the dams.[60] But as Willis Rich and his colleagues had feared in 1944, the situation turned out to be hopeless. On March 6, 1947, the U.S. secretary of the interior signed a memorandum settling the battle between the salmon and the proposed construction program—in favor of the construction program. The memorandum, titled "Columbia River Salmon or Dams," reiterated in unambiguous terms the view that humans had to conquer and control the landscape and rivers of the Northwest. "It is, therefore, the conclusion of all concerned that the overall benefits to the Pacific Northwest from a thorough-going development of the Snake and Columbia are such that the present salmon run must be sacrificed. This means that the department's efforts should be directed toward ameliorating the impact of this development upon the injured interests and not toward a vain attempt to hold still the hands of the clock."[61] With that memorandum, the lower-river plan—the "insurance policy" proposed by the U.S. Fish and Wildlife Service—became the primary strategy for "ameliorating the impacts."

Prior to the "Salmon or Dams" memo, the fisheries agencies had invested heavily in hatcheries and had done little to understand the ecology of the salmon on the Columbia River. Thus in 1947 they faced a massive dam construction program with little information and a single management tool—artificial propagation—that didn't work. To give fish managers time to implement the lower-river plan, the Department of the Interior requested a ten-year delay in the construction program, but the Columbia Basin Inter-Agency Committee recommended against the delay and approved the lower-river strategy for immediate implementation.[62] The plan was renamed the Lower Columbia River Fisheries Development Program, and under it the states of Oregon and Washington became salmon management partners with the federal government. The first major salmon restoration program for the Columbia River was under way.

Different Roads to Restoration

Although major salmon restoration programs were launched on both the Fraser and the Columbia within a ten-year period, the different strategies followed in these two programs produced very different outcomes.

The International Pacific Salmon Fisheries Commission (IPSFC) began its work restoring the Fraser River sockeye in 1937. When Congress ratified the treaty bringing the commission into existence, however, it added to the treaty an understanding that prohibited the enforcement of any fishing regulations for two sockeye life cycles, or eight years. During that time the commission was directed to conduct research and use its findings to establish any necessary regulations, starting in 1945. This in itself was a major departure from the conventional approach to salmon management, which tended to be content with "groping in the dark."

Since the control of the harvest had been put on hold, the IPSFC focused on habitat and natural production. The commission notified the Canadian government that it wanted to be informed of all activities in the Fraser Basin that might modify spawning areas or pollute, dam, or divert water used by adult or juvenile sockeye salmon.

Although salmon fishermen and cannery owners pressured the IPSFC to use hatcheries to quickly rebuild stocks, the commission had other priorities and put off the hatchery issue. According to John Roos, the IPSFC's director of investigations in its waning years, "From the very beginning, a strong commitment was made to protect natural propagation from the detrimental effects of dams, logging, erosion, dredging, pollution and other human-induced changes in the environment."[63]

Long before the commission started its work, biologists suspected that there were many distinct races of sockeye in the Fraser watershed. Hatcherymen, who handled large numbers of salmon, noticed consistent differences in appearance and biology among salmon from different areas in the Fraser watershed.[64] Because W. F. Thompson, the commission's first director of investigations, had been one of Charles Gilbert's students, he knew about Gilbert's earlier discovery of distinct sockeye salmon stocks in the rivers of British Columbia. As a result, Thompson designed harvest management and habitat preservation programs around protection of the individual Fraser River sockeye stocks.[65] Regarding the stock as the basic management unit, Thompson immediately initiated research to learn more about the individual populations of sockeye, including the timing and route of their migration through marine and freshwater habitats and the location of their natural spawning areas. The commission also launched studies on the potential delays or blockages in migration that might still exist at Hells Gate.

The IPSFC's approach to salmon restoration reflected the scientific understanding of the salmon's biology at the time. For example, making hatcheries a low priority was consistent with Russell Foerster's 1936 study of the Cultus Lake hatchery, which had eliminated hatcheries as a scientifically valid option and led to the closure of all British Columbia hatcheries.[66] Instead, the commission emphasized natural production and the protection of the river reaches where spawning took place. The commission's focus on the individual populations of sockeye in the Fraser River was also consistent with the latest understanding of the Pacific salmon's stock structure.[67]

The Lower Columbia River Fisheries Development Program (LCRFDP), meanwhile, took a very different approach to salmon restoration. The program had six parts: removing obstructions to salmon migration in tributaries of the lower Columbia; cleaning up pollution in major tributaries such as the Willamette River; screening water diversions to prevent the loss of juvenile salmon in irrigation ditches and construct fishways over dams in the tributaries; transplanting salmon stocks from above McNary Dam to the lower river; revamping the Columbia River hatchery program by remodeling existing hatcheries and building new facilities; and, finally, creating salmon refuges by setting aside most of the tributaries below McNary Dam exclusively to maintain salmon and steelhead runs.[68]

As it was originally designed, the program contained strong provisions for habitat protection and restoration, but in implementation many of those provisions were given a low priority. For example, to create salmon sanctuaries in all Columbia River tributaries below McNary Dam, Oregon and Washington had to pass enabling legislation. The state of Washington passed its sanctuary bill by overwhelming majorities: 42 to 1 in the Senate and 79 to 18 in the House. But the Oregon House of Representatives voted 41 to 18 to build power dams on the Deschutes River, effectively killing the sanctuary provisions of the LCRFDP. Washington's sanctuary bill was later overturned in the state supreme court.[69]

The hatchery program was only one of the six parts of the LCRFDP, but within a few years it became the dominant part. In 1951, the fourth year of the program, the states spent 49 percent of the budget on hatcheries and used only 5 percent on habitat projects.[70] In 1986, hatcheries received 79 percent of the program's budget. The plan to transfer commercially valuable spring chinook salmon from the upper to the lower river, and then to maintain those fish through artificial propagation, ignored the stock concept and instead treated salmon management on the Columbia as a large agricultural enterprise, a massive extension of the earlier attempts to "till the waters" for the benefit of humans.[71]

On the Fraser, the IPSFC took advantage of the eight-year delay in enforcing harvest regulations and used the time to undertake research and gather the information needed to manage the sockeye on a stock-by-stock basis. Biologists working for the commission studied the migration of individual sockeye stocks through the Strait of Juan de Fuca and the San Juan Islands. Research at Hells Gate revealed that the obstruction there still impaired passage of specific stocks whose migration timing coincided with particular detrimental flow levels. The commission also developed methods to accurately estimate the number of spawning salmon, began collecting more-accurate estimates of the total harvest, and undertook studies to identify and describe the freshwater biology of individual races.[72]

While W. F. Thompson launched major research programs on the Fraser River to strengthen salmon management through sound science, the Columbia managers gave scientific research a low priority. This occurred in part because the salmon management agencies in Oregon and Washington were still dominated by fish culturists who continued to deny the need for research and critical evaluation. Instead, they promoted the LCRFDP as an "action" program that could be implemented immediately, without delay for lengthy research. Their key word was "action."[73] Research was eventually initiated later, but it focused on hatchery operations and fish passage through dams, putting little emphasis on salmon ecology.

The state fisheries managers did not have the same level of concern as the federal biologists regarding the ability of hatcheries to sustain the salmon fishery. Even in the face of massive development of the river, the managers believed they knew how to maintain the salmon through artificial propagation. The fact that their hatcheries had been in use for more than seventy years without documented success did not appear to concern them. They simply assumed that continued implementation of the same unsuccessful programs would somehow in the future produce different results.[74] The failure to use the latest science and to invest in research on salmon ecology was in part a tragic consequence of the failure to create an independent commission to manage and restore the Columbia River salmon.

The bargain that salmon managers had struck with the industrial economy three decades earlier on the Elwha River—wild salmon for dams and hatcheries—was about to pay off with big increases in agency budgets. But the resulting massive increase in spending on salmon only exacerbated the problem. As long as management remained fragmented among several institutions, too much attention would be focused on the competition for funds generated by turbines. The fate of the salmon would not be the center of attention until the Endangered Species Act shocked the institutions out of their complacency.

By 1951, it was apparent that the LCRFDP was in trouble. Funds for the program had been slow to reach the Northwest, and even the available funds were not being used on schedule. The original LCRFDP was designed to be completed in ten years—a schedule that was critical because the construction program was rapidly changing the river. Hatcheries had to be built or remodeled before the dams eliminated upriver populations of salmon altogether. The Korean War added another degree of urgency with calls to accelerate the construction of dams for national security reasons.

On June 14, 1951, Albert Day, the director of the Fish and Wildlife Service, came to Seattle to put the LCRFDP back on track. At a "meeting of great potential importance to the preservation . . . of the salmon fisheries," Day did most of the talking and delivered a strange mixed message of threats, pleas, science, and bribes. He told the assembled managers and biologists that the Northwest's rivers were going to be dammed and "we should not kid ourselves" about it. He did not limit that prediction to American rivers. In his effusive monologue, Day assumed that the Canadians shared American standards and values and that the Fraser River would suffer the same fate as the Columbia. He urged the group to "come up with something that may lead to some concrete results on this problem of dams versus salmon." Day was not bothered by the half-century of failure to stop the decline of salmon, which started before the mainstem dams had been constructed. He wanted the state and federal salmon managers and biologists to come up with a workable program before the end of the day. To give the audience an added incentive, Day laid his bribe on the table. The federal government was willing to speed up and increase funding for the LCRFDP, from $2 million to $5.5 million a year. He also proposed an additional $12.5 million for research on salmon passage at the dams. Sounding like a cheerleader, Day urged the group to "all pull together as one team."[75] The subtext of his speech was clear: The dams are inevitable; take the money, boys; join the team and quit complaining. The group established committees to work out a program.

In hindsight, it's evident that the meeting with Day was the first step in a voracious process that would trap the Northwest in costly salmon programs and actually accelerate the fish's downward spiral.[76] Optimism that a technological solution could ensure salmon survival in "engineered" rivers has not diminished—not even in the face of persistent failure. For the past forty years, the response to continued salmon declines has been to invest in more technology. Through the years, more and more money has been thrown at the problem, but the approach remains rooted in the nineteenth-century myth of universal benefit from human control over natural ecological processes—in the myth that humans could have salmon without healthy rivers.

Columbia River salmon managers became trapped in what historian Paul Hirt has called the "conspiracy of optimism." The goals of stewardship and conservation were displaced by techno-optimism—the belief that it was possible to maximize salmon production through technology, even in a watershed simplified and degraded by a series of continuous impoundments. Salmon managers stuck with their technological fixes even in the face of obvious failure. As Hirt pointed out, when programs based on techno-fixes fail, the technology is never blamed for the failure; rather, the fault is said to lie with the politicians who did not provide sufficient funds to buy enough of it.[77] On the Columbia, the result of this kind of thinking has been a spiral of escalating costs with few tangible results (Figure 8.3).

At first it may appear that Albert Day, an American, erroneously included the Fraser River in his prediction that the hydroelectric potential in every Northwest river would be developed. Day was not just superimposing American values on Canadian rivers, however; the Fraser did not escape the dam-building engineers. In a book on the Fraser River published in 1950, a year before Day's meeting, Canadian Bruce Hutchison left no doubt that the Fraser would be fully developed. Hutchison argued that the government had to weigh the value of the salmon against the value of electricity and the industrial development it would promote, and to him, "The choice is obvi-

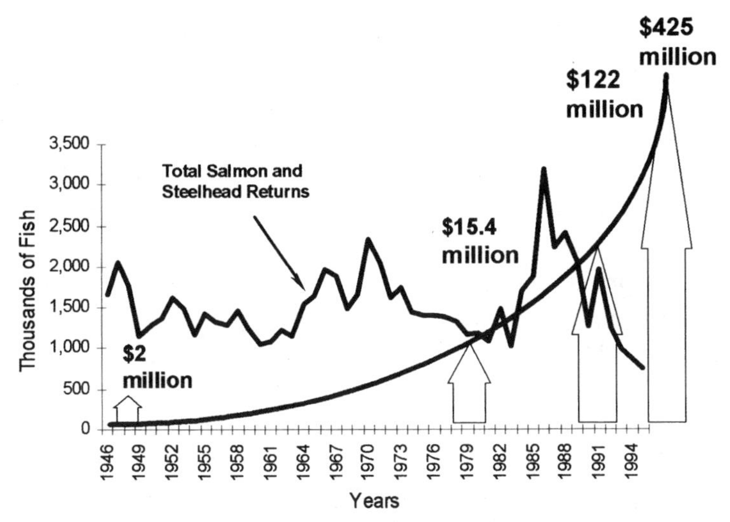

Figure 8.3. Total adult salmon and steelhead returns to the Columbia River and the cost of the salmon restoration program. Initially the program expended $2 million per year. From 1949 to 1981 the average annual cost was $15.4 million per year; from 1982 to 1991 it was $122 million per year; and it could reach $425 million per year in the near future. Note that the first data on the chart are from 1946, when the number of salmon entering the river was about one-tenth their level of historical abundance; the total salmon abundance then was less than half the region's current goal of 5 million fish. (Sources: Laythe 1948, General Accounting Office 1992, Northwest Power Planning Council 1994)

ous." The Fraser would follow the same path as the Columbia. Artificial propagation would be used to maintain the salmon if engineering could not make dams and salmon compatible.[78]

Although the engineers had plans for dams on the Quesnel, Adams, Chilcotin, and Chilko Rivers, the biggest threat to the Fraser River sockeye came in the 1950s in the form of Moran Dam, a 720-foot-high structure that would have been built on the mainstem about twenty miles above Lillooet, British Columbia—roughly 200 miles from the river's mouth. Canadian biologists, together with the IPSFC, concluded that the dam would eliminate all the salmon and steelhead above it. Furthermore, it would eliminate half the salmon in the lower river by changing the river's ecology.[79] This dire prediction was similar to those made by American biologists regarding McNary Dam on the Columbia River. Like the Americans, the Canadians looked for opportunities to have the best of both development and salmon—and they too looked for ways to mitigate the effects of the dams using hatcheries.

Consistent with its commitment to maintaining a scientific basis for its plans and programs, the IPSFC undertook an exhaustive scientific review of salmon mitigation programs. The resulting report concluded, "At the present time, artificial propagation is not a proven method of maintaining even small localized stocks of Fraser River sockeye and pink salmon."[80] The commission also pointed out that the economic cost of destroyed salmon runs would make the power produced by Moran Dam too expensive. Commissioner Roderick Haig-Brown asserted that the salmon's value went beyond economics: the whole attempt to have an "objective" discussion of salmon and dams, he said, "was a betrayal of the salmon and their meaning." To balance the salmon against human "needs" that could be met in other ways revealed an "insensate arrogance that has no place in modern thinking," he argued, that preserving the salmon was "an act of faith in the future."[81]

As a consequence of the commission's research, Moran Dam was not built. The empty claims of success made by hatchery programs did not convince the Canadians to risk their Fraser River salmon. Science clearly showed that intensive artificial propagation did not make dams and salmon compatible. As Canadian biologist Ferris Neave explained, "A natural stream is a complex organization . . . not a specialized farm for the intensive production of one or a few species."[82]

While American biologists turned toward technology to find a way to accommodate the "inevitable" Columbia River dams, commission biologists working on the Fraser effectively used science to halt the construction of mainstem dams. Although development did take place on the Fraser River, the construction of dams was largely limited to upper tributaries that were beyond the limits of salmon migration, as the Canadian biologist John Babcock had once predicted. The mainstem of the Fraser remained free flowing.[83]

Although the commission resisted the hatcheries' false promise of a quick fix, it did experiment with the use of spawning channels adjacent to the river. These artificial stream channels provided a seminatural, protected environment where adult sockeye could spawn naturally and eggs could incubate. The resulting fry left the channel to rear in nearby lakes. Biologists located spawning channels in areas where there was excess rearing capacity in a nearby lake. Of the three spawning channels constructed, the Weaver Creek channel was an immediate success, but the Gates and Nadina channels showed disappointing results.[84]

Inspired by the success of the Weaver Creek channel, the commission developed a comprehensive program to construct twelve additional spawning channels to increase the production of weak stocks. A major purpose of this enhancement program was to avoid additional harvest reductions. When the IPSFC submitted its proposal to the American and Canadian governments for funding approval, the United States quickly agreed, but Canada refused, thereby killing the program. Canada's refusal stemmed from its dissatisfaction with the U.S. position in ongoing negotiations about regulating U.S. fisheries that caught Canadian salmon.[85] This action by Canada was a blessing in disguise, forcing the commission to increase production by cutting the harvest, despite the short-term hardship on the fishermen, and allowing the weak stocks to rebuild naturally. Techno-fixes would not play a large role in the recovery of Fraser River sockeye salmon.

This situation contrasted starkly with that of the Columbia River, where the long-term health of the ecosystem and its salmon were repeatedly sacrificed for short-term economic gains from development or excessive harvest. The sacrifice of long-term ecological health for short-term economics could be made acceptable only by the unfounded optimism invested in techno-fixes—the "conspiracy of optimism." Deprived of the opportunity to rely on technology, the IPSFC was forced to manage the Fraser based on the fundamentals of salmon ecology and good stewardship. And it worked. Except for a small contribution from the three spawning channels, all of the increases in production on the Fraser River occurred through natural processes.[86]

While the IPSFC was concluding that hatcheries could not mitigate for habitat destroyed by dams, the U.S. government was having second thoughts about its own hatchery program. By then twenty hatcheries had been rebuilt or newly constructed with LCRFDP funds. When Ross Leffler, assistant secretary of the interior, reviewed the program, he concluded that habitat improvement needed more attention, especially above McNary Dam. He further recommended that new spending on hatcheries be deferred until the existing program was evaluated and treatments were developed for the diseases that were attacking the salmon in hatcheries.[87] After eighty-seven years of undocumented claims of success, federal authorities finally mandated an evaluation of the effectiveness of the hatchery program.

To carry out the evaluation, federal and state biologists marked hatchery-reared fall chinook salmon from thirteen hatcheries in the Columbia Basin from 1961 to 1964. Then they intensively monitored fisheries from southeast Alaska to northern California to recover the marked fish. Analysis of the data showed that 14 percent of the chinook salmon caught in all the fisheries were hatchery-reared fall chinook from the Columbia River. This level of contribution from hatcheries was a relatively recent phenomenon. Research on nutrition, disease control, and fish-culture practices through the 1950s had yielded positive results in the 1960s, increasing the survival rate of juvenile salmon after release from the hatchery. Changing ocean conditions also contributed to the boost in survival. The results were encouraging: the economic benefits of the surviving hatchery-reared chinook outweighed the costs of operating the hatcheries by as much as seven to one.[88] The evaluation came up with positive results only because it examined a very narrow set of questions, however, and avoided the most important question of all.

The researchers asked two primary questions: Do hatcheries make a contribution to the fisheries (number of fish caught), and is the economic value of the catch greater than the cost of producing it (cost of hatchery operations)? Although these were important questions, they were too narrow to serve as the sole criteria for evaluating the success of hatcheries. Not only were hatcheries supposed to add fish to the fishery at a reasonable price, but they were also supposed to maintain the total level of production by replacing the wild salmon lost as a result of habitat degradation. So the researchers should have also asked, Did hatcheries replace the production lost as a result of habitat destruction in the tributaries and construction of mainstem dams? The obvious answer to that question would have been a resounding "no." As habitat in the tributaries was degraded by grazing, irrigation, mining, and logging, and as the mainstem salmon habitat was engineered into a series of reservoirs, hatcheries clearly failed to compensate for the loss of wild salmon.

Thus the success of the hatchery program was measured by a flawed statistic. As the total number of wild salmon in the Columbia Basin declined, salmon from the hatchery program began to make up a larger and larger percentage of the total run. Today, as proof of their success, hatchery advocates note that artificially propagated salmon make up 80 percent or more of the total number of salmon on the Columbia, but they fail to mention that the total run has crashed to less than 5 percent of its historical abundance. Measuring success by the percentage of hatchery fish in a shrinking production base not only was scientifically invalid but also insidiously enhanced the illusion of hatchery success. At the same time that the percentage of hatchery fish in the run increased, hatcheries were contributing to the decline of wild salmon—by inappropriate stock transfers, egg mining, encouraging overharvest, and allowing managers to ignore habitat degradation.[89] This situation created a deadly extinction spiral that managers, with their myopic evaluation methods, failed to detect.

Nevertheless, the positive results of the federal study gave a green light for further expansion of the hatchery program on the Columbia River. Between 1960 and 1990, the number of juvenile chinook salmon released from hatcheries in the basin nearly tripled, from about 60 million to over 150 million juveniles a year. At the same time, the total abundance showed no sustained increases, failing to even partially recapture historical production levels.

The philosopher and sociologist Max Weber once remarked that man is an animal suspended in a web of ideas he himself has spun.[90] That certainly describes how people have clung to the use of hatcheries to compensate for the effects of dams in the Columbia Basin. Salmon managers became entangled in a web of their own assumptions about nature, the human manipulation of it, and the ability of technology to replace ecological processes. They reinforced their assumptions with studies that were designed not to objectively evaluate but to support their myopic worldview. Caught in the trap of techno-optimism, salmon managers became so committed to their worldview that they never questioned its basic assumptions.

The eminent physicist Freeman Dyson offers an explanation for the increasing commitment to a failed technology. According to Dyson, ideologically driven technology is never allowed to fail; instead, its proponents will ignore signs of failure until it is too late and great damage has been done.[91] And throughout their history, hatcheries have been ideologically driven. Beginning with Spencer Baird's recommendation in 1875 and continuing with the Columbia River program in the 1940s, hatcheries have been promoted not because they were scientifically justified but because they supported the prevailing development ideology. They enabled canneries to have unrestrained access to fish, and they permitted nearly complete development and control of the rivers—that is, nearly complete destruction of the salmon's habitat. To acknowledge that artificial propagation has failed would call into question not only the hatcheries but also the ideology from which they were derived. It has been far easier to construct and maintain a "conspiracy of optimism."

As a result, management costs have rapidly accelerated, with no tangible results; salmon runs on the Columbia have continued to decline and have reached record lows in recent years. About $3 billion has been spent attempting to restore salmon on the Columbia River over the past fifty years, and management institutions are now calling for an additional $50 million for new hatcheries and over $1 billion to improve the passage of juvenile salmon over the dams. In contrast, over roughly the same period (from 1937 to 1985), the International Pacific Salmon Fisheries Commission spent $21.3 million.

More important, the commission's program was successful. From 1917 to 1949 the average Fraser River sockeye salmon run was 3.3 million fish. Under commission management, the average annual run gradually increased to 5.6 million from 1949 to 1982, then to 7.8 million from 1983 to 1986, and finally to 10.2 million in recent years.[92] In 1990, 22 million sockeye returned to the Fraser River to spawn. On the Columbia, in contrast, the 1993 harvest of prime chinook salmon dwindled to 50,000 fish out of a run of about 450,000, a pitiful fraction of the 8 to10 million chinook that once returned to the river.

Why did the management of salmon runs on the Columbia and Fraser Rivers turn out so differently? For one, the focused authority of the IPSFC on the Fraser carried more weight in protecting the river's habitat and preventing the construction of mainstem dams than could the fragmented management practiced on the Columbia. The restoration of salmon on the Columbia River never did come under a central authority, fulfilling the prediction made in 1943 by the Washington State Senate that the states would be "hopelessly defeated in obtaining any solution to the Columbia River fisheries unless we simplif[ied] our administration over the resource." Instead of the five or so agencies the Washington senators were worried about, today a salmon leaving the Lochsa River in Idaho will make the round trip to the ocean and back while passing through a tangled web of seventeen management jurisdictions.[93] Had a central authority been established to manage the Columbia, the development of the river's hydroelectric potential probably would have proceeded anyway, but such an authority might have relied more on science and tempered the worst destruction. Some of the most damaging dams might not have been built, or their design might have been more fish-friendly.

The Americans may have defeated themselves from the beginning. As early as 1925, at the Seattle meeting, American biologists were resigned to the idea that their rivers would be fully developed. Their mood was in sharp contrast to that of the Canadian, John Babcock. There was no question in his mind that no dams would be built on the Fraser's mainstem. Did Babcock's sentiment reflect a broader cultural difference between the American and Canadian attitudes toward the management of natural resources? Maybe, but the Canadians have destroyed far too many salmon streams through logging and development for me to claim that they are fundamentally better conservationists. And recently, Canadian biologists have argued that the problems in the Pacific salmon and Atlantic cod fisheries stem from political or economic influences that interfere with research and dissemination of scientific information.[94]

But from 1930 to 1980, the Canadians and later the IPSFC certainly sought out and accepted the findings of science more readily than the Americans. The Canadians closed all their hatcheries in British Columbia in 1936 after Russell Foerster's study found that they were no less effective than natural propagation. This decision removed the option of mitigating for dams with hatcheries, an option that seems to have been so readily and unjustifiably accepted by the Americans. Even when faced with the threat of Moran Dam in the 1950s, the Canadians still relied on science and did not allow the hatcheries' promise of a quick fix to lure them into trading away the Fraser's mainstem. The commission's restoration program was based on the latest science, which stressed the importance of the salmon's stock structure and the importance of habitat.

On the Columbia, this scientific understanding was ignored until recently. Instead, the Columbia River restoration program invested in a "conspiracy of optimism," clinging to the unfounded hope that hatcheries could restore the salmon. As a result, wild coho in the lower Columbia River have disappeared, and populations of salmon and steelhead in other parts of the basin have become severely depressed. The Canadians and the IPSFC did not do everything right; the Fraser does have its share of environmental tragedies, and some salmon stocks have declined in recent years. But for the period of time when it counted most—the major dam-building era of the 1940s to the 1970s—the International Pacific Salmon Fisheries Commission and its biologists used the science they had to ask the right questions, and then mustered the courage to live with the answers. That, I would argue, made all the difference.

CHAPTER

The Road to Extinction

Chinook and sockeye salmon on the Snake River and coho salmon in Oregon's coastal watersheds spend their freshwater life in very different ecological regions. The Snake River's tributaries flow through the high country of the Rocky Mountains and then drop into an arid landscape, still several hundred miles from the ocean. Oregon's coastal streams are short, usually less than 100 miles from the sea to headwaters, and they flow through a forested landscape watered by large amounts of rainfall. The Snake River chinook and sockeye and the Oregon coastal coho salmon do have one thing in common, though: all three are protected under the federal Endangered Species Act (ESA).

The ESA was authorized by Congress in 1973 in response to the realization that extensive human development was swallowing up the habitat of many types of plants and animals and threatening them with extinction. The winter chinook salmon in the Sacramento River were the first to be protected under the ESA. They were listed as threatened in 1989 and endangered in 1994. But the Sacramento's winter chinook run was not the first salmon to be considered for listing. In 1978, the National Marine Fisheries Service began a formal review of the status of the Pacific salmon in the Snake River. Then, in 1980, Congress enacted the Northwest Electric Power Planning and Conservation Act, creating the Northwest Power Planning Council and giving it the responsibility to protect, mitigate, and enhance the salmon affected by hydroelectric development in the Columbia Basin. With the council defending the salmon, there appeared to be no urgent need to proceed with a listing, so the National Marine Fisheries Service put its status review on hold.

Meanwhile, the Northwest Power Planning Council turned to the basin's state, federal, and tribal salmon managers for advice in crafting its salmon restoration program. Consistent with their history, the managers focused on hatcheries, leaving little funding for habitat. Finally, in 1991, it became apparent the program was not working and that several runs of salmon were indeed headed for extinction. In that year the Snake River sockeye were listed as endangered, and the following year the Snake River fall, spring, and summer chinook were listed as threatened. Those actions started a cascade of listings that is still in progress.

On the Snake River, depletion and extinction came as no surprise. As early as 1944, Willis Rich and his colleagues had predicted that mainstem dams would lead to the extinction of salmon stocks in the upper Columbia and Snake Basins. Since then, biologists and engineers have been engaged in a highly publicized half-century race to build fish-friendly passage devices on the dams before the cumulative mortality inflicted by the hydroelectric system killed off the fish. So far, extinction has a sizable lead in the race.

But in Oregon's coastal watersheds, less than 1 percent of the salmon's habitat has been affected by dams.[1] Unlike the salmon in the highly developed Columbia River, the coastal stocks were not considered as vulnerable to extinction. Yet while everyone focused their attention on the large dams blocking the Columbia, the coastal coho populations were also slipping toward extinction. In 1996, when the National Marine Fisheries Service proposed listing the coho in Oregon's coastal streams under the ESA, their decision was met with a sense of disbelief, especially among some of the state salmon managers.

The collapse of salmon populations outside the highly developed Columbia confirms that the fish are fighting a losing battle throughout the length and breadth of their ecosystems. It is the cumulative effects of many activities in watersheds, not just the highly visible dams, that ultimately are responsible for driving the salmon toward extinction. This means there is no simple solution—no single problem that, if attended to, will bring about recovery. The entire chain of habitats, from headwaters to the ocean, needs attention. Restoration must address entire watersheds, or, to use the terminology popular today, we must begin managing entire ecosystems—managing human activities with ecosystem health in mind.

Overharvest, poor hatchery management, logging, mining, dams, grazing, irrigation, and urban and industrial development are all recognized as proximal causes of the salmon's decline—factors closely associated with the impoverished state of these magnificent fish. Each of these human activities has played a part in the salmon's depletion, but the specific cause-and-effect relationship for each cannot be easily determined. For example, a landslide triggered by logging creates obvious difficulties for salmon, especially if the slide smothers a spawning area, but the specific effect—the number of fish eliminated from the population—is difficult, if not impossible, to measure. The effect of one event cannot be separated from the tangled web of effects of all the other activities in a watershed that degrade habitat or kill salmon. A further complication is that the effects of human activities on the salmon accumulate over the entire watershed, from headwaters to estuary, and through time: the scars of past logging practices such as splash dams, for example, still influence salmon habitat many decades after the practice was stopped. The impossibility of pinpointing the specific effect of any one activity has worked to the advantage of those who destroy salmon habitat and those who overfish the runs: each party contributing to any of the proximal causes, including the management institutions, blames someone else for destroying the salmon.

As a consequence, salmon are now extinct in almost 40 percent of the rivers in which they historically spawned in Oregon, Washington, Idaho, and California. The salmon populations in 44 percent of the remaining streams are at risk.[2] In 1991, I worked with Willa Nehlsen and Jack Williams to inventory the remaining salmon stocks. Our research revealed that of the 214 native, naturally spawning runs of Pacific salmon, steelhead, and sea-run cutthroat trout in Oregon, Washington, Idaho, and California that were at risk, 101 were at high risk of extinction, 58 were at moderate risk, and 54 were of special concern (one stock, the Sacramento River winter chinook, had already been listed under ESA). We also found that at least 106 major stocks had already become extinct. At the time there was some concern that our report, titled "Pacific Salmon at the Crossroads," was too alarming. In retrospect, it turned out to be a timely foreshadowing of the future.[3] Now thirty-four evolutionarily significant units (ESUs) have been or are in the process of being listed under the ESA, including three coho ESUs, five chinook ESUs, and five steelhead ESUs in Oregon, Washington, Idaho, and California (Table 9.1).[4] (An ESU is a population or group of populations that is reproductively isolated from other populations of the same species and that represents an important part of the species' evolutionary legacy.) Since an ESU is usually composed of several populations inhabiting more than one river, the precarious status of the salmon ESUs shown in Table 9.1 covers most of the Northwest.[5]

The massive geographic scale of salmon depletion has been matched by a

Table 9.1. Status of the salmon evolutionarily significant units (ESUs) either listed or proposed for listing under the Endangered Species Act

Evolutionarily Significant Units	*Coho*	*Chinook*	*Chum*	*Sockeye*	*Steelhead*	*Sea-run Cutthroat*
LISTED AS THREATENED OR ENDANGERED						
Central California	Threatened					
Southern Oregon/Northern California coast	Threatened					
Oregon coast	Threatened					
Sacramento River winter run		Endangered				
Snake River fall run		Threatened				
Snake River spring/summer run		Threatened				
Ozette Lake			Threatened			
Snake River			Endangered			
Southern California				Endangered		
South-central California coast				Threatened		
Central California coast				Threatened		
Upper Columbia River				Endangered		
Snake River Basin				Threatened		
Lower Columbia River				Threatened		
California Central Valley				Threatened		
Umpqua River					Endangered	
PROPOSED TO BE LISTED AS THREATENED OR ENDANGERED						
Central Valley spring run		Endangered				
Central Valley fall run		Threatened				
Southern Oregon and California coast		Threatened				
Puget Sound		Threatened				
Lower Columbia River		Threatened				
Upper Willamette River		Threatened				
Upper Columbia River spring run		Endangered				
Deschutes River fall run[a]						
Hood Canal summer run			Threatened			
Columbia River			Threatened			
Ozette Lake					Threatened	
Upper Willamette River					Threatened	
Middle Columbia River					Threatened	
CANDIDATES FOR LISTING						
Puget Sound/Strait of Georgia	Candidate					
Southeast Washington/Lower Columbia River	Candidate					
Baker River				Candidate		
Northern California					Candidate	
Klamath Mountain Province					Candidate	
Oregon coast					Candidate	
All other populations in Washington, Oregon, and California are under review.						Candidate

[a]In addition, the listed Snake River fall run ESU has been proposed to be expanded to include the Deschutes River fall chinook (March 1998).

Source: National Marine Fisheries Service 1998.

massive decline in their numbers. A recent study by Interrain Pacific showed that the drop in adult salmon returning to rivers to spawn has been most pronounced in the southern part of the northeast Pacific. Historically, 56 to 65 percent of the Pacific salmon returned to Alaska streams, 19 to 26 percent returned to streams in British Columbia, and 15 to 16 percent returned to streams in Oregon, Washington, Idaho, and California. Today the picture is markedly different: 81 to 90 percent return to Alaska, 8 to 17 percent return to British Columbia, and only about 1 percent return to the Pacific Northwest. In addition to this dramatic shift and impoverishment of salmon at the southern end of their range, the total number of returning fish declined by 20 to 40 percent over the whole northeast Pacific.[6]

It's clear that the salmon would have slipped into extinction throughout a large part of their range in the Pacific Northwest had the ESA not been in place as a last defense against the failure of management institutions to protect the fish. The ESA has compelled states to finally take inventories of the status of their remaining salmon stocks, to lower their salmon harvest rates, and to reform their forest practices (at least somewhat) to protect salmon habitat. The ESA has also forced some changes in the operation of Columbia River dams in order to improve salmon survival. Yet it is the same institutions that had the responsibility for maintaining the salmon's health, but stood by while they were being exterminated, that now criticize the ESA. It has become common political theater in the Northwest for elected officials and salmon managers to rail against the ESA—to claim it usurps states' rights and to say they want to "save the salmon from the ESA." Yet it was their inaction, and the inaction of their predecessors over many decades, that was responsible for triggering the ESA. In many cases, this political theater has become a dark comedy as the uninformed and self-serving push for still more hatcheries as the quick-fix solution to the salmon crisis.[7]

How did we reach this condition, in which the salmon are teetering on the brink of extinction throughout a large part of their range in the Pacific Northwest? If we look beyond the proximal causes and carefully peel back the layers of history, we can see that our culture's worldview remains as the root cause of this disastrous salmon decline. We assumed that it was possible and desirable to maintain abundant populations of Pacific salmon by simplifying, controlling, and circumventing the ecological processes that created them. We assumed that we were not part of the Northwest's ecosystems but stood apart from them as their managers. We assumed that technology could overcome all problems.

It was these fundamentally flawed assumptions and all the outcomes that flowed from them—including hatcheries, dams, and horrendous habitat destruction—that ultimately led to the salmon's decline throughout the Northwest. From this perspective, the salmon's current state is not the result of a failure of our vision and our management programs; it is the consequence of their success. It is a consequence of making significant progress toward full development of rivers and their surrounding landscape, while at the same time employing technology to eliminate the need for rivers to produce salmon. That was our vision, and the salmon have paid a high price for our success in achieving it.

Even our attempts at restoration have been undermined by the narrow confines of our flawed assumptions about nature, which made us believe we could have salmon without healthy rivers. For example, well-intentioned northwesterners and other Americans funded large salmon restoration programs that cost hundreds of millions of dollars, and then funded management bureaucracies to steward the salmon for present and future generations. Unfortunately, those programs failed, largely because they tried to implement programs based on the same flawed assumptions about nature that had caused depletion in the first place. As environmental historian William Cronon wrote, "Even well-intentioned management can have disastrous consequences if it is predicated on the wrong assumptions."[8]

Why did we so persistently cling to the wrong vision and the wrong assumptions? In part because, for a short time in the 1960s and 1970s, they did appear to be vindicated. Artificially propagated fish finally effected dramatic increases in the supply of salmon, even to the point of producing catches that exceeded peak historical harvests. But in fact, the apparent success of hatcheries, embedded as it was in a faulty worldview, actually accelerated the salmon's decline and set the stage for the latest surge toward extinction. This part of the salmon's story begins in the 1960s, when hatcheries began to enjoy what seemed like phenomenal success.

Hatchery Success and Failure

As with the salmon runs on the Columbia and Fraser Rivers, after the turn of the century the coho in Oregon's coastal streams had been harvested at extremely high rates to support the commercial canning industry. Figure 9.1 shows the pattern of coho harvest in the Oregon Production Index (OPI) area.[9] Although the peak of coho salmon production may have occurred earlier, the first accurate catch records date from 1923, when fish dealers were required to report their fish landings as part of their tax assessments, considerably improving the quality of harvest information.[10] As with other

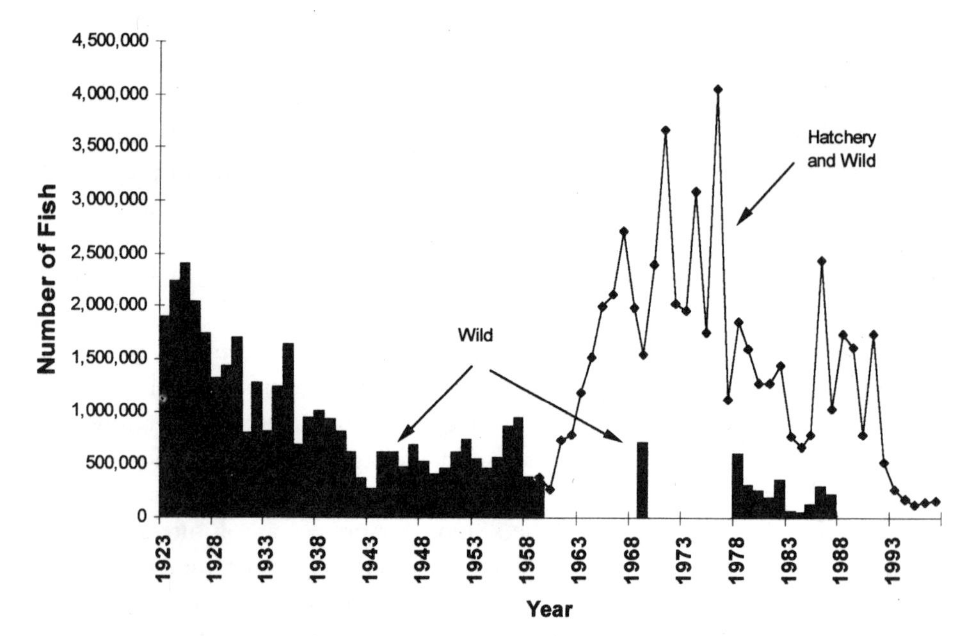

Figure 9.1. Harvest of coho salmon in the Oregon Production Index, partitioned into wild and hatchery fish. Solid bars are catch of wild coho salmon. All coho caught before 1960 are assumed to be wild. (Sources: Unpublished Oregon Department of Fish and Wildlife data [for 1923–1970]; Pacific Fishery Management Council 1992 and 1999 [for 1971–1991 and 1992–1997]; ODFW 1982 [for wild harvest 1959 and 1969]; Borgerson 1992 [for 1978–1987])

salmon runs, the coho catch declined through the 1920s and 1930s and remained at low levels through the 1940s and 1950s. But in the 1960s, it increased rapidly. Because a sizable number of hatchery-reared coho were part of the large catches, this increase was attributed to the success of hatcheries.

By the mid-1950s the Oregon Fish Commission had reported "increasing evidence that the improved methods of operations are resulting in larger returns of adult salmon."[11] And by the 1960s, it appeared that research initiated in the 1930s and 1940s on the treatment of salmon diseases in hatcheries and on the formulation of more nutritious feeds had begun paying off. With better nutrition and disease treatments, hatcheries could hold juvenile salmon until they were ready to migrate to sea. Not only could fish culturists keep juvenile salmon in the protected environment of the hatchery for their entire freshwater life history, but they could also keep the fish healthy until their release. Hatchery managers believed that the longer rearing and improved physiological quality of smolts inside the hatchery led to higher survival in the ocean.[12]

Yet in order to rear juvenile salmon until they smolted and were ready to migrate to sea—a period of six months to a year—most hatcheries needed major renovations. Prior to the 1950s, fish culturists had collected eggs and released fry shortly after hatching. Since the hatcheries were usually empty by late spring, they hadn't needed water during the summer's natural low-flow period. At many facilities the existing supply of water was inadequate to rear juvenile salmon through the summer. And at some stations, logging in the watershed made the naturally, low-summer flows even less reliable.

The renovation of hatcheries to provide water for extended rearing and improved nutrition and disease control led to a dramatic turn of events in 1962. In that year, the Oregon Fish Commission announced that "commission fish cultural operations have become self-sustaining."[13] For the first time in ninety years of operation, hatcheries no longer had to mine wild populations of eggs; the returns of artificially propagated salmon were sufficient to fill the hatcheries to capacity. The number of eggs collected at hatcheries grew dramatically: Oregon Fish Commission hatcheries collected about 6 million coho eggs during the 1953–54 biennium, and by the 1961–62 biennium the egg take had jumped to 45.6 million eggs. During the 1965–66 biennium, 82 million eggs were collected. Instead of searching rivers and streams for wild fish to fill their hatcheries with eggs, managers now had too many hatchery fish returning to their stations. The number of eggs available far exceeded hatchery capacity. Even though salmon biologists recognized that improved ocean survival conditions accounted for part of this phenomenon, they attributed most of the credit to their hatchery program. As the Oregon Fish Commission explained in 1964, "The situation, while most encouraging, was not unplanned or unexpected."[14] The apparent success generated renewed optimism in artificial propagation, but in less than two decades (six coho salmon generations), it would—like all past claims of success—prove to be a costly illusion.

The first indication that the fish culturists' optimism might be premature came in 1975 at the annual conference of fish culturists. Ernie Jeffries, chief of hatcheries for the Oregon Fish Commission, pointed out some disturbing facts: From 1960 to 1966, each increase in the number of coho smolts released from hatcheries resulted in a corresponding increase in the number of returning adults. But after 1966, the abundance of adult coho salmon started to fluctuate wildly and no longer correlated to the release of juveniles from hatcheries. The relationship between hatchery releases and adult returns was breaking down.

Jeffries then went on to express his concerns about the effect of hatchery programs on the health of the ecosystem. He questioned the capacity of the ocean to absorb the increasing numbers of hatchery fish and the pollution of streams by hatchery wastes. Then he raised the issue of genetic and disease impacts of different hatchery release strategies. He even went so far as to "question if building more hatcheries is the correct way to proceed. We might be better off to spend more money learning how to operate those [hatcheries] we have."[15]

But short-term economics crowded out Jeffries's concerns about the future of hatcheries. As one salmon manager once told me, "Fish in the boat is our only concern," and in the boom years of the 1960s and 1970s, there were a lot of fish being harvested by a growing fleet of boats. In 1976, the commercial and sport fishermen harvested a record 3.7 million coho salmon in the ocean fisheries within the OPI area, out of a total of 4.1 million fish. Most of those fish were produced in hatcheries. The harvest exceeded the historical high catches recorded in the 1920s, and it generated a sense of euphoria among salmon managers, who were "enraptured with hatchery fish."[16] In their view, hatchery technology had finally worked. After ninety years of failure, hatcheries were proving that they could be the solution to the problems of salmon supply. Their euphoria, in spite of Jeffries's significant concerns, rapidly spread throughout the Northwest.

The growth and popularity of hatchery-supported sport and commercial fisheries in coastal Oregon did not escape the notice of Canadian biologists. The apparent success of the new hatchery technology caused the Canadians to rethink their policy. British Columbia had closed its salmon hatcheries in 1936, following Foerster's study showing they were no more effective than natural reproduction, but in 1960, after the apparent success of U.S. hatcheries, Canadians began experimenting with the artificial propagation of coho and chinook salmon. Then, in 1970, with the "success" of coho hatcheries in Oregon firmly established by ten years of rapid growth, the Canadians began constructing full-scale hatcheries. The Capilano chinook and coho hatchery was the first full-scale hatchery built in British Columbia in forty years.[17]

At the same time, a shrinking salmon harvest generated intense pressure to rapidly expand the province's hatchery program. By 1970, a century of overharvest and habitat degradation had reduced British Columbia's salmon harvest from 300 million pounds to about half that amount. In this context, as biologist Peter Larkin put it, "Politically, salmon enhancement is a natural."[18] Responding to the pressure, the Canadians launched their Salmonid Enhancement Program (SEP) in 1977, with a budget of about $40 million a year (1978–1991). The program's goal was to double the harvest of salmon in Canadian waters. When SEP began, British Columbia had already built 11 chinook and coho hatcheries, but under the new program, that number exploded. A decade later, 81 facilities were producing chinook salmon and 218 were producing coho salmon. Some of these were small operations run by community involvement programs, but many copied the American approach with large centralized hatcheries that supplemented several different stocks and rivers.[19] By 1994, the program had spent $500 million, largely on hatcheries.[20]

Many saw the growing size of the hatchery-supported salmon runs as evidence that artificial propagation was a sound economic investment.[21] As a result, private entrepreneurs began investing in the new hatchery technology with hopes of creating profits from artificial runs of salmon. The "success" of the new technology made it all seem so simple—easier than growing vegetables, as pioneer fish culturist Seth Green had asserted 100 years earlier. Just build a hatchery near the sea at the mouth of a river, buy some of the state hatcheries' many surplus salmon eggs, hatch the eggs, and release the juveniles into the sea, where they would graze on the open range of the northeast Pacific Ocean. In a year or two, the fat salmon would swim right back into the company hatchery, where they could be easily converted to cash. Because of its resemblance to open-range cattle ranching, this type of operation was termed "sea ranching."

Another version of private salmon ranching more closely resembled feedlots. In this approach, salmon egg incubation and the early rearing of the juveniles occurred in a conventional hatchery. But instead of being released directly to the sea, the juveniles were placed in net-pens suspended in sea water, usually in a bay or nearshore environment. There they were fed until ready for harvest.

While Washington and British Columbia opted for the feedlot approach, in 1971 the Oregon legislature authorized the state fish commission to issue permits for private sea ranching. The initial statute limited these operations to chum salmon, but it was expanded to include coho and chinook salmon in 1973. With the addition of these more profitable species, large corporations such as British Petroleum and Weyerhaeuser became involved in sea ranching. In 1975, Weyerhaeuser launched its sea-ranching operations in Oregon's Yaquina and Coos Bays with an investment of $40 million.[22]

Unlike the publicly funded hatcheries, these corporate hatcheries had to produce tangible benefits and a return on the initial investment. According to the public relations information circulated by the companies at the time, the production and harvest of the Pacific salmon were entering a new and modern era. The corporations would make artificial propagation more efficient and profitable. Some advocates of private sea ranches in Oregon believed their operations would eventually replace commercial fishermen and their primitive salmon hunting in the open ocean.

The apparent success of the Oregon coho hatcheries also encouraged a string of new public salmon enhancement and mitigation programs. For example, the Lower Snake River Compensation Program was authorized by Congress in 1976 to replace the salmon lost due to the construction of Ice Harbor, Lower Granite, Lower Monumental, and Little Goose Dams on the lower Snake River.[23] With twenty-six production, acclimation, and trapping facilities, the program artificially propagates salmon at a cost of about $11 million a year. In 1978, the state of Washington launched another major program to double the salmon harvest largely through hatcheries. The state program was initiated in response to major harvest allocation problems following a federal district court ruling that granted Indians the right to harvest half of the salmon. If hatcheries could double the size of the salmon runs, the reasoning went, then giving half the salmon to the Indians could be accomplished with little disruption to the non-Indian fishermen. By 1987, Washington State had spent $30 million on this program.

To recap, for ninety years, fish culturists and salmon managers had looked for tangible evidence of hatchery success and found none. Again and again they had claimed success based on the flimsiest of evidence or no evidence at all. Then, in the 1960s and 1970s, they saw real, measurable increases in salmon runs that seemed directly attributable to hatcheries. This apparent success unleashed an avalanche of public programs and private investment to enhance salmon runs with artificial propagation. Hatchery success seemed to mean that the region could finally have, in the words of the Washington Department of Fisheries, "salmon without a river."[24]

Then, in 1977, coho production in the OPI area suddenly collapsed. The harvest dropped from a peak of 3.9 million fish in 1976 to a million fish the next year. Less than twenty years later, in 1997, only 28,000 coho were harvested, and there were less than 300,000 in the whole OPI area. By 1997, the Oregon coastal stocks of coho had been listed as threatened under the ESA, and the wild coho in the lower Columbia River were declared extinct.

Within two decades, the promise of a nearly unlimited supply of salmon had vanished, and Oregon's coastal coho were on the brink of extinction.

In the years immediately following the collapse of the coho fishery, salmon managers were perplexed by what had happened but found no clear-cut answers. The chronic lack of monitoring and evaluation that had plagued the hatchery program since the 1870s still persisted in the 1960s, which meant managers could not identify the cause of the coho collapse. Thus they could neither adjust their programs to correct the problem nor even determine if correction was possible.[25]

Consistent with its history since 1872, hatchery enhancement in the 1970s was a house of cards—a house built on the shaky foundation of blind optimism, ideology, and shifting weather patterns. When the collapse came, it was impossible to sort through the cards to determine which one had caused the collapse. A task force of scientists assigned to evaluate the limited information available concluded that there were several possible explanations for the downturn in coho production, but the information was not adequate to determine which was the actual cause.[26]

In British Columbia, SEP had difficulty reaching its goal of doubling salmon harvests. The infectious optimism that had first inspired the program began to erode in the face of economic and biological realities. With regard to SEP's reliance on hatcheries, author Terry Glavin quoted Edgar Birch, a salmon enhancement task force member, as saying, "We should have listened to people who were telling us that all these hatcheries were a bad idea. . . . Hatcheries were a bad idea from the very beginning."[27] In 1994, the Canadian natural resource economist Peter Pearse released his audit of SEP, which concluded that over its lifetime the *economic costs* of the program would *exceed the benefits* by nearly $600 million. Pearse recommended major changes: First, he contended that enhancement and conservation of the wild salmon had to be reconciled so that the wild stocks were protected. Second, he suggested that scientific assessment of the program be strengthened and that there be economic assessment of it as well. Finally, he recommended that those receiving the benefits of salmon enhancement should pay more of the costs.[28]

SEP's economic shortfall matched the program's failure to meet its biological goals. SEP's original goal of doubling the catch meant it had to add about 86,000 metric tons of fish to the harvest. Between 1977 and 1989, the harvest did rise by 31,600 metric tons; however, the increase in catch did not come from the species targeted by SEP. While chinook and coho salmon together received two-thirds of SEP's expenditures, the catch of those species actually declined. By 1992, the failure was severe enough to prompt Pearse to warn Canadians that they faced a tragedy with the Pacific salmon similar to the one they already had on the East Coast with the Atlantic cod.[29]

In their review of the Salmonid Enhancement Program, biologists Ray Hilborn and John Winton reached another conclusion that would be echoed by scientists analyzing other hatchery programs across the Pacific Northwest: the program's lack of monitoring and evaluation, they said, had left biologists unable to determine why the program failed to meet its goals. "Thus it is disconcerting that after 15 yr [years] we still are not sure which technologies will work and under what conditions."[30] In subsequent years in the United States, the National Research Council, the Independent Science Group from the Columbia River, and the National Fish Hatchery Review Panel would all come to the same conclusion.[31]

Private sea ranches fared no better than publicly funded enhancement programs, in spite of claims that corporate hatcheries would cut through the waste and inefficiency of government bureaucracies with their cost-effective salmon production systems. The private sea ranches failed because they depended on two flawed assumptions, in addition to all those that had underlaid artificial propagation since 1872. First, proponents of sea ranches assumed the ocean was a large, stable ecosystem with species-specific niches that existed independently of each other. Since habitat degradation and overharvest had reduced the number of salmon leaving fresh water for the sea, they reasoned that sizable amounts of vacant salmon habitat must exist in the ocean. The goal of private sea ranches was to fill those vacant niches and harvest handsome profits. Second, sea ranchers assumed that state salmon managers would restrict the ocean harvest on mixed stocks of public and private salmon to protect the wild fish. If harvests were cut back to allow adequate numbers of wild salmon to escape the fishery and reach the spawning grounds, enough private sea-ranched fish would also return to make a profit.[32]

Both of these assumptions proved to be false. By 1990, eight out of twelve sea ranches operating in Oregon had gone belly-up after the expected profits failed to materialize, and those still in operation were barely surviving.[33] When Oregon Aqua-Foods, the state's largest sea-ranching operation, faced foreclosure, the company killed the last 3 million juvenile salmon in its hatchery and sold them for fertilizer.[34] That was not the end of OreAqua's hatchery, however; the Oregon Department of Fish and Wildlife and the Port of Newport each contributed funds to keep it operating in order to stimulate sport fishing in the bay. But the program was expensive: each coho caught by a sport fisherman cost roughly $5,000 to produce.[35] Only public hatchery programs, with their chronic lack of accountability, could get away with selling a $10 license to catch a $5,000 fish. The last private sea ranch on the Oregon coast drained its ponds and closed in 1992.

On the Snake River, the Lower Snake River Compensation Plan faced similar problems. In addition to the hatcheries, the U.S. Army Corps of Engineers was applying other forms of technology to mitigate the effects of the dams on the Snake and Columbia Rivers. For example, in the mid-1970s the Corps started barging or trucking juvenile salmon past many of the dams in the mainstem Columbia and Snake Rivers. Then they placed screens in front of the turbine intakes to reduce the entry of juvenile salmon, and installed deflectors on some dams to prevent gas supersaturation of the water below the spillways. When all this technology was in place, one overly optimistic biologist predicted, "It seems possible that we can establish adult runs of steelhead trout and salmon in far greater numbers than existed before" in the Snake River.[36] But as was the case elsewhere, hatcheries and other technological fixes did not work. Today the Snake River coho salmon are extinct, and the steelhead, spring and fall chinook, and sockeye salmon are protected under the ESA.

In 1998, the Lower Snake River Compensation Program held a public review of its accomplishments. Although it generally makes use of the latest information and is implemented with a high degree of professionalism—it is one of the few hatchery programs that conducts extensive monitoring—the program has, like the other efforts begun during the optimistic 1970s, fallen far short of its goals. Congress created the LSRCP to mitigate for the salmon losses created by the lower Snake River dams; however, the published proceedings of the review contained the suggestion that the dams would have to be removed in order to ensure the success of the hatchery program.[37] The program's reviewers apparently forgot that the hatcheries were built to mitigate for the dams.

Like the other enhancement efforts, Washington State's investment in hatcheries in the 1970s failed to live up to expectations. Despite spending $30 million on hatchery enhancement, the state failed to double the harvest; in fact, by 1992 the catch had increased by only 10 percent. Part of the program's failure, however, can be attributed to interception fisheries outside Washington.[38]

The collapse of coho salmon in the OPI area and the other failures of hatchery enhancement programs were, like nearly all salmon depletions, a consequence of the cumulative effects of hundreds of stresses over more than a century. But continuing studies of the general salmon declines in the 1970s and 1980s revealed another previously unrecognized problem: diminished ocean productivity had severely reduced the survival rate of the salmon during their saltwater phase of life.

As pointed out earlier, the harvest of coho salmon in the OPI area collapsed abruptly in 1977. The adult coho harvested that fall had been released from hatcheries in the spring of 1976, and that year proved to be one of monumental change for the ocean. During 1976, as many as forty ocean variables showed a dramatic stepwise change.[39] Studies later showed that the ocean was not a large, stable, and uniform habitat but rather a dynamic ecosystem where changes could exert a strong influence on salmon abundance.[40] In fact, the ocean appeared to change on a forty- to sixty-year cycle.[41] Moreover, scientists realized that the changing productivity of the northeast Pacific Ocean had a zero-sum aspect: during periods of poor productivity in the marine waters off Oregon and Washington, the marine waters off Alaska were in a high-productivity phase, and the reverse was true as well.[42]

Diminished ocean productivity was not the only cause of the coho's precipitous decline, as many would later claim, but it was yet another, admittedly large stress added to all the others. And, more important, increasing hatchery production was as ineffective in making up for changing ocean conditions as it had been in overcoming the effects of dams, logging, and other forms of habitat degradation. In fact, artificially propagated salmon seemed to be particularly vulnerable to changing ocean conditions. During periods of poor oceanic productivity, hatchery-reared coho salmon survived at only about half the rate of their wild cousins.[43]

The change in ocean productivity in 1976 exposed the cumulative effects of decades of habitat degradation, overharvest, and poor management. All of these factors contributed to a narrowing of the salmon's life histories, a loss of their biodiversity that made them vulnerable to natural fluctuations in parts of their ecosystem such as the ocean. Some people claim that the salmon crisis is entirely a consequence of changing ocean conditions, and so there is nothing we can do about it. They are wrong. The performance of salmon in the ocean is not independent of human manipulation of the freshwater phase of their life cycle. The loss of the salmon's biodiversity—their evolutionary legacy—has deepened the effect of changing ocean conditions.[44]

It is now generally accepted that the apparent success of coho hatcheries in the OPI area was largely the result of favorable ocean conditions. All the massive investments in hatchery enhancement programs that followed were based on the assumption that hatcheries could stabilize, maintain, and even enhance salmon abundance. But in reality, the hatchery success of the 1960s and early 1970s was based on the fickle and ephemeral nature of the weather—the climatic patterns that improved ocean conditions.

Because they were under the illusion that hatcheries had solved all the salmon supply problems, managers had continued to allow habitat to degrade and harvest rates to rise.[45] When the survival of hatchery fish dropped abruptly in 1976, salmon managers found it difficult to convince commercial fishermen to reduce their harvest of the dwindling stocks of hatchery and wild salmon.

As the abundance of coho in the OPI area had grown during the 1960s, so too had the number of commercial fishermen. The licenses issued to commercial fishermen in Oregon increased from 2,565 in 1960 to 8,566 in 1978.[46] When the collapse occurred, the large fleet of sport and commercial fishermen exerted pressure to maintain high harvest rates, and Oregon salmon managers relented. Looking back on this period, Oregon salmon manager Jim Martin recalled, "I was the scientist in charge, and I should have seen it coming better than I did." But even if he had, he would have faced intense political pressure to maintain the harvest, "Commercial and sport fishermen clamored for more than we now know we should have been given."[47] By giving in to the pressure from fishermen to maintain high harvest rates, salmon managers set the stage for the regional collapse that followed.

The demographics of hatchery and wild salmon populations meant that wild coho would be the biggest victim of the failure to control the harvest. Under the new artificial rearing regimes, juvenile coho were held in the hatchery until they smolted and were ready to migrate to sea. Remaining in the protected environment of the hatchery throughout their freshwater life increased the egg-to-smolt survival of hatchery-reared fish. But the wild salmon that spawned in streams wrecked by a century of habitat degradation suffered high mortality between egg deposition and smolt migration, so a pair of coho that spawned in a hatchery would send more smolts to the ocean than would a pair of wild coho that spawned in a river. This difference in the survival rates of hatchery and wild fish during their early life stages meant that hatchery adults could sustain a higher harvest rate than wild salmon. But by keeping harvest rates high for mixed catches of hatchery and wild coho, salmon managers permitted the overharvest of the wild fish.

Exploitation rates reached 88 percent, much too high to sustain the wild coho. The number of adult coho salmon reaching the natural spawning areas plummeted in the early 1970s—a clear indication that the harvest rates were already affecting the wild coho.[48] By permitting the rapid growth in the fishery and allowing high harvest rates, salmon managers made it impossible for the wild coho to recover from the lows of the 1920s, 1930s, and 1940s.

In 1980, to shore up the rapidly shrinking wild coho populations and still avoid reduced harvests, the Oregon Department of Fish and Wildlife began planting excess fry from the state's hatcheries in natural rearing and spawning areas. In effect, this was a return to practices carried out in the earliest years of the hatchery program. This time, however, the fry plantings were carefully evaluated; sadly, the attempt to use hatcheries to enhance natural spawning populations not only failed but actually caused a further reduction.[49] The domesticated hatchery stocks spawned a month or two earlier than the wild coho. With this head start, when juvenile hatchery coho were planted in the streams, they were larger than their wild cousins and displaced them. But when the adults of hatchery origin returned, they spawned earlier, during the fall freshets and floods, and their eggs and fry suffered much higher mortality than the later-spawning wild fish. Early spawning, a positive attribute in the hatchery, was a deadly trait in the wild.

As we have seen, the overharvest of wild salmon stocks is a common consequence of hatchery enhancement programs. In 1993, Sam Wright, a senior fisheries biologist for Washington State, published a hard-hitting essay about this problem. He argued against the practice of maximizing the harvest of hatchery fish without regard to the devastation of wild populations it inevitably caused. Management agencies that permitted such high harvest rates were actually causing the extinction of wild populations, he wrote, and "most people do not believe that a fish management agency should condone extinctions." Wright identified 3,500 miles of underused stream habitat in the lower Columbia River where wild fish could have spawned, but which are vacant because of excessive harvest.[50] In effect, expensive hatchery programs were merely replacing the natural reproduction eliminated by overfishing. Even worse, the hatcheries that did create increases in abundance generated fisheries that ended up destroying as much or more of the salmon's biodiversity as had the prior loss of habitat.

In Canada, salmon managers allowed sport and commercial fishermen to harvest as much as 95 percent of the fish, including wild salmon mixed with hatchery fish. These high harvest rates eliminated the smaller wild populations, causing the fishery to become dependent on fewer and fewer stocks.[51]

According to Canadian biologist Carl Walters, "About 50% of the fish we take now come from a very small number of hatchery populations, and marine survival rates of these hatchery fish have been declining in a way that does not bode well for long-term sustainability."[52] The reliance on hatcheries had masked a staggering loss of salmon biodiversity. In British Columbia, about 30 percent of the spawning populations of salmon have been lost since 1950.[53]

Just before his death in 1965, biologist W. F. Thompson published a paper that summarized the insidious effects of a simplified hatchery-harvest management system. "We regulate our fisheries," he wrote, "but we concentrate them on the best races, and one by one these shrink or vanish, and we do not even follow their fate . . . knowing only that our total catches diminish, as one by one small populations disappear unnoticed from the greater mixtures that we fish." He went on to conclude: "So we greatly underestimate what is needed or when it is needed and feel self-righteous about our conservation."[54] With each small population that disappears, with every run that becomes extinct, biodiversity—the very quality that has enabled the salmon to withstand ice ages, mountain uplifts, lava flows, changing ocean conditions, and the whole onslaught of the industrial economy—is lost. Tragically, it is this biodiversity that habitat destruction, overharvest, and the technology of hatcheries have systematically whittled away for over 100 years. If the Pacific salmon are to survive, they will need all of their evolutionary biodiversity that remains. And they will need healthy rivers where that biodiversity is nurtured and maintained.

Today we are faced with a legacy of more than a century of salmon management based on a faulty set of assumptions. Natural salmon habitats have been wrecked while we have spent hundreds of millions of dollars on hatcheries, chasing the foolish dream of producing salmon without rivers. Every independent scientific review of the current management system has called for a major overhaul, but bureaucratic salmon managers still cling to the status quo, defend their hatchery programs, and embrace without thinking the outmoded worldview from which hatcheries first emerged in 1872.

Hatcheries play a unique dual role in the current salmon crisis. They have been clearly identified as part of the problem, yet many consider them an important part of the solution. Salmon managers deal with this apparent contradiction in a straightforward way: they claim that the problems with hatcheries were all in the past and have now been resolved. It's true that hatcheries have changed, and their technology today is very advanced. High-tech equipment, improved disease control, better feeds, new marking technology, and standard operating practices that include animal husbandry based on genetic principles—these have all improved the health of the fish in the hatcheries and allowed us to better estimate their contributions to fisheries.

But what happens after the hatchery fish are released—how do they affect the ecosystem and the wild salmon? This question remains a major concern. Answers are still lacking and will likely remain so in the near future, since managers vigorously protect the status quo and attack scientists who ask questions about fish culture or even conduct research on unanswered questions.[55] In apparent response to the salmon crisis and all the questions it raises, the Oregon Department of Fish and Wildlife recently dismantled its research section and consolidated the power of harvest and hatchery managers—the Lords of Yesterday within the institution—to control policy, thus strengthening the status quo and ensuring that any changes will be superficial.

In fact, the changes made to date are mere tinkering around the edges of a program in need of a serious overhaul. Hatcheries are not the only issue in need of entirely new approaches if the salmon are to survive. For instance, Judge Malcolm Marsh of the federal district court recently chastised the National Marine Fisheries Service for being too timid in using its authority under the Endangered Species Act to change the operation of the federal hydropower system to protect listed salmon stocks. Marsh told the fisheries agency that their efforts were "too heavily geared toward the status quo" and therefore have resulted in "relatively small steps, minor improvements and adjustment—when the situation literally cries out for major overhaul." Judge Marsh was right, but the same criticism could be leveled at the salmon managers and their incremental changes in hatchery programs.[56]

The oft-repeated statement that the problems associated with hatcheries are all in the past rings hollow when the historical record is carefully examined. Hatcheries have changed over the past century, but those changes have been chameleonlike—superficial changes to match the prevailing political and funding environment. The flawed fundamental assumptions remain. It's important to examine those assumptions and their historical roots because, in the words of environmental historian Robert Bunting, "The past is never past."[57] The threads of past assumptions, beliefs, and values are still woven through the fabric of salmon management today.

Conservation biologist Gary Meffe recently examined the fundamental assumptions upon which hatchery-based salmon recovery programs rest. He concluded that current salmon management efforts were based solely on artificial propagation and restocking, even though the salmon had declined as a result of environmental degradation and overharvest. "Without correcting the *cause* of the decline," Meffe contended, salmon management was "not facing biological reality." He concluded, "Salmonid management based largely on hatchery production, with no overt and large-scale ecosystem-level recovery program, is doomed to failure."[57] Meffe was certainly correct. But as I see it, habitat degradation has not simply been a long-overlooked by-product of our industrial economy. It has been the direct result of the large-scale ecosystem simplification that is a central and guiding vision of that economy—the same vision and worldview that engendered hatcheries.

Until we carefully examine our core beliefs and assumptions about nature, our rivers, and our salmon, and change our deeper motivations, salmon management efforts and their institutional Lords of Yesterday—hatcheries and harvest—will continue to threaten the existence of the species they are supposed to protect.

EPILOGUE

Building a New Salmon Culture

I remember the day I saw Grand Coulee Dam for the first time. The day was very warm, and the drive had been long and tiring. There it was, one of the largest blocks of concrete in the world, damming one of the greatest salmon rivers in the world. Grand Coulee has rightfully been termed a salmon killer: it permanently cut off 1,400 miles of habitat where the giant June hogs, the big summer chinook, once thrived. In 1939, biologists estimated the size of the salmon run above Grand Coulee—which had already been decimated by overharvest—at 21,000 to 25,000 fish. Those estimates were very conservative.

Before that day, I had seen the dam only in pictures, where it always looked clean and neat. It seemed to naturally fit—air-brushed into the landscape. But viewed in person, it was an ugly mass of concrete with dark stains down its front face. The nearby canyon walls were gashed and raw. In real life it looked foreign, out of place: an appropriate image for a salmon killer.

I spent a couple of hours on the balcony of my motel room, gazing at the dam and thinking about what it symbolized. In so many ways, this block of concrete has enriched the United States and its citizens. It made us feel good and gave us a sense of accomplishment in the 1930s, when the whole country was down and needed a lift. It gave us jobs, farms and food, labor-saving devices, lights, aluminum, and our great standard of living. It also killed off a significant part of the Northwest. Could there have been another way, I wondered. Did the industrial economy and Grand Coulee Dam, one of its greatest symbols, have to subjugate and then kill so much of the natural economy? Could the two have found a balance, a way to exist in the same ecosystem—the five houses of salmon and the houses and factories of Euro-Americans?

As I sat there wondering and swatting mosquitoes, the face of the dam lit up. It was the start of the nightly laser light show. (The utilitarian engine of the region's economy is also an entertainer.) Appropriately, the lasers sent a series of large green dollar signs floating through the darkness. Then a series of laser salmon swam across the face of the dam. Here were the ideal salmon, I thought, the fish that fit perfectly into our worldview. We have complete control over them—press a button and they appear; press another and they change from green to red; press another and they swim over the dam. Salmon and dams are compatible—as long as you are not particular about the kind of salmon. These laser salmon swimming across the face of Grand Coulee Dam symbolized our culture's story of the relationship between us and the magnificent fish. It has been a story with a very simple plot: control, simplify, dominate, and artificialize salmon and their rivers. If we are to save the real salmon, we will have to change our story. The one we have now does not have a happy ending.

In his book *The Dream of the Earth,* theologian Thomas Berry explains the importance of stories. According to Berry, stories help us to understand and give meaning and value to the often confusing signals we find in the world.[1] Our stories have deep roots in our culture because they have been passed from generation to generation. They do not change easily. Stories sit deep down in human consciousness, quietly shaping reactions to new ideas and information, guiding decisions and ordering expectations for the future.[2] They both form and reflect our worldview. Although they are powerful shapers of what we see, how we interpret what we see, and how we act, our stories usually remain hidden from view. They are rarely examined or challenged. The purpose of this book has been to examine the stories that have guided our relationship with Pacific salmon.

We must pay attention to those stories and not take them for granted. Above all, we must not leave our stories unexamined. They are, after all, human constructs that we impose upon the earth. So far, our culture's stories of conquest over nature have enabled us to convert rivers into reservoirs and irrigated crops, and to shift salmon from their homes to hatcheries. According to Berry, such stories can be the source of deep crisis when they no longer help us to sustain the things we value. We have reached that crisis point in our relationship with the Pacific salmon. For a long time, the story of controlling nature seemed to offer simple solutions to complex problems; but like many such solutions, they turned out to be only simple-minded.[3]

The belief that we could control the production of salmon without healthy rivers has proven to be false. Regardless of how much money we spend on salmon restoration programs, unless we change the story of our relationship with these fish, we face the real possibility of losing them. We need a new story to guide our behavior, one that is in harmony with the ecosystems of the Northwest.

There are hopeful signs that the precarious state of the salmon may be bringing about positive changes in our story. In 1996, a panel of independent scientists acknowledged the failure of our current approach to salmon management and restoration and called for a new conceptual foundation—essentially, a new story. In that same year, another comprehensive study of the salmon crisis, this one by the National Research Council, reached a similar conclusion, especially with regard to the heavy reliance on hatcheries. Key elements of the old story, particularly the assumption that we can improve on nature and simplify the salmon's ecosystem, are finally beginning to lose credibility among scientists.[4]

Salmon are still an economic asset to many local communities, but the strong interest in these fish cannot be explained by economic measures alone. The Pacific salmon are a significant part of the evolutionary and cultural heritage of the Northwest, and one of the most powerful symbols of that heritage. According to a poll taken in Washington State, three out of four voters believe that the salmon are an important part of the Northwest's identity, and 77 percent believe that the salmon are an important indicator of overall environmental health.[5] According to an *Oregonian* poll, 85 percent of Oregon's citizens believe it is important to preserve salmon and steelhead runs on the Columbia and Snake Rivers, and a surprising 60 percent think the salmon should come *before* commerce in managing the Columbia River system. The same poll asked Oregonians why they wanted the salmon to be saved: 35 percent said because they are part of the Northwest's heritage, 36 percent said because they are a measure of the region's environmental health, and only 15 percent cited their commodity value (9 percent for sport fishing and 6 percent for commercial fishing).[6] Thus the salmon management institutions that continue to claim success by reporting only commodity statistics—and they all do—have lost touch with the values held by the public.

After a century of unbridled development of the Northwest's rivers, the people of the region are looking back and asking, What have we done? More important, a growing number of people are asking, Is it possible to undo what we have done? Recently the *Idaho Statesman,* a Boise newspaper, concluded that breaching four dams in the lower Snake River to help restore the salmon made good economic sense.[7] Such analysis and reporting—the kind that critically evaluates technology—would have been unthinkable just five years ago.

Local communities have also taken more interest in their rivers and their salmon runs, and they are coming together to work on grassroots restoration programs. "Save the salmon!" has become a rallying cry that is reviving the spirit of many communities. For example, near my home a group called Wild Olympic Salmon has adopted a small stream, Chimacum Creek, that flows through the community. The group has completed about twenty improvement projects ranging from tree planting and gravel clearing to rebuilding the stream's habitat structure. Each fall, members also host a First Salmon Ceremony that brings the community together to celebrate the salmon and strengthen their ties to the fish and the stream. These are all signs of a growing realization of the need for change. People sense that the relationship between salmon and humans needs mending.

We can look to wildlife biologist Aldo Leopold for guidance on how to reshape the salmon's story. When Leopold began laying a foundation for a new land ethic more than fifty years ago, he realized that it was going to be difficult to change the old ways. "Timidity, optimism, or unbending insistence on old grooves of thought and action," he wrote, "will surely either destroy the remaining resources, or force the adoption of policies which will limit their use to a few."[8] Although Leopold did not write that statement specifically for salmon, it does a nice job of summing up what happened.

We've come to expect that big problems are always solved by big programs, handled entirely by experts with optimistic predictions. Programs do help to organize and keep track of efforts, and experts do play important roles, but the key to fundamental change, and ultimately the key to the survival of the salmon, lies within each of us. Leopold told us there are some decisions that we cannot leave to the experts alone, "which every man must judge for himself, and on which the intuitive conclusion of the non-expert is perhaps as likely to be correct as that of the professional. One of these is what is right. The other is what is beautiful."[9]

In the context of the current debates about our wild salmon, I believe that "what is right" refers to the way we balance today's needs with our moral and legal obligation to manage the salmon for future generations. "What is beautiful" refers to the ecological wonders of the Northwest's rivers, their tremendous natural productivity, their amazing biological, geological, and climatic diversity, and the salmon's biodiversity. Our judgment of what is right and what is beautiful will determine the quality of the world that future generations will live in. It will determine whether future generations see real salmon or laser salmon. The experts, with their technical information, can certainly review the status of the various salmon stocks, but we play a part in deciding whether those stocks will continue to exist.

What is right and what is beautiful are questions of individual values. That is why each individual's opinion carries as much weight as the expert's opinion. However, environmental debates that result from differing values often devolve into choices between economic and ecological values, between short-term economic needs and long-term obligations to future generations. There can be no satisfactory outcome to debates that pit humans against the ecosystems they must live in, or that set this generation against future generations. In the long run, everyone will lose. We must avoid this self-destructive course.

Instead, we must seek a balance between the natural and the industrial economies in the Northwest. Within the framework of the industrial economy and governmental hierarchies that dominate the watersheds of the Pacific Northwest, the struggle to build a sustainable relationship between salmon and humans has been a complicated mixture of politics, science, and myths. For the past several decades, sustainability has eluded our grasp, and the salmon have continued their slide toward extinction. All the power of science, all the organizational power of modern government, and all the wealth of the industrial economy have failed to find a way for humans and salmon to coexist in the watersheds of the Northwest.

If we are to have any hope of restoring the salmon to sustainable levels of abundance, we will have to begin by paying more attention to the salmon's world—not as it exists today, but as it existed before we simplified, controlled, and "improved" it. Although some of our damage to the rivers cannot be undone, much can be changed. Unnecessary dams can be removed or breached. Riparian zones can be protected from grazing by erecting fences. Most important, we can let the rivers heal themselves: we can allow the natural riverine processes to rebuild the natural physical structure of the rivers, so that the salmon can then regain their diverse life histories. And in our restoration efforts, we must address entire watersheds—the whole chain of salmon habitat, from headwaters to estuaries—and then beyond, to the ocean.

Perhaps the series of articles by the *Idaho Statesman* about the economic viability of breaching the lower Snake River dams, together with the polls showing that people are now willing to award salmon survival a higher priority than commerce, are indications that we are starting to inch toward a balance between the natural and industrial economies. But going beyond these first steps and actually achieving the balance will be very difficult. It will require altering fundamental and long-standing beliefs—although it does not, as some fear, mean returning to the standard of living of the 1850s. Historian Arthur McEvoy has written that giving up our attempts to dominate nature "need not entail the exchange . . . of reason for animism or industrialism for a hunting-gathering economy." He explained, "It only means learning to care for other living things," and that is a special talent humans have.[10]

When I talk with my friend Tom Jay, a poet, sculptor, and salmon lover, the conversation often turns to saving the salmon. What can we do, and what must we do, to save the remnants of the shattered populations? Tom believes restoring the salmon will require nothing short of remaking our culture. I agree in principle. But at the same time, change of that magnitude is difficult for me to comprehend. I am convinced that the first step in building a culture capable of coexisting with the salmon is the cultivation of attentiveness—encouraging people to listen to the world they live in. We need to reconnect with the natural world. As individuals and as a society, we need to pay attention to the land and rivers—and especially to the consequences of all the things we do.

In the Pacific Northwest, political leaders and government administrators often speak about the salmon and the current crisis. I always listen to what they say, but I have yet to hear one speak as though he or she had actually paid attention to the salmon. They talk about things that connect humans to humans, not things that connect humans to salmon. They ask questions like Who's in charge? How large is the budget and how is it distributed? How can we avoid the Endangered Species Act? Who is responsible for the crisis? Why do I have to pay to save the salmon? They talk, and they write plans, but they do not listen to what the land, the rivers, and the salmon have to say. As author Nancy Langston has written, "There are ways of living on the land that pay attention to the land, and ways that do not."[11] And we have not been paying attention.

Our first step to achieving Tom Jay's vision of a culture capable of nurturing the salmon must be to follow Langston's advice and find ways to pay attention to the land and water we share with the salmon. How do we learn to pay attention? We can study the natural economy and how other cultures have lived in harmony with it. We can learn more about how rivers work naturally and about how human developments affect them. We can start or join groups that work to protect salmon habitat. We can celebrate and honor the return of the salmon. Or we can watch and listen and maybe pay attention to a single wild fish.

A Single Wild Fish

This is the end of the line, I thought when I let my backpack-electrofisher slide to the ground; not even a cutthroat can climb *this* high. The April air cooled my sweat-soaked shirt. I was standing next to a small tributary of the Pysht River on the north side of the Olympic Peninsula, looking for cutthroat trout. The presence or absence of this salmonid determines how much protection the stream will get when the surrounding timber is cut.

Few people will ever see this stream; it will never appear on the cover of a fishing magazine. At this point, I could plant a foot on each bank and watch the stream flow between. The last few hundred yards were rough going. As the gradient got steeper and steeper, I had to work my way through a tangle of wind-thrown trees. Yet every place where I activated the probes of the electrofisher, a cutthroat trout rolled out from under a log or an overhanging bank. Past experience told me that the upper limit of trout in this small creek should have been farther downstream, closer to the entrance to the canyon. Although I didn't expect to find cutthroat this high, I wasn't surprised. When working with wild trout and salmon, I always expect the unexpected.

The logjam in front of me had to be the end of the line—the upper limit of these colorful wild trout. Rotting logs from the last timber harvest were stacked helter-skelter about four or five feet above the streambed. Over the decades the stream had worked its way through the tangle of wood and then cascaded down the face of the jam in three small rivulets, each dropping through a series of small pools. Each moss-lined pool held a few cups of water. I was convinced there would be no trout above this barrier, but still I felt the need to see for myself.

As I sat on a log, trying to plot the easiest way to drag my gear over the jam, I heard a splashing sound distinctly different from the steady bubbling of the stream. It took me a few seconds to locate the source, but there it was—a trout about seven to eight inches long. The beautiful little fish appeared to be trapped in one of the small pools halfway up the face of the jam.

My first impulse was to capture it: I grabbed the net and headed toward the trout, fully intending to catch it and carry it to the stream above the jam. But then I stopped. I realized that I was witnessing something special. It was time to suppress my desire to intrude and "fix" nature, so I sat back down and watched to see what this little fish would do next. What it did was amazing. Slowly but surely, it leaped and struggled through the tangle of logs and moss-lined pools. With great effort, it eventually reached the stream above the jam.

In this little microcosm of a stream, I was privileged to witness an example of how the Pacific salmon had clung so tenaciously to life in the rivers of the Pacific Northwest, how they had survived mountain uplifts, volcanoes, massive lava flows, centuries of drought, and thousands of years of glacial ice. The genes of that cutthroat trout contained lessons for survival learned over thousands of years. In those lessons lie the strongest hope for the salmon's recovery, if only humans would stop creating insurmountable obstacles.

Preserving wild fish and their genetic resources is just as important to salmon recovery as operating large, expensive hatchery programs. In many cases, those programs are like my first impulse to help the trout over the logjam. Given the size of the obstacles we have placed in the salmon's path, they *do* need our help. But the most important help we can give them is to remove or reduce the obstacles, not to continuously carry the fish over the top.

This point is well illustrated on the Columbia River. The healthiest salmon population remaining in the basin is the wild fall chinook, which spawns in the 55-mile Hanford Reach of the mainstem Columbia, the only free-flowing part of the river left above Bonneville Dam and within the range of salmon. Wild juvenile salmon from the Hanford Reach must pass four dams on their way to the sea, but they are thriving because their spawning and early rearing habitat retains enough of its natural integrity to support them. Moreover, biologists and engineers have successfully reduced the obstacles to salmon survival here. River flows through the Hanford Reach are now regulated to prevent unnatural fluctuations, which for years had dewatered redds before the eggs could hatch. The Hanford Reach is not entirely natural and still has flow problems that need correcting, but it has been improved enough to enable the chinook to survive.

Downstream of Bonneville Dam, there is more evidence that the salmon can bring about their own recovery if given a chance. While the Northwest Power Planning Council's salmon restoration program—probably the largest fisheries-restoration program in the world—emphasizes recovery of salmon in the upper basin, the number of salmon entering the Columbia has continued to decline and is now at a historically low level. While all the human efforts and funds have been concentrated on the upper river, below Bonneville Dam the salmon have been quietly following their own genetic guidelines. Immediately downstream from the dam is an island with extensive gravel bars around its perimeter. In 1993, a few stray hatchery fall chinook were observed spawning in those gravel bars. The fish were probably from the upper-river stock, propagated just upstream at the Bonneville hatchery; the gravel bars are similar to the mainstem bars where the salmon once spawned before they were inundated by reservoirs. By 1997 the spawning population had reached 2,000 to 5,000 fish, a sizable number that shows us what could happen if one or more of the dams that has flooded the historic spawning areas in the upper basin were breached.

As I see it, the lesson is clear: if the habitat is available and healthy, the salmon know how to recover. Should we breach one or more dams to save the salmon? That's not a question for a biologist to answer. The citizens of the Pacific Northwest will have to make that difficult choice after weighing the value they place on the salmon against the value they place on the dams. But if we are to recover the salmon, we need healthy rivers. We must give back to the salmon some of their habitat.

I could have interfered in the trout's struggle and helped it over the logjam, but I didn't. I left the stream that afternoon having learned an important lesson. Anything I do to help the salmon recovery has to work in concert with the salmon's own capacities and capabilities, and those capabilities are tremendous, if only we would pay attention to them. We cannot restore the salmon if the obstacles we have put in their path are beyond the capability of their own genetic resources. All the money in the world will not produce sustainable recovery as long as those obstacles remain in place.

We simply cannot have salmon without healthy rivers. But it's not just the salmon that need healthy rivers. We do too. We live in the same ecosystems as the salmon, so we cannot stand apart, manipulate, control, and simplify those ecosystems without at some fundamental level diminishing ourselves. As biologist Fred Allendorf concluded, "We need to recognize and feel at a deep level that ultimately we are not conservation biologists trying to save other species. Rather, we are one emergence of life on this planet trying to save itself."[12]

Appendix

Classification of anadromous forms of salmon

In the Linnean system of classification for plants and animals, the anadromous Pacific salmon fall into the genus *Oncorhynchus.* Within the genus are seven species exhibiting various degrees of anadromy, or movement from freshwater spawning grounds to saltwater rearing areas and back to fresh water for spawning. The seven species of anadromous Pacific salmon are pink, coho, chinook, chum, sockeye, steelhead, and cutthroat. Two additional species of Pacific salmon are found in the eastern Pacific, the masu and amago salmon; those two species are not discussed in this book. (From Moyle and Cech 1982.)

Infraclass	Teleostei
Division	Euteleostei
Superorder	Protacanthopterygii
Order	Salmoniformes
Suborder	Salmonoidei
Family	Salmonidae
Genus	*Oncorhynchus*
Species	*O. gorbuscha* (pink salmon)
	O. kisutch (coho salmon)
	O. tshawytscha (chinook salmon)
	O. keta (chum salmon)
	O. nerka (sockeye salmon)
	O. mykiss (steelhead)
	O. clarkii (cutthroat)

APPENDIX

B Comparison of the life histories of seven species of Pacific salmon and trout

Life History Details	*Chinook Salmon* (O. tshawytscha)	*Coho Salmon* (O. kisutch)	*Pink Salmon* (O. gorbuscha)
Common Name	Tyee, spring, king, Quinnat, blackmouth.	Silver salmon, coho.	Humpback, humpy, pink.
External Identification	Dark spotting on back and both lobes of the tail. Black flesh along base of teeth in lower jaw.	Small black spots on back and upper lobe of tail. Gums at base of teeth light—not black as in chinook salmon.	Large black spots on back, large oval spots on tail. After entering rivers to spawn, males develop a large hump behind the head.
North American Distribution and Abundance	Ventura River in southern California to Point Hope, Alaska. In that range there are probably more than 1,000 spawning populations. Chinook salmon are the least abundant salmon in North America.	Monterey Bay, California, to Point Hope, Alaska.	Sacramento River, California, north to Bering Strait and east to the Mackenzie River, Canada.
Size	33–36 inches up to 58 inches, 10–50 pounds up to 126 pounds.	18–24 inches up to 38 inches, 8–12 pounds up to 31 pounds.	17–24 inches up to 30 inches, 3–5 pounds up to 14 pounds.

(*continues*)

Life History Details	*Chinook Salmon* (O. tshawytscha)	*Coho Salmon* (O. kisutch)	*Pink Salmon* (O. gorbuscha)
Age at Maturity	Usually matures at 4–5 years but 2–9 years at maturity is observed range.	2–4 years, usually 3 years.	2 years. All adults mature at the same age, so there are no overlapping generations.
Spawning Migration	Mature adults enter rivers from spring through fall. Adults migrate and spawn in lower river reaches above tidewater to as far as 1,200 miles from the sea in the Yukon. Chinook migrate in four distinct runs identified by their timing: spring, fall, summer, and winter. The winter run is rare.	Mature adults enter the rivers in July and August in northern Alaska, September and October in British Columbia, and November and December in California. Coho may migrate as early as April in specific streams.	Mature adults enter rivers between June and September..
Spawning Habitat	Chinook salmon spawn in the middle and upper mainstems of rivers and in the larger tributaries.	Coho spawn throughout smaller coastal tributaries, usually penetrating to the upper reaches to spawn. Spawning takes place from October to March, usually November to January.	Pink salmon may spawn just above or in tidal areas up to 300 miles upstream in larger rivers. Length of upstream migration is usually about 40 miles.
Juvenile Rearing and Migration	Juveniles exhibit two general life history types: Ocean type fish migrate to sea in their first year, usually within six months of hatching. Ocean type juveniles may rear in the estuary for extended periods. Stream type fish migrate to sea in the spring of their second year, although in northern rivers juveniles may remain in fresh water for 2 years or more.	Juveniles establish territories and rear in smaller streams until the spring of their second year, when they rapidly migrate to sea.	Juvenile pink salmon migrate to sea within a few days after emerging from the gravel.

Life History Details	*Chum Salmon* (O. keta)	*Sockeye Salmon* (O. nerka)	*Steelhead Trout* (O. mykiss)	*Coastal Cutthroat Trout* (O. clarkii)
Common Name	Dog salmon.	Anadromous forms: red salmon, blueback salmon. Nonanadromous forms: kokanee, silver trout, little redfish.	Rainbow trout, Kamloops trout.	Cutthroat trout, red-throated trout.
External Identification	No distinct black spots on back and fins. In the rivers, adults develop dirty green vertical bars or blotches.	No black spots on back or tail. At spawning, males are bright red on back and darker red on sides with a light hump behind the head.	Black spots on back dorsal fin and caudal fin, 12 or fewer rays in the anal fin.	Numerous black spots on back and sides, usually extending below the lateral line. Two yellow to red lines in the skin folds of the lower jaw.
North American Distribution and Abundance	Sacramento River, California, to Bering Strait and east to MacKenzie River, Canada.	Klamath River, California, to Point Hope, Alaska.	Northwestern Mexico to the Kuskokwim River, Alaska.	Eel River, California, to Seward, southeastern Alaska.
Size	23–32 inches up to 40 inches, 3–18 pounds up to 45 pounds.	24–28 inches up to 33 inches, 5–7 pounds up to 16 pounds.	20–30 inches up to 48 inches, 8–9 pounds up to 36 pounds.	12–15 inches up to 30 inches, 1 1/2–4 pounds up to 17 pounds.
Age at Maturity	2–7 years, usually 3–4 years.	3–7 years, usually 4–5 years.	3–5 years.	2–4 years.
Spawning Migration	Mature adults enter rivers as early as July, with arrival on the spawning grounds occurring from September to January. Chum salmon may spawn just above tidewater and rarely penetrate more than 100 miles, although in larger rivers they may migrate over great distances. In the Yukon River, chum salmon migrate 1,200 miles.	Mature adults begin entering rivers from May to October.	Steelhead may enter rivers any month of the year, and they spawn in late winter or spring. Migration in the Columbia River extends up to 900 miles from the ocean in the Snake River.	Mature cutthroat trout leave the sea and enter rivers in late autumn or early winter. Spawning takes place from February to May.

Primary sources: Scott and Crassman 1973; Clemens and Wilby 1961.

APPENDIX

Geologic epochs mentioned in the text

Name of Epoch	*Time, Million Years Ago (m.y.a.)*	*Important Events in the Northwest*
	present	Decline and extinction of *Oncorhynchus*
Holocene	began 0.01 m.y.a.	Lake Missoula released floods Cordilleran ice sheet and Wisconsin ice age
Pleistocene	began 2 m.y.a.	Current river systems were established Salmon ancestral lines diverged
Pliocene	began 5 m.y.a.	Catchment basins were forming
Miocene	began 24 m.y.a.	Lava flowed across Northwest Genus *Oncorhynchus* emerged Saber-toothed salmon lived Oceans began cooling Northwest coast began to take shape
Oligocene	began 36 m.y.a.	Anadromy probably developed
Eocene	began 58 m.y.a.	*Eosalmo driftwoodenis* Teleosts achieved dominance in western North America Ocean temperatures warmer and less productive Volcanic islands broke offshore—future Klamath Mountains
Paleocene	began 63 m.y.a.	

Sources: Alt and Hyndman 1995; Kruckeberg 1991.

Endnotes

Introduction

1. D. R. Croes and S. Hackenberger, Hoko River Archaeological Complex: Modeling Prehistoric Northwest Coast Economic Evolution, in *Research in Economic Anthropology,* edited by B. L. Isaac (Greenwich, CT: JAI Press, 1988), pp. 19–85.
2. H. Fish, *Tracks, Trails, and Tales in Clallam County, State of Washington* (Port Angeles, WA: Olympic Printers, 1983).
3. For an excellent description of the historical roots of this point, see Tim Smith's *Scaling Fisheries* (New York: Cambridge University Press, 1994). Even notable salmon biologists such as W. F. Thompson advocated a narrow approach to research and management. Thompson equated the broader study of the fish in its environment to "tunneling a mountain by removing it in its entirety." W. F. Thompson, The Marine Fisheries, the State and the Biologist, *The Scientific Monthly* 15: 542–550 (1922).
4. B. Dietrich, A Clearcut Case, *Seattle Times,* May 10, 1990.
5. Dietrich, A Clearcut Case.
6. The Merrill-Ring Corporation also owns land in the Pysht watershed. During the time that my work took me to that river, the company genuinely tried to incorporate fish and wildlife concerns into the management of its forestlands.
7. D. Worster, *Rivers of Empire* (New York: Pantheon Books, 1985).
8. T. Egan, *The Good Rain: Across Time and Terrain in the Pacific Northwest* (New York: Alfred A. Knopf, 1990).

Chapter 1

1. E. Gunther, A Further Analysis of the First Salmon Ceremony, *Anthropology* 2(5): 129–173 (1928).

2. E. Mayr, *Toward a New Philosophy of Biology: Observations of an Evolutionist* (Cambridge: Harvard University Press, 1988).
3. P. B. Moyle and J. J. Cech, Jr., *Fishes: An Introduction to Ichthyology* (Englewood Cliffs, NJ: Prentice-Hall, 1982); also see E. H. Colbert, *Evolution of the Vertebrates: A History of the Backboned Animals through Time* (New York: Science Edition, John Wiley and Sons, 1966).
4. W. L. Minckley, D. A. Hendrickson, and C. E. Bond, Geography of Western North American Freshwater Fishes: Description and Relationships to Intracontinental Tectonism, in *The Zoogeography of North American Freshwater Fishes,* edited by C. H. Hocutt and E. O. Wiley (New York: John Wiley and Sons, 1986), pp. 519–613.
5. R. J. Behnke, Native Trout of Western North America, American Fisheries Society Monograph 6, Bethesda, MD, 1992.
6. T. M. Cavender, Review of the Fossil History of North American Freshwater Fishes, in *The Zoogeography of North American Freshwater Fishes,* edited by C. H. Hocutt and E. O. Wiley (New York: John Wiley and Sons, 1986), pp. 699–725; also see Behnke, Native Trout of Western North America.
7. R. F. Stearley, Historical Ecology of Salmonidae, with Special Reference to *Oncorhynchus,* in *Systematics, Historical Ecology, and North American Freshwater Fishes,* edited by R. L. Mayden (Stanford, CA: Stanford University Press, 1992), pp. 622–658; also see J. E. Thorpe, Migration in Salmonids, with Special Reference to Juvenile Movements in Freshwater, in *Proceedings of the Salmon and Trout Migratory Behavior Symposium,* edited by E. L. Brannon and E. O. Salo (Seattle: University of Washington, 1982), pp. 86–97.

   Stearley studied the evolution of migratory behavior in salmonids and made a convincing case for a freshwater origin in the Pacific salmon. He concluded that the Atlantic salmon and the Pacific salmon developed anadromy separately rather than inheriting it from a common ancestor. However, as recently as 1981 Thorpe argued for a saltwater origin of the Pacific salmon.
8. Stearley, Historical Ecology of Salmonidae.
9. R. M. Love, *The Chemical Biology of Fishes: With a Key to the Chemical Literature* (London: Academic Press, 1970).
10. Love, The Chemical Biology of Fishes.
11. Thorpe, Migration in Salmonids.
12. M. R. Gross, R. M. Coleman, and R. M. McDowall, Aquatic Productivity and the Evolution of Diadromous Fish Migration, *Science* 239: 1291–1293 (1988); also see R. M. McDowall, The Occurrence and Distribution of Diadromy among Fishes, in *Common Strategies of Anadromous and Catadromous Fishes,* edited by M. J. Dadswell and five others, American Fisheries Society Symposium 1, Bethesda, MD, 1987, pp. 1–13.
13. Stearley, Historical Ecology of Salmonidae.
14. R. E. Bilby, B. R. Fransen, and P. A. Bisson, Incorporation of Nitrogen and Carbon from Spawning Coho Salmon into the Trophic System of Small Streams: Evidence from Stable Isotopes, *Canadian Journal of Fisheries and*

Aquatic Sciences 53(1): 164–173 (1996); also see G. A. Larkin and P. A. Slaney, Implications of Trends in Marine-Derived Nutrient Influx to South Coastal British Columbia Salmonid Production, *Fisheries* 22(11): 16–24 (1997).

15. Stearley, Historical Ecology of Salmonidae.
16. C. A. Frissell, Evolution of the Salmonid Fishes: Zoogeography and the Fossil Record, Department of Fisheries and Wildlife, Oregon State University, Corvallis, 1989.
17. Cavender, Review of the Fossil History.
18. Stearley places the saber-toothed salmon in the genus *Oncorhynchus* (Stearley, Historical Ecology of Salmonidae).
19. Cavender, Review of the Fossil History; also see T. M. Cavender and R. R. Miller, *Smilodonichthys rastrosus*: A New Pliocene Salmonid Fish from the Western United States, Bulletin No. 18, Museum of Natural History, University of Oregon, Eugene, 1972.
20. Minckley et al., Geography of Western North American Freshwater Fishes; also see E. L. Orr, W. N. Orr, and E. M. Baldwin, *Geology of Oregon,* Fourth Edition (Dubuque, IA: Kendall-Hunt Publishing Company, 1992); and B. McKee, *Cascadia: The Geologic Evolution of the Pacific Northwest* (New York: McGraw-Hill Book Company, 1972).
21. McKee, *Cascadia;* also see Orr et al., *Geology of Oregon,* and Minckley et al., Geography of Western North American Freshwater Fishes.
22. Orr et al., *Geology of Oregon;* also see McKee, *Cascadia.*
23. W. K. Thomas, R. E. Withler, and A. T. Beckenbach, Mitochondrial DNA Analysis of Pacific Salmonid Evolution, *Canadian Journal of Zoology* 64: 1058–1064 (1986).
24. McKee, *Cascadia;* also see E. C. Pielou, *After the Ice Age: The Return of Life to Glaciated North America* (Chicago: University of Chicago Press, 1991). Glaciation recurs every 100,000 years on what is known as the Milankovitch cycle—glaciers advance for 60,000 to 90,000 years separated by interglacial periods lasting 10,000 to 40,000 years. The current interglacial peaked about 10,000 years ago, so we have passed the halfway point toward the next glacial period; also see J. D. McPhail and C. C. Lindsey, Freshwater Fishes of Northwestern Canada and Alaska, Bulletin No. 173, Fisheries Research Board of Canada, Ottawa, 1970; and M. C. Healey, Variation in the Life History Characteristics of Chinook Salmon and Its Relevance to Conservation of the Sacramento Winter Run of Chinook Salmon, *Conservation Biology* 8(3): 876–877 (1994).
25. J. E. Allen, M. Burns, and S. C. Sargent, *Cataclysms on the Columbia: A Layman's Guide to the Features Produced by the Catastrophic Bretz Floods in the Pacific Northwest* (Portland, OR: Timber Press, 1991).
26. Allen, Burns, and Sargent, *Cataclysms on the Columbia.*
27. Orr et al., *Geology of Oregon.*
28. C. L. Camp, *Earth Song: A Prologue to History* (Palo Alto, CA: American West Publishing Company, 1970).

29. I have inferred the low levels of productivity during the hot, dry climate cycle from the record of salmon usage by Native Americans. See Chapter 3.
30. C. C. Lindsey and J. D. McPhail, Zoogeography of Fishes of the Yukon and Mackenzie Basins, in *The Zoogeography of North American Freshwater Fishes,* edited by C. H. Hocutt and E. O. Wiley (New York: John Wiley and Sons, 1986), pp. 639–674; also see J. D. McPhail and C. C. Lindsey, Zoogeography of Freshwater Fishes of Cascadia (the Columbia System and Rivers North to Stikine), in *The Zoogeography of North American Freshwater Fishes,* edited by C. H. Hocutt and E. O. Wiley (New York: John Wiley and Sons, 1986), pp. 615–637. The organic machine comes from historian Richard White's book of that title on the Columbia. R. White, *The Organic Machine* (New York: Hill and Wang, 1995).
31. L. Benda and three others, Morphology and Evolution of Salmonid Habitats in a Recently Deglaciated River Basin, Washington State, USA, *Canadian Journal of Fisheries and Aquatic Sciences* 49: 1246–1256 (1992).
32. Benda et al., Morphology and Evolution of Salmonid Habitats.
33. W. Nehlsen, J. Lichatowich, and J. Williams, Pacific Salmon and the Search for Sustainability, *Renewable Resources Journal* 10(2): 20–26 (1992); R. Bakkala, Synopsis of Biological Data on the Chum Salmon, *Oncorhynchus keta* (Walbaum) 1792, FAO Fisheries Symposium No. 41, U.S. Fish and Wildlife Service Circular 315, Washington, DC, 1970; H. Gunnarsdottir, Scale Patterns Indicate Changes in Use of Rearing Habitat by Juvenile Coho Salmon, *Oncorhynchus kisutch,* from 1955 to 1984 in the Ten Mile Lakes, Oregon, M.S. thesis, Oregon State University, Corvallis, 1992; and R. Burgner, Life History of the Sockeye Salmon (*Oncorhynchus nerka*), in *Pacific Salmon Life Histories,* edited by C. Groot and R. Margolis (Vancouver: University of British Columbia Press, 1991), pp. 3–117.
34. M. C. Healey and A. Prince, Scales of Variation in Life History Tactics of Pacific Salmon and the Conservation of Phenotype and Genotype, in *Evolution and the Aquatic Ecosystem: Defining Unique Units in Population Conservation,* edited by J. L. Nielsen and D. A. Powers, American Fisheries Society Symposium 17, Bethesda, MD, 1995, pp. 176–184.
35. W. F. Thompson, An Approach to Population Dynamics of the Pacific Red Salmon, *Transactions of the American Fisheries Society* 88(3): 206–209 (1959).

Chapter 2

1. This description is fiction, constructed from my reading of the archaeological literature and the postglacial environmental history of the Columbia River to illustrate several ideas and events discussed in Chapters 1 and 2.
2. B. M. Fagan, *The Great Journey: The Peopling of Ancient America* (New York: Thames and Hudson, 1987), p. 144; and K. R. Fladmark, Times and Places: Environmental Correlates of Mid-to-late Wisconsinan Human Population Expansion in North America, in *Early Man in the New World,*

edited by R. Shutler, Jr. (Beverly Hills: Sage Publications, 1983), pp. 13–41.

3. E. C. Pielou, *After the Ice Age: The Return of Life to Glaciated North America* (Chicago: University of Chicago Press, 1991).
4. P. S. Martin and H. E. Wright, Jr. (eds.), *Pleistocene Extinctions: The Search for a Cause* (New Haven: Yale University Press, 1967), Table 1, p. 82; and T. F. Flannery, *The Future Eaters: An Ecological History of the Australasian Lands and Peoples* (New South Wales, Australia: Reed Books, 1994), p. 236.
5. A. J. Jelinek, Man's Role in the Extinction of Pleistocene Faunas, in *Pleistocene Extinctions: The Search for a Cause,* edited by P. S. Martin and H. E. Wright, Jr. (New Haven: Yale University Press, 1967), pp. 193–200; also see the other papers in Martin and Wright, *Pleistocene Extinctions;* also see Flannery, *The Future Eaters.* Archaeologists disagree over the cause of the large mammal extinctions, with most arguments falling under two hypotheses. One postulates that the cause was the postglacial climate change; the other points the finger at human overkill. The climate hypothesis is weakened by the timing of megafauna extinctions throughout the world. Although the postglacial climate change occurred at approximately the same time throughout the world, the extinctions occurred at different times: about 35,000 years ago in Australia, as recent as 800 years ago in New Zealand, and about 12,000 years ago in North America and northern Europe. One consistent pattern in the megafauna extinctions, however, is their occurrence shortly after the arrival of modern man. In general, the overkill hypothesis has more evidence to support it, but the truth is probably not a simple either/or answer. The large mammals in North America may have become more vulnerable to overharvest because of the stress associated with the rapid shift to a hotter and drier climate around 11,000 to 12,000 years ago.
6. Quoted in Flannery, *The Future Eaters,* p. 177.
7. C. Darwin, *The Voyage of the Beagle* (New York: P. F. Collier and Son, 1969), p. 405.
8. R. L. Carlson, The Far West, in *Early Man in the New World,* edited by R. Shutler, Jr. (Beverly Hills: Sage Publications, 1983), p. 93.
9. R. G. Matson and G. Coupland, *The Prehistory of the Northwest Coast* (New York: Academic Press, 1995), p. 56.
10. Matson and Coupland, *The Prehistory of the Northwest Coast,* pp. 76, 79, 80, and 81.
11. R. F. Schalk, Land Use and Organizational Complexity Among Foragers of Northeastern North America, in *Affluent Foragers,* edited by S. Koyama and D. H. Thomas, Senri, Osaka, Japan, 1981, pp. 53–75; and R. F. Schalk, The Structure of an Anadromous Fish Resource, in *For Theory Building in Archaeology: Essays on Faunal Remains, Aquatic Resources and Spatial Analysis and Systemic Modeling,* edited by L. R. Binford (New York: Academic Press, 1977), pp. 107–249.

12. K. R. Fladmark, *British Columbia Prehistory* (Ottawa: The Runge Press, 1986).
13. Fladmark, *British Columbia Prehistory.*
14. P. Thomison, When Celilo Was Celilo: An Analysis of Salmon Use During the Past 11,000 Years in the Columbia Plateau, M.A. thesis, Oregon State University, Corvallis, 1987.
15. P. A. Berringer, Northwest Coast Traditional Salmon Fisheries Systems of Resource Utilization, M.A. thesis, University of British Columbia, Vancouver, 1982.
16. Fladmark, *British Columbia Prehistory.*
17. Matson and Coupland, *The Prehistory of the Northwest Coast.*
18. Matson and Coupland, *The Prehistory of the Northwest Coast.*
19. Matson and Coupland, *The Prehistory of the Northwest Coast.*
20. D. R. Croes and S. Hackenberger, Hoko River Archaeological Complex: Modeling Prehistoric Northwest Coast Economic Evolution, in *Research in Economic Anthropology,* Supplement 3, edited by B. L. Isaac (Greenwich, CT: JAI Press, 1988), pp. 19–85.
21. Two excellent studies of the gift in ancient economies are M. Mauss, *The Gift: The Form and Reason for Exchange in Archaic Societies* (New York: W. W. Norton, 1990); and L. Hyde, *The Gift: Imagination and the Erotic Life of Property* (New York: Vintage Books, 1983).
22. W. Suttles, *Coast Salish Essays* (Seattle: University of Washington Press, 1987), p. 8.
23. Suttles, *Coast Salish Essays;* and D. Cole and I. Caikin, *An Iron Hand upon the People: The Law Against the Potlatch on the Northwest Coast* (Seattle: University of Washington Press, 1990).
24. Suttles, *Coast Salish Essays;* Cole and Caikin, In *Iron Hand upon the People;* and S. Piddocke, The Potlatch System of the Southern Kwakiutl: A New Perspective, in *Economic Anthropology: Readings in Theory and Analysis,* edited by E. E. LeClair, Jr., and H. K. Schneider (New York: Holt, Rinehart and Winston, 1968).
25. C. Martin, *Keepers of the Game: Indian-Animal Relationships and the Fur Trade* (Berkeley: University of California Press, 1978), p. 39; and Hyde, *The Gift,* pp. 26–27.
26. P. Drucker, *Indians of the Northwest Coast* (Garden City, NY: Natural History Press, 1963), p. 155; and Hyde, *The Gift,* pp. 26–27.
27. An example of a speech given at a First Salmon Ceremony comes from Erna Gunther's paper, in which she reproduced this Kwakiutl prayer: "Swimmer, I thank you because I am still alive, for the reason why you come is that we may play together with my fishing tackle, Swimmer. Now, go home and tell your friends that you had good luck on account of your coming here, and that they shall come with their wealth bringer, that I may get some of your wealth, Swimmer; also take away my sickness, friend, supernatural one, Swimmer." E. Gunther, An Analysis of the First Salmon Ceremony, *American Anthropologist* 28: 605–617 (1926).

28. Erna Gunther's two papers on the first salmon ceremony give complete descriptions of the ceremony. The First Salmon Ceremony was not confined to this side of the Pacific. On the other side of the ocean, the Ainu people of northern Japan also fished for salmon, principally chum salmon. Many Ainu myths and ceremonies were similar to those practiced in the Pacific Northwest, including their first catch ceremony. The Ainu believed that the salmon took on human form and lived in houses under the sea. The salmon king led his people up certain rivers to visit the Ainu, bringing meals of salmon meat. To keep the salmon people coming back, the Ainu had to treat them with respect. Gunther, An Analysis of the First Salmon Ceremony; E. Gunther, A Further Analysis of the First Salmon Ceremony, *Anthropology* 2(5): 129–173 (1928); and H. Watanabe, *The Ainu Ecosystem: Environment and Group Structure* (Seattle: University of Washington Press, 1972).
29. C. J. Cederholm, D. B. Houston, D. L. Cole, and W. J. Scarlett, Fate of Coho Salmon (*Oncorhynchus kisutch*) Carcasses in Spawning Streams, *Canadian Journal of Fisheries and Aquatic Sciences* 46: 1347–1355 (1989); T. C. Kline, Jr., and four others, Recycling of Elements Transported Upstream by Runs of Pacific Salmon: I. Evidence in Sashin Creek, Southeastern Alaska, *Canadian Journal of Fisheries and Aquatic Sciences* 47: 136–144 (1990); R. E. Bilby, B. R. Fransen, and P. A. Bisson, Incorporation of Nitrogen and Carbon from Spawning Coho Salmon into the Trophic System of Small Streams: Evidence from Stable Isotopes, *Canadian Journal of Fisheries and Aquatic Sciences* 53(1): 164–173 (1996); M. F. Willson and K. C. Halupka, Anadromous Fish as Keystone Species in Vertebrate Communities, *Conservation Biology* 9(3): 489–497 (1995); and J. E. Richey, M. A. Perkins, and C. R. Goldman, Effects of Kokanee Salmon (*Oncorhynchus nerka*) Decomposition on the Ecology of a Subalpine Stream, *Journal of the Fisheries Research Board of Canada* 32: 817–820 (1975).
30. H. Stewart, *Indian Fishing: Early Methods on the Northwest Coast* (Seattle: University of Washington Press, 1977).
31. J. C. Swan, *The Northwest Coast or, Three Years' Residence in Washington Territory* (Seattle: University of Washington Press, 1857). Cannery crews seined for salmon in the lower Columbia River in the late nineteenth century. The seines may have been larger and horses were used to haul in the nets, but it was the same basic technique. In 1974, I used the same technique to seine spring chinook salmon in the lower Rogue River as part of a research program. A frame for holding the net was attached to the gunwale of a fiberglass boat instead of a canoe, and we didn't have to paddle because we had the benefit of an outboard motor.
32. T. T. Waterman and A. L. Kroeber, The Kepel Fish Dam, *University of California Publications in American Archaeology and Ethnology (1934–1943)* 35: 49–79 [New York: Kraus Reprint Corporation (reprint), 1965].
33. R. K. Nelson, A Conservation Ethic and Environment: The Koyukon of Alaska. In *Resource Managers: North American and Australian Hunter-Gather-*

ers, edited by N. M. Williams and E. S. Hunn, American Association for the Advancement of Science Selected Symposium 67, Boulder, CO, 1982, pp. 211–228.

34. A. F. McEvoy, *The Fisherman's Problem: Ecology and Law in the California Fisheries,* 1850–1980 (New York: Cambridge University Press, 1986).
35. C. A. Simenstad, J. A. Estes, and K. W. Kenyon, Aleuts, Sea Otters, and Alternate Stable-State Communities, *Science* 200(28): 403–411 (1978).
36. McEvoy, *The Fisherman's Problem.*
37. Stories of starvation, particularly among the more northerly tribes, likely reflect natural fluctuation in the size of the salmon runs. However, we cannot discount the possibility that such stories are echoes from the prehistoric trial and error that contributed to the coevolution of a sustainable economy. For example, see Piddocke, The Potlatch System of the Southern Kwakiutl.
38. Gunther, A Further Analysis of the First Salmon Ceremony.
39. Gunther, A Further Analysis of the First Salmon Ceremony.
40. Martin, *Keepers of the Game.*
41. M. G. Beck, *Shamans and Kushtakas: North Coast Tales of the Supernatural* (Anchorage: Alaska Northwest Books, 1991).

Chapter 3

1. For a discussion of viable and non-viable salmon stocks, see Salmon and Steelhead Advisory Commission, A New Management Structure for Anadromous Salmon and Steelhead Resources and Fisheries of the Washington and Columbia River Conservation Areas, National Marine Fisheries Service, Seattle, 1984. For a discussion of intentional management for extinction of "nonviable" wild salmon, see S. Wright, Fishery Management of Wild Salmon Stocks to Prevent Extinction, *Fisheries* 18: 3–4 (1993).
2. W. G. Robbins, *Landscapes of Promise: The Oregon Story, 1800–1940* (Seattle: University of Washington Press, 1997).
3. J. R. McGoodwin, *Crisis in the World's Fisheries: People, Problems, and Policies* (Stanford, CA: Stanford University Press, 1990).
4. A. F. McEvoy, *The Fisherman's Problem: Ecology and Law in the California Fisheries, 1850–1980* (New York: Cambridge University Press, 1986); and G. Hardin, The Tragedy of the Commons, *Science* 192: 1243–1248 (1968).
5. S. P. Hays, *The Response to Industrialism* (Chicago: University of Chicago Press, 1957); also see W. G. Robbins, *Colony and Empire: The Capitalist Transformation of the American West* (Lawrence: University Press of Kansas, 1994).
6. Oregon Business Council, A New Vision for Pacific Salmon, Portland, 1996.
7. McEvoy, *The Fisherman's Problem.*
8. J. A. Lichatowich, Evaluating the Performance of Salmon Management Institutions: The Importance of Performance Measures, Temporal Scales and Production Cycles, in *Pacific Salmon and Their Ecosystems: Status and Future*

Options, edited by D. J. Stouder, P. A. Bisson, and R. J. Naiman (New York: Chapman and Hall, 1996), pp. 69–87.

9. McEvoy, *The Fisherman's Problem.*

Chapter 4

1. G. E. Moulton, *The Journals of the Lewis and Clark Expedition, July 28–November 1, 1805,* volume 5 (Lincoln: University of Nebraska Press, 1989), p. 74.
2. Moulton, *The Journals of the Lewis and Clark Expedition,* pp. 71–83.
3. Moulton, *The Journals of the Lewis and Clark Expedition,* pp. 71–83.
4. J. K. Mumford, *John Ledyard's Journal of Captain Cook's Last Voyage* (Corvallis: Oregon State University Press, 1964), p. 70.
5. R. Fisher, *Contact and Conflict: Indian-European Relations in British Columbia, 1774–1890* (Vancouver: University of British Columbia Press, 1992).
6. J. Nisbet, *Sources of the River: Tracking David Thompson across Western North America* (Seattle: Sasquatch Books, 1994), p. 76.
7. C. F. Wilkinson, *Crossing the Next Meridian: Land, Water, and the Future of the West* (Washington, DC: Island Press, 1992), p. 17.
8. J. R. Gibson, *Farming the Frontier: The Agricultural Opening of the Oregon Country, 1786 to 1846* (Seattle: University of Washington Press, 1985).
9. Cited in R. Bunting, *The Pacific Raincoast: Environment and Culture in an American Eden, 1778–1900* (Lawrence: University Press of Kansas, 1997), p. 32.
10. D. R. Johnson and D. H. Chance, Presettlement Overharvest of Upper Columbia River Beaver Populations, *Canadian Journal of Zoology* 52: 1519–1521 (1974).
11. C. H. Carey, *A General History of Oregon* (Portland, OR: Binford and Mort, 1971).
12. R. J. Naiman, J. M. Melillo, and J. E. Hobbie, Ecosystem Alteration of Boreal Forest Streams by Beaver (*Castor canadensis*), *Ecology* 67(5): 1254–1269 (1986); also see Gibson, *Farming the Frontier.*
13. National Research Council, *Upstream: Salmon and Society in the Pacific Northwest* (Washington, DC: National Academy Press, 1996).
14. Naiman et al., Ecosystem Alteration of Boreal Forest Streams by Beaver; and R. J. Naiman, C. A. Johnston, and J. C. Kelley, Alteration of North American Streams by Beaver: The Structure and Dynamics of Streams Are Changing as Beaver Recolonize Their Historic Habitat, *Bioscience* 38(11): 753–762 (1988).
15. National Research Council, *Upstream.*
16. C. Huntington, W. Nehlsen, and J. Bowers, Healthy Native Stocks of Native Salmonids in the Pacific Northwest and California, Oregon Trout, Portland, 1994.
17. G. Ontko, *Thunder over the Ochoco: The Gathering Storm,* volume 1 (Bend, OR: Maverick Publications, 1993).

18. Interview with Errol Claire, retired employee of the Oregon Department of Fish and Wildlife, April 3, 1997.
19. T. E. Nickelson and three others, Seasonal Changes in Habitat Use by Juvenile Coho Salmon (*Oncorhynchus kisutch*) in Oregon Coastal Streams, *Canadian Journal of Fisheries and Aquatic Sciences* 49: 783–789 (1992).
20. A. F. McEvoy, *The Fisherman's Problem: Ecology and Law in the California Fisheries, 1850–1980* (New York: Cambridge University Press, 1986).
21. McEvoy, *The Fisherman's Problem,* p. 84.
22. California Division of Fish and Game, Thirtieth Biennial Report, 1926–1928, Sacramento, 1928.
23. J. W. Moffet and S. H. Smith, Biological Investigations of the Fishery Resources of California, Special Scientific Report 12, U.S. Fish and Wildlife Service, Washington, DC, 1950.
24. H. Oliver, *Gold and Cattle Country* (Portland, OR: Binford and Mort, 1967); and D. Johansen and C. Gates, *Empire of the Columbia* (New York: Harper and Brothers, 1957).
25. C. Rivers, *History of the Rogue River Fisheries.* Undated draft manuscript covering the history of the Pacific salmon fishery in the Rogue River, Oregon. Oregon Department of Fisheries and Wildlife, Portland.
26. Oregon State Board of Fish Commissioners, Second Report to the Governor, Salem, 1888, p. 15; also see Oregon Department of Fisheries, Annual Report to the Governor, Salem, 1903, p. 16.
27. Oregon State Board of Fish Commissioners, Fourth Annual Report, Salem, 1890, p.19.
28. H. Oliver, *Gold and Cattle Country,* p. 30.
29. J. Leethem, The Western Gold Dredge Company of John Day, Oregon, *Oregon Geology* 41(6): 91–95 (1979).
30. J. C. Swan, *The Northwest Coast or, Three Years' Residence in Washington Territory* (Seattle: University of Washington Press, 1857).
31. J. Crawford, First Annual Report of the Washington State Fish Commissioner, Olympia, 1890.
32. L. Stone, Report of Operations at the Salmon-Hatching Station on the Clackamas River, Oregon, in 1877, Report of the U.S. Fish Commissioner for 1877, Washington, DC, 1879.
33. S. Johnson, Freshwater Environmental Problems and Coho Production in Oregon, Information Report 84-11, Oregon Department of Fish and Wildlife, Corvallis, 1984.
34. R. Bunting, Abundance and the Forests of the Douglas-fir Bioregion, 1840 to 1920, *Environmental History Review* 18(4): 41–62 (1994).
35. T. Cox, *Mills and Markets* (Seattle: University of Washington Press, 1974); also see S. Holbrook, *The Columbia* (New York: Rinehart, 1956).
36. Cox, *Mills and Markets.*
37. A. Beckman, *Swift Flows the River* (Coos Bay, OR: Arago Books, 1970); and Cox, *Mills and Markets.*

38. H. Wendler and G. Deschamps, Logging Dams on Coastal Washington Streams, Washington Department of Fisheries, *Fisheries Research Papers* 1(3): 27–38 (1955).
39. Wendler and Deschamps, Logging Dams on Coastal Washington Streams.
40. N. Langston, *Forest Dreams, Forest Nightmares* (Seattle: University of Washington Press, 1995); Wendler and Deschamps, Logging Dams on Coastal Washington Streams; Beckman, *Swift Flows the River;* and J. R. Sedell and K. J. Luchessa, Using the Historical Record as an Aid to Salmonid Habitat Enhancement, in *Symposium on Acquisition and Utilization of Aquatic Habitat Inventory Information,* October 23–28, 1981, edited by N. B. Armantrout (Portland, OR: American Fisheries Society, Western Division, 1981), pp. 210–223.
41. J. Lichatowich, Habitat Alteration and Changes in Abundance of Coho (*Oncorhynchus kisutch*) and Chinook (*O. tshawytscha*) Salmon in Oregon's Coastal Streams, in *Proceedings of the National Workshop on Effects of Habitat Alteration on Salmonid Stocks,* edited by C. D. Levings, L. B. Holtby, and M. A. Henderson, Canadian Special Publication of Fisheries and Aquatic Sciences 105, Ottawa, 1989, pp. 92–99.
42. Beckman, *Swift Flows the River.*
43. P. Benner, Historical Reconstruction of Coquille River and Surrounding Landscape, in *Near Coastal Waters National Pilot Project: The Coquille River, Oregon,* Action Plan for Oregon Coastal Watersheds, Estuary and Ocean Waters, 1988–1991, Grant X-000382-1, Prepared by the Oregon Department of Environmental Quality for the U.S. Environmental Protection Agency, 1991; Sedell and Luchessa, Using the Historical Record as an Aid to Salmonid Habitat Enhancement; and Swan, *The Northwest Coast.* Swan gives a firsthand account of the amount of debris in undisturbed rivers around 1850.
44. R. Schoettler, Sixty-second Annual Report of the Washington Department of Fisheries, Olympia, 1953.
45. Sedell and Luchessa, Using the Historical Record.
46. Bunting, *The Pacific Raincoast.*
47. M. A. Grainger, *Woodsmen of the West* (Toronto: McClelland and Stewart, 1964).
48. Sedell and Luchessa, Using the Historical Record.
49. B. J. Bradbury, J. Brenneman, and C. Hosticka, Report of the Interim Task Force on Forest Practices, State of Oregon, Salem, 1986.
50. J. Brinckman, U.S. Salmon Plan Would Be Ruinous, Forestry Officials, Companies Say, *Oregonian* (Portland), March 5, 1998, p. D1.
51. Bunting, *The Pacific Raincoast.*
52. D. Ferguson and N. Ferguson, *Sacred Cows at the Public Trough* (Bend, OR: Maverick Publications, 1983).
53. W. Platts, Livestock Grazing, *American Fisheries Society Special Publication* 19: 389–423 (1991).

54. Oliver, *Gold and Cattle Country.*
55. A. McGregor, *Counting Sheep: From Open Range to Agribusiness on the Columbia Plateau* (Seattle: University of Washington Press, 1982).
56. Ferguson and Ferguson, *Sacred Cows at the Public Trough.*
57. J. Oliphant, The Cattle Trade from the Far Northwest to Montana, *Agricultural History* 6: 1 (1932).
58. G. Buckley, Desertification of Camp Creek Drainage in Central Oregon, 1826–1905, M.S. thesis, University of Oregon, Eugene, 1992.
59. Buckley, Desertification of Camp Creek.
60. Buckley, Desertification of Camp Creek.
61. Wilkinson, *Crossing the Next Meridian.*
62. J. Kuhler, A History of Agriculture in the Yakima Valley, Washington, from 1880 to 1900, M.S. thesis, University of Washington, Seattle, 1940.
63. W. Rowley, *U.S. Forest Service Grazing and Rangelands: A History* (College Station: Texas A&M University Press, 1985).
64. H. Rees, *Shaniko: From Wool Capital to Ghost Town* (Portland, OR: Binford and Mort, 1982).
65. Rowley, *U.S. Forest Service Grazing and Rangelands.*
66. J. Smith, Retrospective Analysis of Changes in Stream and Riparian Habitat Characteristics Between 1935 and 1990 in Two Eastern Cascade Streams, M.S. thesis, University of Washington, Seattle, 1993.
67. F. Plummer, Forest Conditions in the Cascade Range, Washington, between the Washington and Mount Rainier Forest Reserves, U.S. Geological Survey, Washington, DC, 1902.
68. S. P. Hays, *Conservation and the Gospel of Efficiency: The Progressive Conservation Movement, 1890–1920* (New York: Atheneum Press, 1969).
69. R. Wissmar and five others, A History of Resource Use and Disturbance in the Riverine Basin of Eastern Oregon and Washington (Early 1880s–1990s), *Northwest Science* 69, Special Issue (1994).
70. S. P. Hays, *Beauty, Health and Permanence: Environmental Politics in the United States, 1955–1985* (New York: Cambridge University Press, 1987).
71. Ferguson and Ferguson, *Sacred Cows at the Public Trough.*
72. Interview with Homer Campbell, April 1995. Forty years later Lester Campbell's son Homer, a biologist with the Oregon Game Commission, would be assigned the job of reducing the loss of juvenile salmon to irrigation ditches in northeast Oregon.
73. R. Van Cleve and R. Ting, The Condition of Salmon Stocks in the John Day, Umatilla, Walla Walla, Grande Ronde, and Imnaha Rivers as Reported by Various Fisheries Agencies, Department of Oceanography, University of Washington, Seattle, 1960; and Western Editorial on Important Fishery Questions, *Pacific Fisherman* 2(6): (1904).
74. D. L. Bottom, To Till the Water: A History of Ideas in Fisheries Conservation, in *Pacific Salmon and Their Ecosystems: Status and Future Options,* edited by D.

J. Stouder, P. A. Bisson, and R. J. Naiman (New York: Chapman and Hall, 1997), pp. 569–597.

75. D. Worster, *Nature's Economy: A History of Ecological Ideas* (New York: Cambridge University Press, 1977).
76. C. Scofield, The Settlement of Irrigated Lands, *U.S. Department of Agriculture Yearbook, 1912,* Government Printing Office, Washington, DC, 1912, pp. 483–494; also see Johansen and Gates, *Empire of the Columbia.*
77. W. Nehlsen, Historical Salmon Runs of the Upper Deschutes and Their Environments, Report to Portland General Electric Company, Portland, OR, 1993.
78. Oregon State Board of Fish Commissioners, Fourth Annual Report, Salem, 1890, p. 21.
79. Washington State Fish Commissioner, Fourteenth and Fifteenth Annual Reports of the Washington Department of Fisheries, Seattle, 1904.
80. J. Wales, California's Fish Screen Program, *California Fish and Game* 34 (2): 45–51 (1948).
81. California Division of Fish and Game, Thirty-first Biennial Report, Sacramento, 1930.
82. W. D. Lyman, *History of the Yakima Valley, Washington* (S. J. Clarke Publishing Company, 1919).
83. G. Fite, *The Farmer's Frontier, 1865–1900* (New York: Holt, Rinehart and Winston, 1966).
84. D. Worster, *Rivers of Empire* (New York: Pantheon Books, 1985).
85. W. D. Rowley, *Reclaiming the Arid West: The Career of Francis G. Newlands* (Bloomington: Indiana University Press, 1996).
86. Hays, *Conservation and the Gospel of Efficiency.*
87. Hays, *Conservation and the Gospel of Efficiency.*
88. Federal Reclamation Bureau Arraigned for Ignoring State Fishway Laws, *Pacific Fisherman* 29(4): (1931).
89. B. E. Stoutemyer, Letter to Henry O'Malley, Commissioner of Fisheries, January 28, 1931, National Archives, Washington, DC, Record Group 22.
90. For example, see Washington State Fish Commissioner, Fourteenth and Fifteenth Annual Reports; also see H. Van Dusen, Annual Report of the Oregon Department of Fisheries, Salem, 1903.
91. Loss of Salmon Fry in Irrigation, *Pacific Fisherman* 18(2): (1920).
92. R. Cogwill, Notes on salmon losses in irrigation ditches, 1930, Cogwill records, Oregon Historical Society, Portland, OR.
93. F. Davidson, The Development of Irrigation in the Yakima River Basin and Its Effect on the Migratory Fish Population in the River, Manuscript report, Ephrata, WA, 1965.
94. M. R. Delarm, E. Wold, and R. Z. Smith, Columbia River Fisheries Development Program Annual Report for FY 1986, NOAA Technical Memorandum, NMFS F/NWR-21, Seattle, 1987.
95. National Research Council, *Upstream.*

96. D. Lavier, Major Dams on Columbia River and Tributaries, Investigative Reports of Columbia River Fisheries Project, Pacific Northwest Regional Commission, Portland, OR, 1976.
97. R. Dunfield, Atlantic Salmon in the History of North America, Canadian Special Publication of Fisheries and Aquatic Sciences 80, Ottawa, 1985.
98. Johnson, Freshwater Environmental Problems and Coho Production.
99. B. Brown, *Mountain in the Clouds: A Search for the Wild Salmon* (New York: Simon and Schuster, 1982).
100. Rivers, History of the Rogue River Fisheries; and G. Dodds, *The Salmon King of Oregon: R. D. Hume and the Pacific Fisheries* (Chapel Hill: University of North Carolina Press, 1959).
101. Rivers, History of the Rogue River Fisheries.
102. Rivers, History of the Rogue River Fisheries.
103. Oregon Fish Commission, Biennial Report of the Fish Commission of the State of Oregon for 1931 and 1932, Salem, 1933.
104. L. A. Fulton, Spawning Areas and Abundance of Chinook Salmon (*Oncorhynchus tshawytscha*) in the Columbia River Basin—Past and Present, Special Scientific Report: Fisheries No. 571, U.S. Fish and Wildlife Service, Bureau of Commercial Fisheries, Washington, DC, 1968.
105. R. R. Miller and R. G. Miller, The Contribution of the Columbia River System to the Fish Fauna of Nevada: Five Species Unrecorded from the State, *Copeia* 3: 174–187 (1948).

Chapter 5

1. J. R. Gibson, *Farming the Frontier: The Agricultural Opening of the Oregon Country, 1786 to 1846* (Seattle: University of Washington Press, 1985), p. 25. Gabriel Franchere also mentions the laxative effect of salmon in his account of his travels in the Northwest from 1811 to 1814. See G. Franchere, *A Voyage to the Northwest Coast of North America* (New York: Citadel Press, 1968), p. 179.
2. J. A. Craig and R. L. Hacker, The History and Development of the Fisheries of the Columbia River, *Bulletin of the Bureau of Fisheries* 49(32): 133–216 (1940).
3. Gibson, *Farming the Frontier,* p. 24.
4. Gibson, *Farming the Frontier,* p. 48.
5. R. T. Tetlow and G. J. Barbey, *Barbey: The Story of a Pioneer Columbia River Salmon Packer* (Portland, OR: Binford and Mort, 1990), p. 3.
6. M. V. Hayden, History of the Salmon Industry in the Pacific Northwest, M.S. thesis, University of Oregon, Eugene, 1930; also see Craig and Hacker, The History and Development of the Fisheries of the Columbia River; see S. E. Morrison, New England and the Opening of the Columbia River Salmon Trade, *Oregon Historical Quarterly* 28(2):111–132 (1927).
7. Craig and Hacker, The History and Development of the Fisheries of the Columbia River.

8. D. B. DeLoach, The Salmon Canning Industry, Oregon State College Economic Studies No. 1, Corvallis, 1939, p. 9.
9. A. Lufkin (ed.), *California's Salmon and Steelhead: The Struggle to Restore an Imperiled Resource* (Berkeley: University of California Press, 1991); also see G. H. Clark, Sacramento–San Joaquin Salmon (*Oncorhynchus tshawytscha*) Fishery of California, Fish Bulletin No. 17, California Division of Fish and Game, Sacramento, 1929.
10. A. F. McEvoy, *The Fisherman's Problem: Ecology and Law in the California Fisheries, 1850–1980* (New York: Cambridge University Press, 1986).
11. Craig and Hacker, The History and Development of the Fisheries of the Columbia River, p. 150.
12. E. C. May, *The Canning Clan: A Pageant of Pioneering Americans* (New York: Macmillan, 1937).
13. R. D. Hume, *Salmon of the Pacific Coast* (San Francisco: Schmidt Label & Lithographic Company, 1893).
14. May, *The Canning Clan.*
15. R. D. Hume, History of the Salmon Industry of the Pacific, *Pacific Fisherman* 1: 9 (1903).
16. Hume, History of the Salmon Industry of the Pacific.
17. Hayden, History of the Salmon Industry in the Pacific Northwest, p. 10.
18. DeLoach, The Salmon Canning Industry, p. 15.
19. G. Meggs, *Salmon: The Decline of the British Columbia Fishery* (Vancouver, BC: Douglas and McIntyre, 1991).
20. D. A. Stacey, Sockeye and Tinplate: Technological Change in the Fraser River Canning Industry, 1871–1912, British Columbia Provincial Museum Heritage Record No. 15, Victoria, 1982, p. 2.
21. W. G. Robbins, *Colony and Empire: The Capitalist Transformation of the American West* (Lawrence: University Press of Kansas, 1994), p. 63.
22. J. N. Cobb, Pacific Salmon Fisheries, Appendix 13 to the Report of the U.S. Commissioner of Fisheries for 1930, Bureau of Fisheries Document No. 1092, Washington, DC, 1930.
23. Stacey, Sockeye and Tinplate; also see Cobb, Pacific Salmon Fisheries.
24. D. S. Jordan and C. H. Gilbert, The Salmon Fishing and Canning Interests of the Pacific Coast, in *The Fisheries and Fishing Industry of the United States,* sec. 5, *History and Methods of the Fisheries,* volume I, part 13, edited by G. B. Goode, Washington, DC, 1887, pp. 729–751; and M. McDonald, The Salmon Fisheries of the Columbia Basin, *Bulletin of the U.S. Fish Commission,* 14: 153–167 (1895).
25. C. L. Smith, *Salmon Fishers of the Columbia* (Corvallis: Oregon State University Press, 1979).
26. Fourth Annual Salmon Day Celebrated, *Pacific Fisherman* 15(3): 1 (1916).
27. Fourth Annual Salmon Day, *Pacific Fisherman.*
28. Fourth Annual Salmon Day, *Pacific Fisherman.*

29. D. Newell, *The Development of the Pacific Salmon-Canning Industry: A Grown Man's Game* (Montreal: McGill–Queen's University Press, 1989).
30. H. E. Gregory and K. Barnes, North Pacific Fisheries with Special Reference to Pacific Salmon, Studies of the Pacific No. 3, American Council Institute of Pacific Relations, New York, 1939, pp. 93–96.
31. G. Dodds, *The Salmon King of Oregon: R. D. Hume and the Pacific Fisheries* (Chapel Hill: University of North Carolina Press, 1959).
32. Cork was a term used by gillnetters and seiners; to "cork" another fisherman meant to set one's net immediately in front of his gear in order to rob him of any catch.
33. A. Netboy, *The Salmon: Their Fight for Survival* (Boston: Houghton Mifflin, 1974), p. 277; also see Tetlow and Barbey, *Barbey: The Story of a Pioneer Columbia River Salmon Packer,* p. 17; and H. W. McKervill, *The Salmon People: The Story of Canada's West Coast Salmon Fishing Industry* (Sidney, BC: Gray's Publishing, 1967), p. 58.
34. M. F. McKeown, *The Trail Led North: Mont Hawthorne's Story* (New York: Macmillan, 1949), p. 13.
35. W. Ricker, Effects of the Fishery and of Obstacles to Migration on the Abundance of Fraser River Sockeye Salmon (*Oncorhynchus nerka*), Canadian Technical Report of Fisheries and Aquatic Sciences No. 1522, Nanaimo, BC, 1987.
36. Ricker, Effects of the Fishery and of Obstacles to Migration.
37. Cited in P. Thompson, T. Wailey, and T. Lummis, *Living the Fishing* (London: Routledge and Kegan Paul, 1983).
38. I. Martin, *Legacy and Testament: The Story of the Columbia River Gillnetters* (Pullman: Washington State University Press, 1994).
39. Smith, Salmon Fishers of the Columbia, p. 28.
40. Robbins, *Colony and Empire,* p.180.
41. Interview with A. L. Flesher, retired fisherman, April 20, 1995, Newport, OR.
42. E. Beard, Biography, Oregon Historical Society, Manuscript 1509, Portland.
43. Hayden, History of the Salmon Industry, p. 45.
44. Martin, *Legacy and Testament.*
45. McEvoy, *The Fisherman's Problem,* p. 96.
46. Jordan and Gilbert, The Salmon Fishing and Canning Interests, p. 736.
47. Smith, *Salmon Fishers of the Columbia,* p. 25.
48. For an account of the discrimination against the Chinese fishermen in California, see McEvoy, *The Fisherman's Problem.* Also, Smith discusses the Chinese fishermen as well as other ethnic fishers in *Salmon Fishers of the Columbia;* Meggs covers the problems of Indian and Japanese fishermen in the salmon fishery in British Columbia in *Salmon: The Decline of the British Columbia Fishery.* Accounts of attacks on the Austrian purse seine fishermen are contained in Washington State Fish Commissioner, Twenty-eighth and Twenty-ninth Annual Reports, Olympia, 1920. Also see Aliens Threaten the Salmon Industry, *Oregon Journal* (Portland), August 8, 1919, p. 10. For a dis-

cussion of the Indian fishery, see Washington State Senate, Minutes of the Washington State Columbia River Salmon Interim Committee, October 20, 1943.

49. B. Leibhardt, Law, Environment, and Social Change in the Columbia River Basin: The Yakima Indian Nation as a Case Study, 1840–1933, Ph.D. dissertation, University of California, Berkeley, 1990.
50. American Friends Service Committee, *Uncommon Controversy* (Seattle: University of Washington Press, 1970), p. 41.
51. McEvoy, *The Fisherman's Problem.*
52. M. A. Baumhoff, *Ecological Determinants of Aboriginal California Populations* (Berkeley: University of California Press, 1963).
53. McEvoy, *The Fisherman's Problem,* pp. 34–35 and 51–60.
54. McEvoy, *The Fisherman's Problem,* pp. 58–60; and Dodds, *The Salmon King of Oregon,* pp. 174–179.
55. For a summary of various reports that describe the decline of the Yakima River salmon, see J. A. Lichatowich and L. E. Mobrand, Analysis of Chinook Salmon in the Columbia River from an Ecosystem Perspective, Bonneville Power Administration, DOE/BP-25105-2, Portland, OR, 1995, p. 36.
56. Leibhardt, Law, Environment, and Social Change.
57. Leibhardt, Law, Environment, and Social Change.
58. Leibhardt, Law, Environment, and Social Change, p. 346.
59. Leibhardt, Law, Environment, and Social Change, p. 361.
60. T. Roosevelt, Eighth Annual Message, in *The State of the Union Messages of the Presidents, 1790–1966,* volume III, edited by F. Israel (New York: Chelsea House Publishers, 1967).
61. H. McAllister, Annual Report of the Oregon Department of Fisheries, Portland, 1908.
62. Cited in McAllister, Annual Report of the Oregon Department of Fisheries.
63. S. G. Kohlstedt, *The Origins of Natural Science in America: The Essays of George Brown Goode* (Washington, DC: Smithsonian Institution Press, 1991), pp. 164–165.
64. J. F. Reiger, *American Sportsmen and the Origins of Conservation* (Norman: University of Oklahoma Press, 1986).
65. G. B. Goode, The Status of the U.S. Fish Commission in 1884, part XII: Report of the Commissioner for 1884, Government Printing Office, Washington, DC, 1886.
66. C. Cranston, The Fish and Game Laws of Oregon, *Transactions of the American Fisheries Society,* Forty-second Annual Meeting, September 3–5, 1912, Denver (Washington, DC, 1913), pp. 75–87.
67. J. Crutchfield and G. Pontecorvo, *The Pacific Salmon Fisheries: A Study of Irrational Management* (Baltimore: Resources for the Future/Johns Hopkins Press, 1969); and B. Mace, Chronological History of Fish and Wildlife Administration in Oregon, Oregon Department of Fish and Wildlife, Portland, 1960.
68. H. O. Wendler, Regulation of Commercial Fishing Gear and Seasons on the

Columbia River from 1859 to 1963, Washington Department of Fisheries, *Fisheries Research Papers* 2(4): 19–31 (1966).
69. J. Gharrett, Summary Coastal River Regulations—1878–1950, Unpublished manuscript, Oregon Fish Commission, Portland, 1953; Crutchfield and Pontecorvo, *The Pacific Salmon Fisheries;* and Meggs, *Salmon: The Decline of the British Columbia Fishery;* also see R. Cooley, *Politics and Conservation: The Decline of the Alaska Salmon* (New York: Harper and Row, 1963).
70. Oregon State Board of Fish Commissioners, Second Report to the Governor, Salem, 1888.
71. J. L. Riseland, Twenty-second and Twenty-third Annual Reports of the State Fish Commissioner to the Governor of the State of Washington, Washington Department of Fisheries and Game, Olympia, 1912.
72. Stacey, Sockeye and Tinplate.
73. Smith, *Salmon Fishers of the Columbia.*
74. Jordan and Gilbert, The Salmon Fishing and Canning Interests.
75. M. McDonald, The Salmon Fisheries of the Columbia Basin.
76. W. Rich, The Salmon Runs in the Columbia River in 1938, Fishery Bulletin No. 37, U.S. Fish and Wildlife Service, Washington, DC, 1942.
77. Wendler, Regulation of Commercial Fishing Gear and Seasons on the Columbia River.
78. Dodds, *The Salmon King of Oregon,* p. 145.
79. Senator I. H. Bingham, Letter to H. G. Van Dusen, April 26, 1909, Oregon Historical Society Archives, Portland, Warren Packing Company, Record 1141, Box 1; also see Master Fish Warden Van Dusen Dismissed, *Oregon Journal* (Portland), March 26, 1908, Oregon Historical Society Archives, Seufert Canning Company, Record 1102.
80. H. G. Van Dusen, Letter to Senator I. H. Bingham, April 19, 1909, Oregon Historical Society Archives, Portland, Warren Packing Company, Record 1141, Box 1.
81. Senator I. H. Bingham, Letter to H. G. Van Dusen, April 26, 1909.
82. Vanity of Salmon Theories (editorial), *Oregonian* (Portland), November 13, 1906.
83. Rich, Salmon Runs of the Columbia River in 1938.
84. D. Johnson, W. Chapman, and R. Schoning, The Effects on Salmon Populations of the Partial Elimination of Fixed Fishing Gear on the Columbia River in 1935, Contribution No. 11, Oregon Fish Commission, Portland, 1948.
85. Increasing Use of Gasoline Engines in Fishing Industry, *Pacific Fisherman* 5(6): (1906); also see Sixty Years Ago: When a Fish Boat Engine Was a Dream of Pioneers, *Pacific Fisherman* 56(13): (1958).
86. F. Caldwell, *Pacific Troller* (Anchorage: Alaska Northwest Publishing Company, 1978).
87. Cobb, Pacific Salmon Fisheries.
88. Martin, *Legacy and Testament.*

89. E. Smith, The Taking of Immature Salmon in Waters of the State of Washington, Washington Department of Fisheries, Olympia, 1920.
90. D. Milne, Recent British Columbia Spring and Coho Salmon Tagging Experiments and a Comparison with Those Conducted from 1925 to 1930, Bulletin No. 113, Fisheries Research Board of Canada, Ottawa, 1957.
91. California Fish and Game Commission, Twenty-sixth Biennial Report, Sacramento, 1921; and also California Fish and Game Commission, Twenty-seventh Biennial Report, Sacramento, 1923.
92. Smith, The Taking of Immature Salmon.
93. Washington State Fisheries Board, Meeting of Fisheries Executives of the Pacific Coast, Seattle, March 16–17, 1925, General stacks, Fisheries Library, University of Washington, Seattle.
94. Salmon Fisheries in Oregon, *Oregonian* (Portland), March 3, 1875.
95. Remonstrance, *Oregonian* (Portland), January 1, 1877.

Chapter 6

1. International Fisheries Exhibition, *An Illustrated Description of the Electric Light Machinery in the Exhibition, with Elementary Notes on the Production of Electric Currents* (London: William Clowes and Sons, 1883).
2. H. R. MacCrimmon, The Beginning of Salmon Culture in Canada, *Canadian Geographical Journal* 71(3): 96–103 (1965).
3. International Fisheries Exhibition, *Fisheries Exhibition Literature,* volume VI, *Conferences,* part III (London: William Clowes and Sons, 1883).
4. W. H. Fry, *Complete Treatise on Artificial Fish-Breeding* (New York: D. Appleton and Company, 1854). This is the first book on artificial propagation published in the United States, and it describes all the early European work on hatcheries; also see G. B. Goode, Pisciculture, in *The Encyclopaedia Britannica,* volume 19 (Chicago: Werner Company, 1898).
5. S. G. Kohlstedt, *The Origins of Natural Science in America: The Essays of George Brown Goode* (Washington, DC: Smithsonian Institution Press, 1991); also see L. White, Jr., The Historical Roots of Our Ecologic Crisis, *Science* 155(3767): 1203–1206 (1967).
6. Fry, *Complete Treatise on Artificial Fish-Breeding,* p. 27.
7. For example, see these two articles in the January–February 1992 issue of *Fisheries*: J. Martin, J. Webster, and G. Edwards, Hatchery and Wild Stocks: Are They Compatible? and R. Hilborn, Hatcheries and the Future of Salmon in the Northwest, *Fisheries* 17(1): 4–8 (1992).
8. Fry, *Complete Treatise on Artificial Fish-Breeding,* p. 11; also see Fish-culture in America, *Harper's New Monthly Magazine* 37(222): 721–739 (1868).
9. Independent Scientific Group, Return to the River: Restoration of Salmonid Fishes in the Columbia River Ecosystem, Northwest Power Planning Council, Portland, OR, 1996; and National Research Council, *Upstream: Salmon and Society in the Pacific Northwest* (Washington, DC: National Academy Press, 1996).

10. R. McIntosh, *The Background of Ecology: Concept and Theory* (New York: Cambridge Studies in Ecology, Cambridge University Press, 1985); and D. L. Bottom, To Till the Water: A History of Ideas in Fisheries Conservation, in *Pacific Salmon and Their Ecosystems: Status and Future Options,* edited by D. J. Stouder, P. A. Bisson, and R. J. Naiman (New York: Chapman and Hall, 1997), pp. 569–597.
11. Fry, *Complete Treatise on Artificial Fish-Breeding,* pp. 21–22.
12. E. Clark, *The Oysters of Locmariaquer* (New York: Harper Perennial, 1992), pp. 86–87; also see *Harper's New Monthly Magazine,* Fish-culture in America.
13. Fry, *Complete Treatise on Artificial Fish-Breeding,* p. 187.
14. T. Steinberg, *Nature Incorporated: Industrialization and the Waters of New England* (New York: Cambridge University Press, 1991), p. 167.
15. *Harper's New Monthly Magazine,* Fish-culture in America, p. 738.
16. Steinberg, *Nature Incorporated,* pp. 186–187.
17. G. P. Marsh, *Artificial Propagation of Fish: Report Made under the Legislative Authority of Vermont* (Burlington, VT: Free Press Printing, 1857).
18. L. Stone, Some Brief Reminiscences of the Early Days of Fish-Culture in the United States, *Bulletin of the U.S. Fish Commission* 25(22): 338 (1897).
19. D. C. Allard, Jr., *Spencer Fullerton Baird and the U.S. Fish Commission* (New York: Arno Press, 1978).
20. Allard, *Spencer Fullerton Baird*; also see Steinberg, *Nature Incorporated.*
21. W. T. Bower, History of the American Fisheries Society, *Transactions of the American Fisheries Society,* Fortieth Annual Meeting, September 27–29, 1910 (Washington, DC, 1911).
22. U.S. Commission of Fish and Fisheries, Report of the Commissioner for 1872 and 1873, part II, Government Printing Office, Washington, DC, 1874, pp. 757–772.
23. Allard, *Spencer Fullerton Baird,* p. 160.
24. L. Stone, Artificial Propagation of Salmon on the Pacific Coast of the United States: With Notes on the Natural History of the Quinnat Salmon, U.S, Commission of Fish and Fisheries, Washington, DC, 1896; also see J. W. Hedgpeth, Livingston Stone and Fish Culture in California, *California Fish and Game* 27(3): 126–148 (1941).
25. L. Stone, Letter to Professor Spencer Baird, August 31, 1872, National Archives, Beltsville, MD, Record Group 22, Volume 7, Entry 1.
26. L. Stone, Letter to Spencer Baird, February 24, 1873, National Archives, Beltsville, MD, Record Group 22.
27. Stone, Some Brief Reminiscences of the Early Days of Fish-Culture in the United States, p. 339.
28. Hedgpeth, Livingston Stone and Fish Culture in California.
29. L. Stone, Letter to Spencer Baird, February 24, 1873.
30. M. V. Hayden, History of the Salmon Industry in the Pacific Northwest, M.S. thesis, University of Oregon, Eugene, 1930.

31. R. J. Miller and E. L. Brannon, The Origin and Development of Life History Patterns in Pacific Salmon, in *Salmon and Trout Migratory Behavior Symposium,* edited by E. L. Brannon and E. O. Salo, School of Fisheries, University of Washington, Seattle, 1982, pp. 296–309.
32. For reduced survival of transplanted fish, see R. R. Reisenbichler, Relation Between Distance Transferred from Natal Stream and Recovery Rate for Hatchery Coho Salmon, *North American Journal of Fisheries Management* 8: 172–174 (1988); also see Columbia Basin Fish and Wildlife Authority, Review of the History, Development, and Management of Anadromous Fish Production Facilities in the Columbia River Basin, Columbia Basin Fish and Wildlife Authority, Portland, OR, 1990, p. 17; A. M. McGie, Analysis of Relationships Between Hatchery Coho Salmon Transplants and Adult Escapements in Oregon Coastal Watersheds, Information Report Series, Fisheries No. 80-6, Oregon Department of Fish and Wildlife, Charleston, OR, 1980; and L. Shapovalov, An Evaluation of Steelhead and Salmon Management in California, Inland Fisheries Branch, California Department of Fish and Game, Sacramento, 1953, p. 29.
33. G. B. Goode, The Status of the U.S. Fish Commission in 1884, part XII, Report of the Commissioner for 1884, Government Printing Office, Washington, DC, 1886, p. 1162.
34. Hedgpeth, Livingston Stone and Fish Culture in California.
35. National Research Council, *Upstream.*
36. D. Worster, *Nature's Economy: A History of Ecological Ideas* (New York: Cambridge University Press, 1977), p. 53.
37. H. M. Smith, A Review of the History and Results of the Attempts to Acclimatize Fish and Other Water Animals in the Pacific States, *Bulletin of the U.S. Fish Commission for 1895,* volume 15, Washington, DC, 1896; also see Allard, *Spencer Fullerton Baird.*
38. A. Nicols, *The Acclimatisation of the Salmonidae at the Antipodes: Its History and Results* (London: Sampson, Low, Marston, Searle, and Rivington, 1882).
39. H. M. Smith, Fish Acclimatization of the Pacific Coast, *Science* 22(550): 88–89 (1893).
40. L. Stone, The Artificial Propagation of Salmon in the Columbia River Basin, *Transactions of the American Fish-Culture Association,* Thirteenth Annual Meeting, May 13–14, 1884, Washington, DC, p. 21.
41. Oregon State Board of Fish Commissioners, Fourth Annual Report, Salem, 1890, p. 26.
42. A. C. Little, Tenth and Eleventh Annual Reports of the Washington State Fish Commissioner, Olympia, 1901, p. 18.
43. S. Johnson, Freshwater Environmental Problems and Coho Production in Oregon. Information Report 84-11, Oregon Department of Fish and Wildlife, Corvallis.
44. R. Bunting, *The Pacific Raincoast: Environment and Culture in an Eden, 1778–1900* (Lawrence: University Press of Kansas, 1997).

45. Bunting, *The Pacific Raincoast.*
46. Little, Tenth and Eleventh Annual Reports of the State Fish Commissioner.
47. J. Titcomb, Report on the Propagation and Distribution of Food-fishes, Report of the U.S. Commissioners of Fish and Fisheries for 1902, Washington, DC, 1904.
48. Oregon Game Commission, Biennial Report of the Oregon Game Commission, Salem, 1942.
49. U.S. Department of Interior and the Lower Elwha Klallam Tribe, The Elwha Report: Restoration of the Elwha River Ecosystem and Native Anadromous Fisheries, Report Submitted Pursuant to Public Law 102–495, National Park Service, Port Angeles, WA, 1994.
50. B. Brown, *Mountain in the Clouds: A Search for the Wild Salmon* (New York: Simon and Schuster, 1982).
51. P. Johnson, Historical Assessment of Elwha River Fisheries, Draft Report, Olympic National Park, Port Angeles, WA, 1994.
52. Johnson, Historical Assessment of Elwha River Fisheries.
53. Cited in Johnson, Historical Assessment of Elwha River Fisheries.
54. Cited in Johnson, Historical Assessment of Elwha River Fisheries.
55. Brown, *Mountain in the Clouds.*
56. Brown, *Mountain in the Clouds.*
57. There are two dams on the Elwha; the other was built about nine miles upstream from Aldwell's dam, at Glines Canyon.
58. For budgets, see U.S. General Accounting Office, Endangered Species: Past Actions Taken to Assist Columbia River Salmon, GAO/RCED-92-173BR, Washington, DC, 1992. For independent panels, see National Research Council, *Upstream;* Independent Scientific Group, Return to the River; and National Fish Hatchery Review Panel, Report of the National Fish Hatchery Review Panel, The Conservation Fund, Arlington, VA, 1994. For a tenacious defense, see *Oregonian* (Portland), Overhauling Fish Factories, December 7, 1998.
59. L. Stone, A National Salmon Park, *Transactions of the American Fisheries Society* 21: 149–162 (1892).
60. B. Harrison, Afognak Forest and Fish Culture Reserve: A Proclamation, Washington, DC, 1892, National Archives, Washington, DC, Record Group 22, U.S. Fish and Wildlife Service.
61. P. Roppel, Salmon from Kodiak, Alaska Historical Commission Studies in History No. 216, Anchorage, 1986. Willis Rich studied Litnik Lake in 1931 and concluded that it lacked enough plankton to support the juvenile salmon released from the hatchery. The removal of salmon carcasses by the hatchery might have contributed to the lake's low productivity.
62. J. J. Brice, Establishment of Stations for the Propagation of Salmon on the Pacific Coast, Miscellaneous Documents, U.S. House of Representatives, 53rd Congress, Washington, DC, 1895. Thirty years later, the California Fish

and Game Commission sponsored an initiative petition to prevent the construction of dams in the lower Klamath River. The initiative passed. Statement of N. B. Scofield, Minutes of Meeting of Fisheries Executives of the Pacific Coast, 1925, p. 124, General stacks, Fisheries Library, University of Washington, Seattle.

63. H. Ward, The Preservation of the American Fish Fauna, *Transactions of the American Fisheries Society* 42: 157–170 (1912).
64. C. R. Pollock, Fortieth and Forty-first Annual Reports of the Washington State Department of Fisheries and Game, Olympia, 1932.
65. Save Rivers for Fish, *Oregon Voter* (Portland), July 14, 1928, pp. 21–24.
66. J. Cone and S. Ridlington, *The Northwest Salmon Crisis: A Documentary History* (Corvallis: Oregon State University Press, 1996).
67. Oregon State Planning Board, A Study of Commercial Fishing Operations on the Columbia River. Report Submitted to the Governor of Oregon, Salem, 1938.
68. The story of salmon refuges was adapted from J. A. Lichatowich and three others, Sanctuaries for Pacific Salmon, in *Sustainable Fisheries Management: Balancing the Conservation and Use of Pacific Salmon,* edited by E. E. Knutson and four others (Boca Raton, FL: Lewis Publishers, in press).
69. G. B. Goode, The Status of the U.S. Fish Commission in 1884.
70. Goode, The Status of the U.S. Fish Commission in 1884, p. 1157.
71. Stone, Artificial Propagation of Salmon on the Pacific Coast of the United States, p. 319; also see Hedgpeth, Livingston Stone and Fish Culture in California, p. 143.
72. G. H. Clark, Sacramento–San Joaquin Salmon (*Oncorhynchus tshawytscha*) Fishery of California, Fish Bulletin No. 17, California Division of Fish and Game, Sacramento, 1929.
73. Allard, *Spencer Fullerton Baird,* p. 293.
74. International Fisheries Exhibition, *Fisheries Exhibition Literature,* p. 173.
75. Goode, The Status of the U.S. Fish Commission in 1884, p. 1157.
76. E. M. Wood, A Century of American Fish Culture, 1853–1953, *The Progressive Fish-Culturist* 15(4): 147–162 (1953).
77. M. McDonald, Address of the Chairman of the General Committee on the World Fisheries Congress, *Bulletin of the U.S. Fish Commission,* volume 13, Washington, DC, 1894, p. 15.
78. H. Whitaker, Address of the President, *Transactions of the American Fisheries Society,* Twenty-second Annual Meeting, July 15, 1893 (New York, 1893); and H. Whitaker, Some Observations on the Moral Phases of Modern Fish Culture, *Transactions of the American Fisheries Society,* June 12–13, 1895 (New York, 1896).
79. H. Van Dusen, Annual Report of the Department of Fisheries of the State of Oregon to the Legislative Assembly, 1909, Salem, 1908, p. 9. The need to take hatcheries on faith until the industry or the people were willing to pay

for studies analyzing their contribution to the fisheries was also voiced by F. M. Chamberlin, Artificial Propagation, *Pacific Fisherman* 1(11): 10 (1903).

80. J. N. Cobb, Pacific Salmon Fisheries, Appendix 13 to the Report of the U.S. Commissioner of Fisheries for 1930, Bureau of Fisheries Document No. 1092, Washington, DC, 1930.
81. J. Wallis, An Evaluation of the Alsea River Salmon Hatchery, Oregon Fish Commission Research Laboratory, Clackamas, OR, 1963, p. 11.
82. Wallis, An Evaluation of the Alsea River Salmon Hatchery, p. 25.
83. Shapovalov, An Evaluation of Steelhead and Salmon Management in California, p. 29.
84. C. F. Culler, Progress in Fish Culture, *Transactions of the American Fisheries Society,* Sixty-second Annual Meeting, September 21–23, 1932, p. 118.
85. G. Dodds, *The Salmon King of Oregon: R. D. Hume and the Pacific Fisheries* (Chapel Hill: University of North Carolina Press, 1959).
86. H. Van Dusen, Annual Report of the Department of Fisheries of the State of Oregon to the Legislative Assembly, 1909, p. 6; also see J. L. Riseland, Sixteenth and Seventeenth Annual Reports of the State Fish Commissioner and Game Warden to the Governor of the State of Washington (1905–1906). Washington Department of Fisheries and Game, Olympia, 1907, p. 13.
87. Oregon Fish and Game Commission, Biennial Report of the Fish and Game Commission of the State of Oregon, Salem, 1919, p. 16.
88. L. H. Darwin, Twenty-sixth and Twenty-seventh Annual Reports of the State Fish Commissioner to the Governor of the State of Washington, Washington Department of Fisheries and Game, Olympia, 1917, p. 9.
89. Willis Rich suggested that the increase in cannery pack was also due to World War I, which caused a shift from mild cured salmon to canned salmon, and that the total catch did not change at all. W. H. Rich, A Statistical Analysis of the Results of Artificial Propagation of Chinook Salmon, 1922, Manuscript Library, Northwest and Alaska Fisheries Science Center, National Marine Fisheries Service, Seattle.
90. R. Van Cleve and R. Ting, The Condition of Salmon Stocks in the John Day, Umatilla, Walla Walla, Grande Ronde, and Imnaha Rivers as Reported by Various Fisheries Agencies, University of Washington, Department of Oceanography, Seattle, 1960, p. 8.
91. V. Smith, Fish Culture Methods in the Hatcheries of the State of Washington, Washington State Fish Commission, Olympia, 1919; also see Oregon Fish Commission, Biennial Report of the Fish Commission of the State of Oregon for 1931 and 1932, Salem, 1933, p. 15; C. Brown, Adjustment of Environment vs. Stocking to Increase the Productivity of Fish Life, *Transactions of the American Fisheries Society,* Fifty-first Annual Meeting, September 5–7, 1921, Allentown, PA (Washington, DC), volume 51, 1921–1922; P. N. Needham, Natural Propagation Versus Artificial Propagation in Relation to Angling,

Transactions of the Fourth North American Wildlife Conference, February 13–15, 1939 (Washington, DC: American Wildlife Institute, 1939); R. E. Foerster, Sockeye Salmon Propagation in British Columbia, Bulletin No. 53, Biological Board of Canada, Ottawa, 1936; and S. Gordon, Scientific Management: Our Future Fisheries Jobs, *Transactions of the American Fisheries Society,* Sixty-third Annual Meeting, 1933, Columbus, OH.

92. Needham, Natural Propagation Versus Artificial Propagation.
93. R. Hile, The Increase in the Abundance of the Yellow Pike-Perch, *Stizostedion vitreum* (Mitchill), in Lakes Huron and Michigan, in Relation to the Artificial Propagation of the Species, in *The Collected Papers of Ralph Hile, 1928–73,* U.S. Fish and Wildlife Service, volume I, 1928–46, Washington, DC, 1977, p. 154.
94. All Sockeye Hatcheries Closed in British Columbia, *Pacific Fisherman* 34(5): 17 (1936).
95. J. Taylor, Making Salmon: Economy, Culture, and Science in the Oregon Fisheries, Precontact to 1960, Ph.D. dissertation, University of Washington, Seattle, 1996.
96. Rich, A Statistical Analysis of the Results of Artificial Propagation of Chinook Salmon.
97. Cobb, Pacific Salmon Fisheries, p. 493.
98. J. Gottschalk, Reports of Vice-Presidents of Divisions, *Transactions of the American Fisheries Society,* Seventy-first Annual Meeting, August 25–26, 1941, St. Louis, MO (Washington, DC, 1942), p. 22.

Chapter 7

1. Kitzhaber Convenes a Summit on Salmon, *Oregonian* (Portland), June 4, 1997.
2. Washington State Department of Fisheries and Game, Thirty-fourth and Thirty-fifth Annual Reports, Olympia, 1925.
3. Washington State Fisheries Board, Meeting of Fisheries Executives of the Pacific Coast, Seattle, March 16–17, 1925, General stacks, Fisheries and Oceanography Library, University of Washington, Seattle.
4. Washington State Fisheries Board, Meeting of Fisheries Executives of the Pacific Coast, pp. 95–96.
5. W. Hornaday, *Hornaday's American Natural History* (New York: Charles Scribner's Sons, 1935).
6. Washington State Fisheries Board, Meeting of Fisheries Executives of the Pacific Coast, pp. 140–155.
7. Washington State Fisheries Board, Meeting of Fisheries Executives of the Pacific Coast, p. 153.
8. Washington State Fisheries Board, Meeting of Fisheries Executives of the Pacific Coast, p. 102.
9. Washington State Fisheries Board, Meeting of Fisheries Executives of the

Pacific Coast, p. 122. Priest Rapids was a dam proposed for the mainstem Columbia River above the town of Richland, Washington.

10. Washington State Fisheries Board, Meeting of Fisheries Executives of the Pacific Coast, p. 122.
11. Washington State Fisheries Board, Meeting of Fisheries Executives of the Pacific Coast, p. 80.
12. H. O'Malley, Symposium on Fisheries and Fishery Investigations: Our Opportunities, Our Responsibilities, part II: Proceedings of the Divisional Conference, January 4–7, 1927, in Progress in Biological Inquiries 1926, edited by E. Higgins, Appendix 7 to the Report of the U.S. Commissioner of Fisheries for 1927, Bureau of Fisheries Document No. 1029, Washington, DC, 1928.
13. E. Higgins, ed., Progress in Biological Inquiries 1926, Appendix 7 to the Report of the U.S. Commissioner of Fisheries for 1927, Bureau of Fisheries Document No. 1029, Washington, DC, 1928.
14. Higgins, Progress in Biological Inquiries 1926.
15. Fish culturists today have developed a revisionist history of their craft, which denies that hatcheries were ever expected to make up for "all possible destructive influences." See Higgins, Progress in Biological Inquiries 1926.
16. W. Ricker, The Fisheries Research Board of Canada—Seventy-five Years of Achievements, *Journal of Fisheries Research Board of Canada* 32(8): 1465–1490 (1975).
17. R. Van Cleve, The School of Fisheries, *The Progressive Fish-Culturist* 14(4): 159–164 (1952).
18. S. Gordon, Scientific Management: Our Future Fisheries Jobs, *Transactions of the American Fisheries Society,* Sixty-third Annual Meeting, 1933, Columbus, OH.
19. For example, Finley asked Oregon State College professor R. E. Dimick if he was aware of any evidence that the planting of 122 million game fish at a cost of $261,000 actually ever increased the number of fish in Oregon's streams. W. Finley, Letter to R. E. Dimick, June 6, 1935, Oregon State University Archives, Corvallis, Dimick Papers, Correspondence with W. Finley.
20. W. Mathewson, *William L. Finley: Pioneer Wildlife Photographer* (Corvallis: Oregon State University Press, 1986).
21. Investigation of Ousting of Finley Proposed, *Oregon Journal* (Portland), January 12, 1920; also see Make Finley Biologist Is [Governor] Ulcott View, *Oregon Journal,* December 23, 1919; and School Children Getting Up Big Finley Petition, *Oregon Journal,* December 12, 1919, p. 4.
22. The Reason for Finley, *Oregon Journal* (Portland), December 30, 1919, p. 6.
23. Pacific Fishery Management Council, Amendment 13 to the Pacific Coast Salmon Plan: Fishery Management Regime to Ensure Protection and Rebuilding of Oregon Coastal Natural Coho, Portland, OR, 1997.
24. L. Maps, Wildlife Director May Lose His Job, *Seattle Times,* March 10, 1998.
25. Oregon Game Commission, First Progress Report of the Ten-Year Plan, Portland, 1932.

26. E. Moore, Report of the Vice President, Division of Fish Culture, New York State Conservation Commission, *Transactions of the American Fisheries Society,* Fifty-fifth Annual Meeting, 1925, Denver, pp. 17–23.
27. Interview with Phil Snyder, former Oregon Game Commission director, August 10, 1995.
28. Oregon Fish Commission, Biennial Report of the Fish Commission of the State of Oregon to the Governor and the Thirty-ninth Legislative Assembly, 1937, State of Oregon, Salem, 1937.
29. D. Allen, *Our Wildlife Legacy* (New York: Funk and Wagnalls, 1962).
30. J. R. Dunn, Charles Henry Gilbert (1859–1928): An Early Fishery Biologist and His Contribution to Knowledge of Pacific Salmon (*Oncorhynchus* spp.), *Reviews in Fisheries Science* 4(2): 133–184 (1996).
31. D. S. Jordan, The Salmon of the Pacific, *Pacific Fisherman* 2: 1 (1904).
32. In a 1972 paper, Ricker states that up to the 1930s, many biologists thought salmon species were genetically uniform. W. Ricker, Hereditary and Environmental Factors Affecting Certain Salmonid Populations, in *The Stock Concept of Pacific Salmon,* H. R. MacMillan Lectures in Fisheries, edited by R. Simon and P. Larkin (Seattle: U.S. Bureau of Fisheries, 1972).
33. Salmon Marking Experiments on the Pacific Coast, *Pacific Fisherman* 2(6): (1904).
34. This theory was still being advocated in 1928, when Willis Rich and Harlan Holmes conducted studies proving that the timing of migration is hereditary. W. H. Rich and H. B. Holmes, Experiments in Marking Young Chinook Salmon on the Columbia River, 1916 to 1927, *Bulletin of the Bureau of Fisheries,* Document No. 1047, Washington, DC, 1929.
35. C. H. Gilbert, Age at Maturity of the Pacific Coast Salmon of the Genus *Oncorhynchus, Bulletin of the Bureau of Fisheries,* Washington, DC, 1913.
36. Cited in Dunn, Charles Henry Gilbert.
37. See W. H. Rich, Local Populations and Migration in Relation to the Conservation of Pacific Salmon in the Western States and Alaska, Contribution No. 1, Department of Research, Fish Commission of the State of Oregon, Salem, 1939; A. G. Huntsman, "Races" and "Homing" of Salmon, *Science* 85(2216): 582–583 (1937); and W. H. Rich, "Homing" of Pacific Salmon, *Science* 85(2211): 477–478 (1937).
38. Dunn, Charles Henry Gilbert.
39. See the following for statements about the greater efficiency of artificial vs. natural production of salmon. V. Smith, Fish Culture Methods in the Hatcheries of the State of Washington, Washington State Fish Commission, Olympia, 1919, p. 6; Washington State Department of Fisheries and Game, Thirtieth and Thirty-first Annual Reports of the State Fish Commissioner to the Governor of the State of Washington, Olympia, 1921, p. 17; and Oregon State Fish and Game Protector, Third and Fourth Annual Reports of the State Fish and Game Protector of the State of Oregon, 1895–1896, Salem, 1896, p. 33.

40. M. C. Healey, Variation in the Life History Characteristics of Chinook Salmon and Its Relevance to Conservation of the Sacramento Winter Run of Chinook Salmon, *Conservation Biology* 8(5): 876–877 (1994); M. C. Healey and A. Prince, Scales of Variation in Life History Tactics of Pacific Salmon and the Conservation of Phenotype and Genotype, in *Evolution and the Aquatic Ecosystem: Defining Unique Units in Population Conservation,* edited by J. L. Nielsen and D. A. Powers, American Fisheries Society Symposium 17, Bethesda, MD, 1995, pp. 176–184; and P. E. Reimers, The Length of Residence of Juvenile Fall Chinook Salmon in Sixes River, Oregon, *Research Reports of the Fish Commission of Oregon* 4(2): 3–43 (1973). Also see M. Schluchter and J. A. Lichatowich, Juvenile Life Histories of Rogue River Spring Chinook Salmon *Oncorhynchus tshawytscha* (Walbaum), as Determined from Scale Analysis, Information Report Series, Fisheries No. 77-5, Oregon Department of Fish and Wildlife, Corvallis, 1977; and L. M. Carl and M. C. Healey, Differences in Enzyme Frequency and Body Morphology among Three Juvenile Life History Types of Chinook Salmon (*Oncorhynchus tshawytscha*) in the Nanaimo River, British Columbia, *Canadian Journal of Fisheries and Aquatic Sciences* 41: 1070–1077 (1984). And see Independent Scientific Group, Return to the River: Restoration of Salmonid Fishes in the Columbia River Ecosystem, Northwest Power Planning Council, Portland, OR, 1996; and W. F. Thompson, An Approach to Population Dynamics of the Pacific Red Salmon, *Transactions of the American Fisheries Society* 88(3): 206–209 (1959).
41. W. H. Rich, Early History and Seaward Migration of Chinook Salmon in the Columbia and Sacramento Rivers, *Bulletin of the Bureau of Fisheries* No. 37, Washington, DC, 1920; and W. H. Rich, Progress in Biological Inquiries, July 1 to December 31, 1924, Bureau of Fisheries Document No. 990, Washington, DC, 1925.
42. Rich and Holmes, Experiments in Marking Young Chinook Salmon on the Columbia River.
43. F. A. Davidson, "Migration" and "Homing" of Pacific Salmon, *Science* 86(2220): 55–56 (1937).
44. Rich, Local Populations and Migration in Relation to the Conservation of Pacific Salmon.
45. Oregon Fish Commission, Biennial Report of the Fish Commission of the State of Oregon to the Governor and the Fortieth Legislative Assembly, 1939, State of Oregon, Salem, 1939.
46. T. A. Flagg and four others, The Effect of Hatcheries on Native Coho Salmon Populations in the Lower Columbia River, in *Uses and Effects of Cultured Fishes in Aquatic Ecosystems,* edited by H. L. Schramm, Jr., and R. G. Piper, American Fisheries Society Symposium 15, Bethesda, MD, 1995, pp. 366–375.
47. A. C. Benke, A Perspective on America's Vanishing Streams, *Journal of North American Benthological Society* 9(1): 77–88 (1990).

Chapter 8

1. R. Burgner, Life History of the Sockeye Salmon, in *Pacific Salmon Life Histories,* edited by C. Groot and R. Margolis (Vancouver: University of British Columbia Press, 1991), pp. 3–117.
2. Burgner, Life History of the Sockeye Salmon.
3. F. Ward and P. Larkin, Cyclic Dominance in Adams River Sockeye Salmon, International Pacific Salmon Fisheries Commission, Progress Report No. 11, New Westminster, BC, 1964.
4. J. Roos, *Restoring Fraser River Salmon: A History of the International Pacific Salmon Fisheries Commission, 1937–1985,* Pacific Salmon Fisheries Commission, Vancouver, BC, 1991.
5. Catch and escapement data obtained from Ian Guthrie of the Pacific Salmon Commission, August 2, 1995.
6. W. Ricker, Effects of the Fishery and of Obstacles to Migration on the Abundance of Fraser River Sockeye Salmon (*Oncorhynchus nerka*), Canadian Technical Report of Fisheries and Aquatic Sciences No. 1522, Nanaimo, BC, 1987.
7. Roos, *Restoring Fraser River Salmon;* also see A. Netboy, *The Salmon: Their Fight for Survival* (Boston: Houghton Mifflin, 1974); and R. E. Foerster, The Sockeye Salmon, Bulletin No. 162, Fisheries Research Board of Canada, Ottawa, 1968.
8. W. Ricker, Hells Gate and the Sockeye, *Journal of Wildlife Management* 11(1): 10–20 (1947); also see Roos, *Restoring Fraser River Salmon.*
9. Ricker, Effects of the Fishery and of Obstacles to Migration.
10. Roos, *Restoring Fraser River Salmon.*
11. H. W. McKervill, *The Salmon People: The Story of Canada's West Coast Salmon Fishing Industry* (Sidney, BC: Gray's Publishing, 1967); also see Roos, *Restoring Fraser River Salmon.*
12. McKervill, *The Salmon People.*
13. Roos, *Restoring Fraser River Salmon.*
14. Ricker, Effects of the Fishery and of Obstacles to Migration.
15. D. L. Boxberger, *To Fish in Common: The Ethnohistory of Lummi Indian Salmon Fishing* (Lincoln: University of Nebraska Press, 1986).
16. Boxberger, *To Fish in Common.*
17. Boxberger, *To Fish in Common.*
18. G. Meggs, *Salmon: The Decline of the British Columbia Fishery* (Vancouver, BC: Douglas and McIntyre, 1991).
19. Meggs, *Salmon: The Decline of the British Columbia Fishery.*
20. Boxberger, *To Fish in Common;* also see C. Lyons, *Salmon: Our Heritage* (Vancouver, BC: Mitchell Press, 1969).
21. Lyons, *Salmon: Our Heritage.*
22. Roos, *Restoring Fraser River Salmon;* also see Lyons, *Salmon: Our Heritage.*
23. L. Parsons, Management of Marine Fisheries in Canada, National Research Council of Canada and the Department of Fisheries and Oceans, Ottawa, 1993.

24. Boxberger, *To Fish in Common,* p. 99.
25. Lyons, *Salmon: Our Heritage, pp.* 327–328.
26. J. Crutchfield and G. Pontecorvo, *The Pacific Salmon Fisheries: A Study of Irrational Management* (Baltimore: Resources for the Future/Johns Hopkins Press, 1969).
27. L. H. Darwin, Thirtieth and Thirty-first Annual Reports of the State Fish Commissioner, Seattle, 1921, p. 15.
28. R. E. Bilby, B. R. Fransen, and P. A. Bisson, Incorporation of Nitrogen and Carbon from Spawning Coho Salmon into the Trophic System of Small Streams: Evidence from Stable Isotopes, *Canadian Journal of Fisheries and Aquatic Sciences* 53(1): 164–173 (1996); also see G. A. Larkin and P. A. Slaney, Implications of Trends in Marine-Derived Nutrient Influx to South Coastal British Columbia Salmonid Production, *Fisheries* 22(11): 16–24 (1997).
29. Roos, *Restoring Fraser River Salmon.*
30. Roos, *Restoring Fraser River Salmon.*
31. *U.S. Congressional Record,* Seventy-first Congress, Third Session, 1931, volume 74, pp. 392–393.
32. Roos, *Restoring Fraser River Salmon.*
33. W. M. Freeman, The Memoirs of Miller Freeman 1875–1955 (1956), General stacks, Fisheries Library, University of Washington, Seattle.
34. Netboy, *The Salmon: Their Fight for Survival.*
35. Northwest Power Planning Council, Compilation of Information on Salmon and Steelhead Losses in the Columbia River Basin, Portland, OR, 1986.
36. W. Rich, Future of the Columbia River Salmon Fisheries, *Stanford Ichthyological Bulletin* 2(2): 37–47 (1940).
37. W. Thompson, An Outline for Salmon Research in Alaska, Fisheries Research Institute Circular 18, University of Washington, Seattle, 1951.
38. H. McGuire, First and Second Annual Reports of the Fish and Game Protector, Salem, OR, 1894.
39. R. Neuberger, *Our Promised Land* (New York: Macmillan, 1938).
40. Oregon Fish Commission, Biennial Report for 1933 and 1934, Portland, 1935.
41. Oregon Fish Commission, Biennial Report for 1933 and 1934.
42. In 1980, the fishery agencies found it necessary to publish a fifty-four-page directory that simply listed and described the various agencies and committees affecting the Columbia River salmon.
43. Oregon State Planning Board, A Study of Commercial Fishing Operations on the Columbia River, Report Submitted to the Governor of Oregon, Salem, 1938.
44. Washington State Senate, Report on the Problems Affecting the Fisheries of the Columbia River, Columbia River Fisheries Interim Investigating Committee, Olympia, 1943.
45. D. Parman, Inconsistent Advocacy: The Erosion of Indian Fishing Rights in

the Pacific Northwest, 1933–1956, *Pacific Historical Review* 53: 163–189 (1983).
46. Washington State Senate, Report on the Problems Affecting the Fisheries.
47. Address by B. M. Brennan, Director of Fisheries, State of Washington, to the Legislative Meeting of the Columbia River Fisheries Protective Union at Astoria, Oregon, October 1, 1940, National Archives, Washington, DC, Record Group 22.
48. W. Rich, The Present State of the Columbia River Salmon Resources, Contribution No. 3, Oregon Fish Commission, Salem, 1941; also see W. Rich, Fishery Problems Raised by the Development of Water Resources, Contribution No. 2, Oregon Fish Commission, Salem, 1940; W. Rich, Future of the Columbia River Salmon Fisheries, *Stanford Ichthyological Bulletin* 2(2): 37–47 (1940); W. Rich, The Biology of Columbia River Salmon, *Northwest Science* 40(1): 3–14 (1935); and J. Craig, The Effects of Power and Irrigation Projects on the Migratory Fish of the Columbia River, *Northwest Science* 40(1): 19–22 (1935). Also see L. Griffin, Certainties and Risks Affecting Fisheries Connected with Damming the Columbia River, *Northwest Science* 40(2): 25–30 (1935).
49. Griffin, Certainties and Risks Affecting Fisheries Connected with Damming the Columbia River.
50. W. Rich, F. Fish, M. Hanavan, and H. Holmes, Memorandum to Elmer Higgins, Program and Policy for the Fish and Wildlife Service in Respect of the Salmon Fisheries of the Columbia River, February 9, 1944, National Archives, Pacific Northwest Region, Seattle, Record Group 77, Box 34621.
51. Rich, Fish, Hanavan, and Holmes, Program and Policy for the Fish and Wildlife Service.
52. Rich, Fish, Hanavan, and Holmes, Program and Policy for the Fish and Wildlife Service.
53. Rich, Fish, Hanavan and Holmes, Program and Policy for the Fish and Wildlife Service.
54. Rich, The Present State of the Columbia River Salmon Resources.
55. J. Barnaby, Memorandum to Director, Fish and Wildlife Service, Development of the Salmon Fisheries of the Lower Columbia River, 1945, National Archives, Pacific Northwest Region, Seattle, Record Group 22, Box 12.
56. F. Bell, Guarding the Columbia's Silver Horde, *Nature Magazine* 29(1): 43–47 (1937); also see R. Neuberger, The Great Salmon Experiment, *Harper's Magazine* 190(1137): 229–236 (1945).
57. City of Tacoma, Washington, Undated sixteen-page document on the letterhead of the Department of Public Utilities, National Archives, Washington, DC, Record Group 48, Box 865.
58. W. G. Robbins, *Landscapes of Promise: The Oregon Story, 1800–1940* (Seattle: University of Washington Press, 1997).
59. Columbia Basin Fisheries Development Association, Wealth of the River: A Presentation of Fact Concerning the Columbia River Salmon Industry and a

Petition for the Conservation of This Industry Submitted with Reference to Proposals to Construct High Dams on the Columbia River and Its Tributaries, Astoria, OR, 1945.

60. The Columbia River Fisheries Development Association included the U.S. Fish and Wildlife Service, Washington Department of Fisheries, Washington Department of Game, Oregon Fish Commission, Oregon Game Commission, Idaho Fish and Game Commission, Columbia River Fishermen's Protective Union, Columbia River Salmon and Tuna Packers Association, International Fishermen and Allied Workers, Oregon Legislative Fisheries Committee, Washington Legislative Interim Fisheries Committee, Indian Tribal Council, and Northwest Federation of Indians.
61. W. Gardner, Memo to Secretary of Interior, Columbia River Dams and Salmon, March 6, 1947, National Archives, Washington, DC, Record Group 48.
62. R. W. Scheufele, History of the Columbia Basin Inter-Agency Committee, Pacific Northwest River Basins Commission, Vancouver, WA (no date, perhaps 1970).
63. Roos, *Restoring Fraser River Salmon.*
64. A. Robertson, Further Proof of the Parent Stream Theory, *Transactions of the American Fisheries Society* 51: 87–90 (1921).
65. L. Royal, The Effects of Regulatory Selectivity on the Productivity of Fraser River Sockeye Salmon, *The Canadian Fish Culturist* 14: 1–12 (1953).
66. All Sockeye Hatcheries Closed in British Columbia, *Pacific Fisherman* 34(5): 17 (1936).
67. Rich, Local Populations and Migration in Relation to the Conservation of Pacific Salmon.
68. L. L. Laythe, The Fishery Development Program in the Lower Columbia River, *Transactions of the American Fisheries Society,* September 13–15, 1948, Atlantic City, NJ.
69. C. G. Davidson, Fisheries and Other Resources in the Columbia Basin. Statements made by C. Girard Davidson under questioning by the Public Works Committee of the U.S. Senate, 1949, pp. 41–55, National Archives, Pacific Northwest Region, Seattle, Record Group 22.
70. U.S. Fish and Wildlife Service, A Program for the Preservation of the Fisheries of the Columbia River Basin, 1951, National Archives, Washington, DC, Record Group 48.
71. D. L. Bottom, To Till the Water: A History of Ideas in Fisheries Conservation, in *Pacific Salmon and Their Ecosystem: Status and Future Options,* edited by D. J. Stouder, P. A. Bisson, and R. J. Naiman (New York: Chapman and Hall, 1997) pp. 569–597.
72. Roos, *Restoring Fraser River Salmon.*
73. Pacific Northwest Field Committee, Reevaluation of the Columbia River Fishery Program—Interim Report, Portland, OR, 1950, National Archives, Washington, DC, Record Group 22.

74. This attitude still plagues salmon management. Recently, after serving on a scientific review panel for the Columbia River, I discussed the panel's recommendations with a salmon manager. He disagreed with the panel's call for additional research into new approaches to stream restoration and evaluation of existing approaches. Saying he represented the opinion of other salmon managers, he declared that the need was for action and not research.
75. U.S. Fish and Wildlife Service, Transcript of a conference with state, provincial, and federal agencies to develop an engineering-biological research program that would attempt to solve the problems of migratory fish passage over high dams and related problems, 1951, National Archives, Washington, DC, Record Group 48.
76. J. Brinckman, Three Billion Later: Columbia Basin Salmon Dwindle, *Oregonian* (Portland), July 27, 1997, section A.
77. P. Hirt, *A Conspiracy of Optimism* (Lincoln: University of Nebraska Press, 1994).
78. B. Hutchison, *The Fraser* (Toronto: Clarke, Irwin and Company, 1950).
79. Roos, *Restoring Fraser River Salmon.*
80. F. Andrew and G. Green, Sockeye and Pink Salmon Production in Relation to Proposed Dams in the Fraser River System, Bulletin No. 11, International Pacific Salmon Fisheries Commission, New Westminster, BC, 1960.
81. Haig-Brown cited in R. Bocking, *Mighty River* (Vancouver, BC: Douglas and McIntyre, 1997).
82. F. Neave, Stream Ecology and Production of Anadromous Fish, in *The Investigation of Fish-power Problems,* H. R. MacMillan Lectures in Fisheries, edited by P. Larkin (Vancouver: University of British Columbia, 1958), pp. 43–48.
83. Roos, *Restoring Fraser River Salmon.* It was not necessary to dam the mainstem of the Fraser River or the tributaries within the range of salmon to destroy salmon habitat, as is well illustrated by the Nechako-Kemano project. The upper Nechako was dammed above a falls and beyond the range of salmon; however, the impounded water was diverted via a tunnel outside the Fraser River Basin to a coastal stream, the Kemano River, where it was used to generate electricity for aluminum production. The reduced flows in the Nechako River below the project had severe detrimental effects on the salmon. Sockeye died before spawning because of elevated temperatures, and chinook production declined in the area below the project. The Canadian government gave the aluminum company extraordinary latitude and freedom from environmental-protection regulations. The fishery problems created by the project continued to be sources of debate and conflict for thirty-five years, but were never resolved.
84. R. Hilborn, Institutional Learning and Spawning Channels for Sockeye Salmon, *Canadian Journal of Fisheries and Aquatic Sciences* 49: 1126–1136 (1992).
85. Roos, *Restoring Fraser River Salmon.*

86. R. Hilborn and J. Winton, Learning to Enhance Salmon Production: Lessons from the Salmonid Enhancement Program, *Canadian Journal of Fisheries and Aquatic Sciences* 50(9): 2043–2056 (1993).
87. R. Leffler, The Program of the U.S. Fish and Wildlife Service for the Anadromous Fishes of the Columbia River, *Bulletin of the Oregon State Game Commission* 10: 3–7 (1959).
88. R. J. Wahle and R. R. Vreeland, Bioeconomic Contribution of Columbia River Hatchery Fall Chinook Salmon, 1961 through 1964 Broods, to the Pacific Salmon Fisheries, *Fisheries Bulletin* 76(1): 179–208 (1978).
89. T. A. Flagg and four others, The Effect of Hatcheries on Native Coho Salmon Populations in the Lower Columbia River, in *Uses and Effects of Cultured Fishes in Aquatic Ecosystems,* edited by H. L. Schramm, Jr., and R. G. Piper, American Fisheries Society Symposium 15, Bethesda, MD, 1995, pp. 366–375; also see National Research Council, *Upstream: Salmon and Society in the Pacific Northwest* (Washington, DC: National Academy Press, 1996); and Independent Scientific Group, Return to the River: Restoration of Salmonid Fishes in the Columbia River Ecosystem, Northwest Power Planning Council, Portland, OR, 1996.
90. C. Geertz, *The Interpretation of Culture* (New York: Basic Books, 1973).
91. F. Dyson, *Imagined Worlds* (Cambridge: Harvard University Press, 1997).
92. Roos, *Restoring Fraser River Salmon.*
93. C. Wilkinson and D. Conner, The Law of the Pacific Salmon Fishery: Conservation and Allocation of a Transboundary Common Property Resource, *Kansas Law Review* 32(1): 109 (1983).
94. J. Hutchings, C. Walters, and R. Haedrich, Is Scientific Inquiry Incompatible with Government Information Control? *Canadian Journal of Fisheries and Aquatic Sciences* 54(5): 1198–1210 (1997).

Chapter 9

1. Pacific Fishery Management Council, Freshwater Habitat, Salmon Produced, and Escapements for Natural Spawning along the Pacific Coast of the United States, Portland, OR, 1979.
2. National Research Council, *Upstream: Salmon and Society in the Pacific Northwest* (Washington, DC: National Academy Press, 1996); also see Wilderness Society, The Living Landscape, volume 2: Pacific Salmon and Federal Lands, Washington, DC, 1993.
3. W. Nehlsen, J. E. Williams, and J. A. Lichatowich, Pacific Salmon at the Crossroads: Stocks at Risk from California, Oregon, Idaho, and Washington, *Fisheries* 16(2): 4–21 (1991).
4. R. S. Waples, Definition of "Species" Under the Endangered Species Act: Application to Pacific Salmon, NOAA Technical Memorandum, NMFS F/NWC-194, U.S. Department of Commerce, National Marine Fisheries Service, Seattle, 1991. Waples gives the definition of ESU as: "A population

(or group of populations) will be considered 'distinct' (and hence a 'species') for purposes of the ESA if it represents an evolutionarily significant unit (ESU) of the biological species. A population must satisfy two criteria to be considered an ESU: 1) It must be reproductively isolated from other conspecific population units, and 2) It must represent an important component in the evolutionary legacy of the species."

5. National Marine Fisheries Service, Progress of Species Status Reviews in NMFS Northwest Region, www.nwr.noaa.gov, updated August 5, 1998.
6. T. Gresh, J. Lichatowich, and P. Shoonmaker, An Estimation of Historical and Current Levels of Salmon Production in the Northeast Pacific Ecosystem: Evidence of a Nutrient Deficit, Draft Report to Interrain Pacific, Portland, OR, 1998.
7. M. Dawson, Buck: Save Fish from ESA, *Peninsula Daily News* (Port Angeles, WA), August 27, 1997.
8. N. Langston, *Forest Dreams, Forest Nightmares* (Seattle: University of Washington Press, 1995).
9. The Oregon Production Index includes coho salmon production from southwestern Washington to northern California.
10. R. Mullen, Oregon's Commercial Harvest of Coho Salmon *Oncorhynchus kisutch* (Walbaum), 1892–1960, Information Report Series, Fisheries No. 81-3, Oregon Department of Fish and Wildlife, Portland, 1981.
11. Oregon Fish Commission, Biennial Report, Portland, 1955.
12. Oregon Fish Commission, Biennial Report July 1, 1962–June 30, 1964, Portland, 1964.
13. Oregon Fish Commission, Biennial Report July 1, 1960–June 30, 1962, Portland, 1962.
14. Oregon Fish Commission, Biennial Report July 1, 1962–June 30, 1964, p. 16.
15. E. Jeffries, Present: Role and Challenge of Fish Culture in the Northwest, in Proceedings of the Twenty-sixth Annual Northwest Fish Culture Conference, Otter Rock, Oregon, December 3–5, 1975, edited by J. Brophy, Oregon Aqua-Foods, Newport, OR, 1975, 167–174.
16. C. Finley, Dwindling Coho Salmon Runs Leave Oregon in Quandary, *Oregonian* (Portland), November 16, 1993.
17. Resource Development Branch of Department of Fisheries Vancouver, Hatcheries in British Columbia, part 1, *Western Fisheries,* April 1971, pp. 12–41; also see Resource Development Branch of Department of Fisheries Vancouver, Hatcheries in British Columbia, part 2, *Western Fisheries,* May 1971, pp. 72–78.
18. P. Larkin, Play It Again Sam—An Essay on Salmon Enhancement (with appendices by B. Campbell and C. Clay), *Journal of Fisheries Research Board of Canada* 31: 1433–1439 (1974).
19. J. Winton, Supplementation of Wild Salmonids: Management Practices in British Columbia, M.S. thesis, University of Washington, Seattle, 1991.

20. T. Glavin, *Dead Reckoning: Confronting the Crisis in Pacific Fisheries* (Vancouver: Douglas and McIntyre, 1996).
21. Larkin, Play It Again Sam.
22. T. Cummings, Private Salmon Hatcheries in Oregon 1986, Processed Report, Oregon Department of Fish and Wildlife, Portland, 1987.
23. U.S. Army Corps of Engineers, Lower Snake River Fish and Wildlife Compensation Plan, Special Report, Walla Walla [WA] District, 1975.
24. Washington State Department of Fisheries, *Fisheries*, volume III, Olympia, 1960, p. 201.
25. National Research Council, *Upstream.*
26. J. Lichatowich, H. Wagner, and T. Nickelson, Summary of Salmon Task Force Meetings, Oregon Department of Fish and Wildlife, Portland, 1978.
27. Glavin, *Dead Reckoning.*
28. P. Pearse, An Assessment of the Salmon Stock Development Program of Canada's Pacific Coast, Final Report to Department of Fisheries and Oceans, Vancouver, BC, 1994.
29. A. Rose, High-tech Hatcheries a Failure, *Vancouver (BC) Globe and Mail,* August 1, 1992.
30. R. Hilborn and J. Winton, Learning to Enhance Salmon Production: Lessons from the Salmonid Enhancement Program, *Canadian Journal of Fisheries and Aquatic Sciences* 50: 2043–2056 (1993).
31. National Research Council, *Upstream;* also see Independent Scientific Group, Return to the River: Restoration of Salmonid Fishes in the Columbia River Ecosystem, Northwest Power Planning Council, Portland, OR, 1996; and National Fish Hatchery Review Panel, Report of the National Fish Hatchery Review Panel, The Conservation Fund, Arlington, VA, 1994.
32. Cummings, Private Salmon Hatcheries in Oregon.
33. C. Finley, Ocean Ranching Hits a Snag When Profits Turn Belly-up, *Oregonian* (Portland), August 26, 1990.
34. D. Tims, OreAqua Turning Salmon into Fertilizer, *Oregonian* (Portland), October 5, 1990.
35. J. Griffith, Port Takes a Giant Gamble with Tiny Coho, *Oregonian* (Portland), March 25, 1995.
36. W. J. Ebel, Panel 2: Fish Passage Problems and Solutions to Major Passage Problems, in *Columbia River Salmon and Steelhead,* Proceedings of a symposium, March 5–6, 1976, edited by E. Schwiebert, American Fisheries Society, Special Publication 10, Washington, DC, 1977, pp. 33–39.
37. U.S. Fish and Wildlife Service, Proceedings of the Lower Snake River Compensation Plan Status Review Symposium, Boise, ID, 1998.
38. Washington State Department of Fisheries, Salmon 2000 Technical Report, Phase 2: Puget Sound, Washington Coast and Integrated Planning, Olympia, 1992.
39. C. Ebbesmeyer and five others, 1976 Step in the Pacific Climate: Forty Environmental Changes between 1968–1975 and 1977–1984, in Proceedings of the Seventh Annual Pacific Climate (PACLIM) Workshop, April 1990, edited

by J. Betancourt and V. Tharp, Technical Report 26, California Department of Water Resources Interagency Ecological Studies Program, Sacramento, 1991.

40. T. Nickelson, Influences of Upwelling, Ocean Temperature and Smolt Abundance on Marine Survival of Coho Salmon (*Oncorhynchus kisutch*) in the Oregon Production Index, *Canadian Journal of Fisheries and Aquatic Sciences* 43(3): 527–535 (1986).
41. D. Ware and R. Thompson, Link Between Long-term Variability in Upwelling and Fish Production in the Northeast Pacific Ocean, *Canadian Journal of Fisheries and Aquatic Sciences* 48(12): 2296–2306 (1991).
42. R. Francis, Climate Change and Salmonid Production in the North Pacific Ocean, in Proceedings of the Ninth Annual Pacific Climate (PACLIM) Workshop, edited by K. Redmond and V. Tharp, Technical Report 34, California Department of Water Resources Interagency Ecological Studies Program, Sacramento, 1993, pp. 33–43.
43. Nickelson, Influences of Upwelling; also see R. Francis and S. Hare, Decadal-scale Regime Shifts in the Large Marine Ecosystems of the Northeast Pacific: A Case for Historical Science, *Fisheries Oceanography* 3(4): 270–291 (1994); and R. Beamish and D. Bouillion, Pacific Salmon Production Trends in Relation to Climate, *Canadian Journal of Fisheries and Aquatic Sciences* 50(5): 1002–1016 (1993).
44. Independent Scientific Group, Return to the River.
45. As a recently hired biologist at the Oregon Department of Fish and Wildlife in the mid-1970s, I attended an employee orientation workshop that included a session titled "Compromise Is the Name of the Game." In it, a department administrator set out to indoctrinate the young biologists with the attitude that development and habitat destruction were inevitable, and that the department's (our) job was to get the best hatchery deal we could. We were told that everything, including habitat and wild salmon, was negotiable. That meeting was extremely troubling to me, and to some degree, the seeds for this book were sown there.
46. N. Carter, Multi-fishery Activity in Oregon Commercial Fishing Fleets: An Economic Analysis of Short-run Decision-making Behavior, Ph.D. dissertation, Oregon State University, Corvallis, 1981.
47. B. Monroe, State Fisheries Chief Martin Might Take Scientific Approach to Politics, *Oregonian* (Portland), April 9, 1995.
48. Oregon Department of Fish and Wildlife, Comprehensive Plan for Production and Management of Oregon's Anadromous Salmon and Trout, part II, Coho Salmon Plan, Oregon Department of Fish and Wildlife, Portland, 1982.
49. T. E. Nickelson, M. F. Solazzi, and S. L. Johnson, Use of Hatchery Coho Salmon (*Oncorhynchus kisutch*) Presmolts to Rebuild Wild Populations in Oregon Coastal Streams, *Canadian Journal of Fisheries and Aquatic Sciences* 43: 2443–2449 (1986).

50. S. Wright, Fishery Management of Wild Pacific Salmon Stocks to Prevent Extinctions, *Fisheries* 18(5): 3–4 (1993).
51. R. Hilborn, Hatcheries and the Future of Salmon in the Northwest, *Fisheries* 17(1): 4–8 (1992).
52. C. Walters, Fish on the Line: The Future of Pacific Fisheries, A Report to the David Suzuki Foundation, Vancouver, BC, 1995.
53. B. Riddell, Spatial Organization of Pacific Salmon: What to Conserve? In *Genetic Conservation of Salmonid Fishes,* edited by J. Cloud and G. Thorgaard (New York: Plenum Press, 1993).
54. W. F. Thompson, Fishing Treaties and the Salmon of the North Pacific, *Science* 150: 1786–1789 (1965).
55. B. Espenson, Tribes Rip Council's Hatchery Report, *The Columbia Basin Bulletin,* January 11–15, 1999, www.nwppc.org/bulletin.htm.
56. M. Marsh, opinion, civil no. 92–973 MA (lead) in the United States District Court for the District of Oregon, 1994, pp. 36–37.
57. R. Bunting, *The Pacific Raincoast: Environment and Culture in an American Eden, 1778–1900* (Lawrence: University Press of Kansas, 1997).
58. G. Meffe, Techno-arrogance and Halfway Technologies: Salmon Hatcheries on the Pacific Coast of North America, *Conservation Biology* 6(3): 350–354 (1992).

Epilogue

1. T. Berry, *The Dream of the Earth* (San Francisco: Sierra Club Books, 1988).
2. A. Durning, *This Place on Earth* (Seattle: Sasquatch Books, 1996), p. 247. For a history of stories used in salmon management, see J. A. Lichatowich, L. E. Mobrand, R. J. Costello, and T. S. Vogel, A History of Frameworks Used in the Management of Columbia River Chinook Salmon, Report Submitted to Bonneville Power Administration, Portland, OR, 1996.
3. J. Crutchfield and G. Pontecorvo, *The Pacific Salmon Fisheries: A Study of Irrational Management* (Baltimore: Resources for the Future/Johns Hopkins Press, 1969).
4. National Research Council, *Upstream: Salmon and Society in the Pacific Northwest* (Washington, DC: National Academy Press, 1996); and also Independent Scientific Group, Return to the River: Restoration of Salmonid Fishes in the Columbia River Ecosystem, ISG Draft, September 1996, Portland, OR.
5. Emerging Issue: Long Way from Salmon Consensus, *The Elway Poll* 1(5): 3 (1992).
6. J. Brinckman, Salmon Tops Environmental Worries, *Oregonian* (Portland), December 7, 1997, p. A1.
7. Dollars, Sense and Salmon: An Argument for Breaching Four Dams on the Lower Snake River, *Idaho Statesman* (Boise), September 22, 1997.
8. C. Meine, *Aldo Leopold: His Life and Work* (Madison: University of Wisconsin Press, 1987).

9. Meine, *Aldo Leopold: His Life and Work.*
10. A. F. McEvoy, *The Fisherman's Problem: Ecology and Law in the California Fisheries, 1850–1980* (New York: Cambridge University Press, 1986).
11. N. Langston, *Forest Dreams, Forest Nightmares* (Seattle: University of Washington Press, 1995).
12. F. Allendorf, The Conservation Biologist as Zen Student, *Conservation Biology* 11(5): 1045–1046 (1997).

Bibliography

Allard, D. C., Jr. 1978. *Spencer Fullerton Baird and the U.S. Fish Commission.* Arno Press, New York.

Allen, D. 1962. *Our Wildlife Legacy.* Funk and Wagnalls, New York.

Allen, J. E., M. Burns, and S. C. Sargent. 1991. *Cataclysms on the Columbia: A Layman's Guide to the Features Produced by the Catastrophic Bretz Floods in the Pacific Northwest.* Timber Press, Portland, OR.

Allendorf, F. 1997. The Conservation Biologist as Zen Student. *Conservation Biology* 11(5): 1045–1046.

Alt, D., and D. W. Hyndman. 1995. *Northwest Exposures: A Geologic Story of the Northwest.* Mountain Press Publishing Company, Missoula, MT.

American Friends Service Committee. 1970. *Uncommon Controversy.* University of Washington Press, Seattle.

Andrew, F., and G. Green. 1960. Sockeye and Pink Salmon Production in Relation to Proposed Dams in the Fraser River System. Bulletin No. 11. International Pacific Salmon Fisheries Commission, New Westminster, BC.

Baird, S. 1875. The Salmon Fisheries of Oregon. *Oregonian* (Portland), March 3.

Bakkala, R. 1970. Synopsis of Biological Data on the Chum Salmon, *Oncorhynchus keta* (Walbaum) 1792. FAO Fisheries Symposium No. 41, U.S. Fish and Wildlife Service Circular 315, Washington, DC.

Barnaby, J. 1945. Memorandum to Director, Fish and Wildlife Service. Development of the Salmon Fisheries of the Lower Columbia River. National Archives, Pacific Northwest Region, Seattle, Record Group 22, Box 12.

Baumhoff, M. A. 1963. *Ecological Determinants of Aboriginal California Populations.* University of California Press, Berkeley.

Beamish, R., and D. Bouillion. 1993. Pacific Salmon Production Trends in Relation to Climate. *Canadian Journal of Fisheries and Aquatic Sciences* 50(5): 1002–1016.

Beard, E. Biography. Oregon Historical Society, Manuscript 1509, Portland.

Beck, M. G. 1991. *Shamans and Kushtakas: North Coast Tales of the Supernatural.* Alaska Northwest Books, Anchorage.

Beckman, A. 1970. *Swift Flows the River.* Arago Books, Coos Bay, OR.

Behnke, R. J. 1992. Native Trout of Western North America. American Fisheries Society Monograph 6, Bethesda, MD.

Beiningen, K. T. 1976. Fish Runs. In Investigative Reports of Columbia River Fisheries Project. Pacific Northwest Regional Commission, Vancouver, WA, section E.

Bell, F. 1937. Guarding the Columbia's Silver Horde. *Nature Magazine* 29(1): 43–47.

Benda, L., T. J. Beechie, R. C. Wissmar, and A. Johnson. 1992. Morphology and Evolution of Salmonid Habitats in a Recently Deglaciated River Basin, Washington State, USA. *Canadian Journal of Fisheries and Aquatic Sciences* 49: 1246–1256.

Benke, A. C. 1990. A Perspective on America's Vanishing Streams. *Journal of North American Benthological Society* 9(1): 77–88.

Benner, P. 1991. Historical Reconstruction of Coquille River and Surrounding Landscape. In *Near Coastal Waters National Pilot Project: The Coquille River, Oregon.* Action Plan for Oregon Coastal Watersheds, Estuary and Ocean Waters, 1988–1991. Grant X-000382-1. Prepared by the Oregon Department of Environmental Quality for the U.S. Environmental Protection Agency.

Berringer, P. A. 1982. Northwest Coast Traditional Salmon Fisheries Systems of Resource Utilization. M.A. thesis, University of British Columbia, Vancouver.

Berry, T. 1988. *The Dream of the Earth.* Sierra Club Books, San Francisco.

Bilby, R. E., B. R. Fransen, and P. A. Bisson. 1996. Incorporation of Nitrogen and Carbon from Spawning Coho Salmon into the Trophic System of Small Streams: Evidence from Stable Isotopes. *Canadian Journal of Fisheries and Aquatic Sciences* 53(1): 164–173.

Bingham, Senator I. H. 1909. Letter to H. G. Van Dusen, April 26. Oregon Historical Society Archives, Portland, Warren Packing Company, Record 1141, Box 1.

Bocking, R. 1997. *Mighty River.* Douglas and McIntyre, Vancouver, BC.

Borgerson, L. 1992. Memo to Jim Lichatowich, dated December 1992.

Bottom, D. L. 1997. To Till the Water: A History of Ideas in Fisheries Conservation. In *Pacific Salmon and Their Ecosystems: Status and Future Options.* Edited by D. J. Stouder, P. A. Bisson, and R. J. Naiman, pp. 569–597. Chapman and Hall, New York.

Bower, W. T. 1911. History of the American Fisheries Society. *Transactions of the American Fisheries Society,* Fortieth Annual Meeting, September 27–29, 1910, Washington, DC.

Boxberger, D. L. 1986. *To Fish in Common: The Ethnohistory of Lummi Indian Salmon Fishing.* University of Nebraska Press, Lincoln.

Bradbury, B., J. Brenneman, and C. Hosticka. 1986. Report of the Interim Task Force on Forest Practices. State of Oregon, Salem.

Brennan, B. M. 1940. Address by B. M. Brennan, Director of Fisheries, State of Washington, to the Legislative Meeting of the Columbia River Fisheries Protective Union at Astoria, Oregon, October 1. National Archives, Washington, DC, Record Group 22.

Brice, J. J. 1895. Establishment of Stations for the Propagation of Salmon on the Pacific Coast. Miscellaneous Documents, U.S. House of Representatives, 53rd Congress, Washington, DC.

Brinckman, J. 1997. Three Billion Later: Columbia Basin Salmon Dwindle. *Oregonian* (Portland), July 27, section A.

Brinckman, J. 1997. Salmon Tops Environmental Worries. *Oregonian* (Portland), December 7, p. A1.

Brinckman, J. 1998. U.S. Salmon Plan Would Be Ruinous, Forestry Officials, Companies Say. *Oregonian* (Portland), March 5, p. D1.

Brown, B. 1982. *Mountain in the Clouds: A Search for the Wild Salmon.* Simon and Schuster, New York.

Brown, C. 1922. Adjustment of Environment vs. Stocking to Increase the Productivity of Fish Life. *Transactions of the American Fisheries Society,* Fifty-first Annual Meeting, September 5–7, 1921, Allentown, PA. Volume 51, 1921–1922. Washington, DC.

Buckley, G. 1992. Desertification of Camp Creek Drainage in Central Oregon. 1826–1905. M.S. thesis, University of Oregon, Eugene.

Bunting, R. 1994. Abundance and the Forests of the Douglas-fir Bioregion, 1840 to 1920. *Environmental History Review* 18(4): 41–62.

Bunting R. 1997. *The Pacific Raincoast: Environment and Culture in an American Eden, 1778–1900.* University Press of Kansas, Lawrence.

Burgner, R. 1991. Life History of the Sockeye Salmon (*Oncorhynchus nerka*). In *Pacific Salmon Life Histories.* Edited by C. Groot and R. Margolis, pp. 3–117. University of British Columbia Press, Vancouver.

Caldwell, F. 1978. *Pacific Troller.* Alaska Northwest Publishing Company, Anchorage.

California Division of Fish and Game. 1928. Thirtieth Biennial Report. 1926–1928. Sacramento.

California Division of Fish and Game. 1930. Thirty-first Biennial Report. Sacramento.

California Fish and Game Commission. 1921. Twenty-sixth Biennial Report. Sacramento.

California Fish and Game Commission. 1923. Twenty-seventh Biennial Report. Sacramento.

Camp, C. L. 1970. *Earth Song: A Prologue to History.* American West Publishing Company, Palo Alto, CA.

Carey, C. H. 1971. *A General History of Oregon.* Binford and Mort, Portland, OR.

Carl, L. M., and M. C. Healey. 1984. Differences in Enzyme Frequency and Body

Morphology among Three Juvenile Life History Types of Chinook Salmon (*Oncorhynchus tshawytscha*) in the Nanaimo River, British Columbia. *Canadian Journal of Fisheries and Aquatic Sciences* 41: 1070–1077.

Carlson, R. L. 1983. The Far West. In *Early Man in the New World.* Edited by R. Shutler, Jr. Sage Publications, Beverly Hills, CA.

Carter, N. 1981. Multi-fishery Activity in Oregon Commercial Fishing Fleets: An Economic Analysis of Short-run Decision-making Behavior. Ph.D. dissertation. Oregon State University, Corvallis.

Cavender, T. M. 1986. Review of the Fossil History of North American Freshwater Fishes. In *The Zoogeography of North American Freshwater Fishes.* Edited by C. H. Hocutt and E. O. Wiley, pp. 699–725. John Wiley and Sons, New York.

Cavender, T. M., and R. R. Miller. 1972. *Smilodonichthys rastrosus:* A New Pliocene Salmonid Fish from the Western United States. Bulletin No. 18. Museum of Natural History, University of Oregon, Eugene.

Cederholm, C. J., D. B. Houston, D. L. Cole, and W. J. Scarlett. 1989. Fate of Coho Salmon (*Oncorhynchus kisutch*) Carcasses in Spawning Streams. *Canadian Journal of Fisheries and Aquatic Sciences* 46: 1347–1355.

Chamberlin, F. M. 1903. *Pacific Fisherman.* 1(1): 10.

City of Tacoma, Washington. Undated. Sixteen-page document on the letterhead of the Department of Public Utilities. National Archives, Washington, DC, Record Group 48, Box 865.

Clark, E. 1992. *The Oysters of Locmariaquer.* HarperPerennial, New York.

Clark, G. H. 1929. Sacramento–San Joaquin Salmon (*Oncorhynchus tshawytscha*) Fishery of California. Fish Bulletin No. 17. California Division of Fish and Game, Sacramento.

Clemens, W. A., and G. V. Wilby. 1961. Fishes of the Pacific Coast of Canada. Bulletin no. 68, Fisheries Research Board of Canada, Ottawa.

Cobb, J. N. 1930. Pacific Salmon Fisheries. Appendix 13 to the Report of the U.S. Commissioner of Fisheries for 1930. Bureau of Fisheries Document No. 1092. Washington, DC.

Cogwill, R. 1930. Notes on salmon losses in irrigation ditches. Cogwill records, Oregon Historical Society, Portland.

Colbert, E. H. 1966. *Evolution of the Vertebrates: A History of the Backboned Animals through Time.* Science Edition. John Wiley and Sons, New York.

Cole, D., and I. Caikin. 1990. *An Iron Hand upon the People: The Law Against the Potlatch on the Northwest Coast.* University of Washington Press, Seattle.

Columbia Basin Fish and Wildlife Authority. 1990. Review of the History, Development, and Management of Anadromous Fish Production Facilities in the Columbia River Basin. Columbia Basin Fish and Wildlife Authority, Portland, OR.

Columbia Basin Fisheries Development Association. 1945. Wealth of the River: A Presentation of Fact Concerning the Columbia River Salmon Industry and a Petition for the Conservation of This Industry Submitted with Reference to Proposals to Construct High Dams on the Columbia River and Its Tributaries. Astoria, OR.

Cone, J., and S. Ridlington. 1996. *The Northwest Salmon Crisis: A Documentary History.* Oregon State University Press, Corvallis.

Cooley, R. 1963. Politics and Conservation: The Decline of the Alaska Salmon. Harper and Row, New York.

Cox, T. 1974. *Mills and Markets.* University of Washington Press, Seattle.

Craig, J. 1935. The Effects of Power and Irrigation Projects on the Migratory Fish of the Columbia River. *Northwest Science* 40(1): 19–22.

Craig, J. A., and R. L. Hacker. 1940. The History and Development of the Fisheries of the Columbia River. *Bulletin of the Bureau of Fisheries* 49(32):133–216 (1940).

Cranston, C. 1913. The Fish and Game Laws of Oregon. *Transactions of the American Fisheries Society,* Forty-second Annual Meeting, September 3–5, 1912, Denver, pp. 75–87.

Crawford, J. 1890. First Annual Report of the Washington State Fish Commissioner. Olympia.

Croes, D. R., and S. Hackenberger. 1988. Hoko River Archaeological Complex: Modeling Prehistoric Northwest Coast Economic Evolution. In *Research in Economic Anthropology.* Edited by B. L. Isaac, pp. 19–85. JAI Press, Greenwich, CT.

Crutchfield, J., and G. Pontecorvo. 1969. *The Pacific Salmon Fisheries: A Study of Irrational Management.* Resources for the Future/Johns Hopkins Press, Baltimore.

Culler, C. F. 1932. Progress in Fish Culture. *Transactions of the American Fisheries Society,* Sixty-second Annual Meeting, September 21–23.

Cummings, T. 1987. Private Salmon Hatcheries in Oregon, 1986. Processed Report. Oregon Department of Fish and Wildlife, Portland.

Darwin, C. 1969. *The Voyage of the Beagle.* P. F. Collier and Son, New York.

Darwin, L. H. 1917. Twenty-sixth and Twenty-seventh Annual Reports of the State Fish Commissioner to the Governor of the State of Washington. Washington Department of Fisheries and Game, Olympia.

Darwin, L. H. 1921. Thirtieth and Thirty-first Annual Reports of the State Fish Commissioner. Seattle.

Davidson, C. G. 1949. Fisheries and Other Resources in the Columbia Basin. Statements made by C. Girard Davidson, under questioning by the Public Works Committee of the U.S. Senate. National Archives, Pacific Northwest Region, Seattle, Record Group 22.

Davidson, F. A. 1937. "Migration" and "Homing" of Pacific Salmon. *Science* 86(2220): 55–56.

Davidson, F. 1965. The Development of Irrigation in the Yakima River Basin and Its Effect on the Migratory Fish Population in the River. Manuscript report. Ephrata, WA.

Dawson, M. 1997. Buck: Save Fish from ESA. *Peninsula Daily News* (Port Angeles, WA), August 27.

Delarm, M. R., E. Wold, and R. Z. Smith. 1987. Columbia River Fisheries Development Program Annual Report for FY 1986. NOAA Technical Memorandum, NMFS F/NWR-21, Seattle.

DeLoach, D. B. 1939. The Salmon Canning Industry. Oregon State College, Economic Studies No. 1, Corvallis.

Dietrich, B. 1990. A Clearcut Case. *Seattle Times,* May 10.

Dodds, G. 1959. *The Salmon King of Oregon: R. D. Hume and the Pacific Fisheries.* University of North Carolina Press, Chapel Hill.

Drucker, P. 1963. *Indians of the Northwest Coast.* The Natural History Press, Garden City, NY.

Dunfield, R. 1985. Atlantic Salmon in the History of North America. Canadian Special Publication of Fisheries and Aquatic Sciences 80. Ottawa.

Dunn, J. R. 1996. Charles Henry Gilbert (1859–1928): An Early Fishery Biologist and His Contribution to Knowledge of Pacific Salmon (*Oncorhynchus* spp.). *Reviews in Fisheries Science* 4(2): 133–184.

Durning, A. 1996. *This Place on Earth.* Sasquatch Books, Seattle.

Dyson, F. 1997. *Imagined Worlds.* Harvard University Press, Cambridge.

Ebbesmeyer, C., D. R. Cayan, D. R. McClain, F. H. Nichols, D. H. Peterson, and K. T. Redmond. 1991. 1976 Step in the Pacific Climate: Forty Environmental Changes between 1968–1975 and 1977–1984. In Proceedings of the Seventh Annual Pacific Climate (PACLIM) Workshop, April 1990. Edited by J. Betancourt and V. Tharp. Technical Report 26. California Department of Water Resources Interagency Ecological Studies Program, Sacramento.

Ebel, W. J. 1977. Panel 2: Fish Passage Problems and Solutions to Major Passage Problems. In *Columbia River Salmon and Steelhead.* Proceedings of a symposium, March 5–6, 1976. Edited by E. Schwiebert, pp. 33–39. American Fisheries Society, Special Publication 10, Washington, DC.

Egan, T. 1990. *The Good Rain: Across Time and Terrain in the Pacific Northwest.* Alfred A. Knopf, New York.

The Elway Poll. 1992. Emerging Issue: Long Way from Salmon Consensus. 1(5): 3.

Espenson, B. 1999. Tribes Rip Council's Hatchery Report. *The Columbia Basin Bulletin,* January 11–15, www.nwppc.org/bulletin.htm.

Fagan, B. M. 1987. *The Great Journey: The Peopling of Ancient America.* Thames and Hudson, New York.

Ferguson, D., and N. Ferguson. 1983. *Sacred Cows at the Public Trough.* Maverick Publications, Bend, OR.

Finley, C. 1990. Ocean Ranching Hits a Snag When Profits Turn Belly-up. *Oregonian* (Portland), August 26.

Finley, C. 1993. Dwindling Coho Salmon Runs Leave Oregon in Quandary. *Oregonian* (Portland), November 16.

Finley, W. 1935. Letter to R. E. Dimick, June 6. Oregon State University Archives, Corvallis, Dimick Papers, Correspondence with W. Finley.

Fish, H. 1983. *Tracks, Trails, and Tales in Clallam County, State of Washington.* Olympic Printers, Port Angeles, WA.

Fisher, R. 1992. *Contact and Conflict: Indian-European Relations in British Columbia, 1774–1890.* University of British Columbia Press, Vancouver.

Fite, G. 1966. *The Farmer's Frontier, 1865–1900.* Holt, Rinehart and Winston, New York.

Fladmark, K. R. 1986. *British Columbia Prehistory.* The Runge Press, Ottawa.

Fladmark, K. R. 1983. Times and Places: Environmental Correlates of Mid-to-late Wisconsinan Human Population Expansion in North America. In *Early Man in the New World.* Edited by R. Shutler, Jr., pp. 13–41. Sage Publications, Beverly Hills, CA.

Flagg, T. A., F. W. Waknitz, D. J. Maynard, G. B. Milner, and C. V. W. Mahkhen. 1995. The Effect of Hatcheries on Native Coho Salmon Populations in the Lower Columbia River. In *Uses and Effects of Cultured Fishes in Aquatic Ecosystems.* Edited by H. L. Schramm, Jr., and R. G. Piper, pp. 366–375. American Fisheries Society Symposium 15, Bethesda, MD.

Flannery, T. F. 1994. *The Future Eaters: An Ecological History of the Australasian Lands and Peoples.* Reed Books, New South Wales, Australia.

Foerster, R. E. 1936. Sockeye Salmon Propagation in British Columbia. Bulletin No. 53. Biological Board of Canada, Ottawa.

Foerster, R. E. 1968. The Sockeye Salmon. Bulletin No. 162. Fisheries Research Board of Canada, Ottawa.

Franchere, G. 1968. *A Voyage to the Northwest Coast of North America.* Citadel Press, New York.

Francis, R. 1993. Climate Change and Salmonid Production in the North Pacific Ocean. In Proceedings of the Ninth Annual Pacific Climate (PACLIM) Workshop. Edited by K. Redmond and V. Tharp, pp. 33–43. Technical Report 34. California Department of Water Resources Interagency Ecological Study Program, Sacramento.

Francis, R., and S. Hare. 1994. Decadal-scale Regime Shifts in the Large Marine Ecosystems of the Northeast Pacific: A Case for Historical Science. *Fisheries Oceanography* 3(4): 270–291.

Freeman, W. M. 1956. The Memoirs of Miller Freeman 1875–1955. General stacks, Fisheries Library, University of Washington, Seattle.

Frissell, C. A. 1989. Evolution of the Salmonid Fishes: Zoogeography and the Fossil Record. Department of Fisheries and Wildlife, Oregon State University, Corvallis.

Fry, W. H. 1854. *Complete Treatise on Artificial Fish-Breeding.* D. Appleton and Company, New York.

Fulton, L. A. 1968. Spawning Areas and Abundance of Chinook Salmon (*Oncorhynchus tshawytscha*) in the Columbia River Basin—Past and Present. Special Scientific Report: Fisheries No. 571. U.S. Fish and Wildlife Service, Bureau of Commercial Fisheries, Washington, DC.

Gardner, W. 1947. Memorandum to Secretary of Interior. Columbia River Dams and Salmon. March 6. National Archives, Washington, DC, Record Group 48.

Geertz, C. 1973. *The Interpretation of Culture.* Basic Books, New York.

Gharrett, J. 1953. Summary Coastal River Regulations—1878–1950. Unpublished manuscript. Oregon Fish Commission, Portland.

Gibson, J. R. 1985. *Farming the Frontier: The Agricultural Opening of the Oregon Country, 1786 to 1846.* University of Washington Press, Seattle.

Gilbert, C. H. 1912. Age at Maturity of the Pacific Coast Salmon of the Genus *Oncorhynchus. Bulletin of the Bureau of Fisheries,* Washington, DC.

Glavin, T. 1996. *Dead Reckoning: Confronting the Crisis in Pacific Fisheries.* Douglas and McIntyre, Toronto.

Goode, G. B. 1886. The Status of the U.S. Fish Commission in 1884. Part XII: Report of the Commissioner for 1884. Government Printing Office, Washington, DC.

Goode, G. B. 1898. Pisciculture. In *The Encyclopaedia Britannica.* Volume 19. Werner Company, Chicago.

Gordon, S. 1933. Scientific Management: Our Future Fisheries Jobs. *Transactions of the American Fisheries Society,* Sixty-third Annual Meeting, Columbus, OH.

Gottschalk, J. 1942. Reports of Vice-Presidents of Divisions. *Transactions of the American Fisheries Society.* Seventy-first Annual Meeting, August 25–26, 1941, St. Louis, MO.

Grainger, M. A. 1964. *Woodsmen of the West.* McClelland and Stewart, Toronto.

Gregory, H. E., and K. Barnes. 1939. North Pacific Fisheries with Special Reference to Pacific Salmon. Studies of the Pacific No. 3. American Council Institute of Pacific Relations, New York.

Gresh, T., J. Lichatowich, and P. Shoonmaker. 1998. An Estimation of Historical and Current Levels of Salmon Production in the Northeast Pacific Ecosystem: Evidence of a Nutrient Deficit. Draft Report to Interrain Pacific. Portland, OR.

Griffin, L. 1935. Certainties and Risks Affecting Fisheries Connected with Damming the Columbia River. *Northwest Science* 40(2): 25–30.

Griffith, J. 1995. Port Takes a Giant Gamble with Tiny Coho. *Oregonian* (Portland), March 25.

Gross, M. R., R. M. Coleman, and R. M. McDowall. 1988. Aquatic Productivity and the Evolution of Diadromous Fish Migration. *Science* 239: 1291–1293.

Gunnarsdottir, H. 1992. Scale Patterns Indicate Changes in Use of Rearing Habitat by Juvenile Coho Salmon, *Oncorhynchus kisutch,* from 1955 to 1984 in the Ten Mile Lakes, Oregon. M.S. thesis. Oregon State University, Corvallis.

Gunther, E. 1926. An Analysis of the First Salmon Ceremony. *American Anthropologist* 28: 605–617.

Gunther, E. 1928. A Further Analysis of the First Salmon Ceremony. *Anthropology* 2(5): 129–173.

Hardin, G. 1968. The Tragedy of the Commons. *Science* 1912: 1243–1248.

Harper's New Monthly Magazine. 1868. Fish-culture in America. 37(222): 721–739.

Harrison, B. 1892. Afognak Forest and Fish Culture Reserve: A Proclamation, Washington, DC. National Archives, Washington, DC, Record Group 22, U.S. Fish and Wildlife Service.

Hayden, M. V. 1930. History of the Salmon Industry in the Pacific Northwest. M.S. thesis. University of Oregon, Eugene.

Hays, S. P. 1957. *The Response to Industrialism.* University of Chicago Press, Chicago.

Hays, S. P. 1969. *Conservation and the Gospel of Efficiency: The Progressive Conservation Movement, 1890–1920.* Atheneum Press, New York.

Hays, S. P. 1987. *Beauty, Health, and Permanence: Environmental Politics in the United States, 1955–1985.* Cambridge University Press, New York.

Healey, M. C. 1994. Variation in the Life History Characteristics of Chinook Salmon and Its Relevance to Conservation of the Sacramento Winter Run of Chinook Salmon. *Conservation Biology* 8(3): 876–877.

Healey, M. C., and A. Prince, 1995. Scales of Variation in Life History Tactics of Pacific Salmon and the Conservation of Phenotype and Genotype. In *Evolution and the Aquatic Ecosystem: Defining Unique Units in Population Conservation.* Edited by J. L. Nielsen and D. A. Powers, pp. 176–184. American Fisheries Society Symposium 17, Bethesda, MD.

Hedgpeth, J. W. 1941. Livingston Stone and Fish Culture in California. *California Fish and Game* 27(3): 126–148.

Higgins, E., ed. 1928. Progress in Biological Inquiries 1926. Appendix 7 to the Report of the U.S. Commissioner of Fisheries for 1927. Bureau of Fisheries Document No. 1029. Washington, DC.

Hilborn, R. 1992. Institutional Learning and Spawning Channels for Sockeye Salmon. *Canadian Journal of Fisheries and Aquatic Sciences* 49: 1126–1136.

Hilborn, R. 1992. Hatcheries and the Future of Salmon in the Northwest. *Fisheries* 17(1): 4–8.

Hilborn, R., and Winton, J. 1993. Learning to Enhance Salmon Production: Lessons form the Salmonid Enhancement Program. *Canadian Journal of Fisheries and Aquatic Sciences* 50(9): 2043–2056.

Hile, R. 1977. The Increase in the Abundance of the Yellow Pike-Perch, *Stizostedion vitreum* (Mitchill), in Lakes Huron and Michigan, in Relation to the Artificial Propagation of the Species. In *The Collected Papers of Ralph Hile, 1928–73.* U.S. Fish and Wildlife Service. Volume I, 1928–46. Washington, DC.

Hirt, P. 1994. *A Conspiracy of Optimism.* University of Nebraska Press, Lincoln.

Holbrook, S. 1956. *The Columbia.* Rinehart, New York.

Hornaday, W. 1935. *Hornaday's American Natural History.* Charles Scribner's Sons, New York.

Hume, R. D. 1893. *Salmon of the Pacific Coast.* Schmidt Label & Lithographic Company, San Francisco.

Hume, R. D. 1903. History of the Salmon Industry of the Pacific. *Pacific Fisherman* 1: 9.

Huntington, C., W. Nehlsen, and J. Bowers. 1994. Healthy Native Stocks of Native Salmonids in the Pacific Northwest and California. Oregon Trout, Portland.

Huntsman, A. G. 1937. "Races" and "Homing" of Salmon. *Science* 85(2216): 582–583.

Hutchings, J., C. Walters, and R. Haedrich. 1997. Is Scientific Inquiry Incompatible with Government Information Control? *Canadian Journal of Fisheries and Aquatic Sciences* 54 (5): 1198–1210.

Hutchison, B. 1950. *The Fraser.* Clarke, Irwin and Company, Toronto.

Hyde, L. 1983. *The Gift: Imagination and the Erotic Life of Property.* Vintage Books, New York.

Idaho Statesman (Boise). 1997. Dollars, Sense, and Salmon: An Argument for Breaching Four Dams on the Lower Snake River. September 22.

Independent Scientific Group. 1996. Return to the River: Restoration of Salmonid Fishes in the Columbia River Ecosystem. Northwest Power Planning Council, Portland, OR.

International Fisheries Exhibition. 1883. *An Illustrated Description of the Electric Light Machinery in the Exhibition, with Elementary Notes on the Production of Electric Currents.* William Clowes and Sons, London.

International Fisheries Exhibition. 1884. *Fisheries Exhibition Literature.* Volume VI, *Conferences,* part III. William Clowes and Sons, London.

Jeffries, E. 1975. Present: Role and Challenge of Fish Culture in the Northwest. In Proceedings of the Twenty-sixth Annual Northwest Fish Culture Conference. Otter Rock, Oregon, December 3–5. Edited by J. Brophy, pp. 167–174. Oregon Aqua-Foods, Newport, OR.

Jelinek, A. J. 1967. Man's Role in the Extinction of Pleistocene Faunas. In *Pleistocene Extinctions: The Search for a Cause.* Edited by P. S. Martin and H. E. Wright, Jr., pp. 193–200. Yale University Press, New Haven, CT.

Johansen, D., and C. Gates. 1957. *Empire of the Columbia.* Harper and Brothers, New York.

Johnson, D., W. Chapman, and R. Schoning. 1948. The Effects on Salmon Populations of the Partial Elimination of Fixed Fishing Gear on the Columbia River in 1935. Contribution No. 11. Oregon Fish Commission, Portland.

Johnson, D. R., and D. H. Chance. 1974. Presettlement Overharvest of Upper Columbia River Beaver Populations. *Canadian Journal of Zoology* 52: 1519–1521.

Johnson, P. 1994. Historical Assessment of Elwha River Fisheries. Draft Report. Olympic National Park, Port Angeles, WA.

Johnson, S. 1984. Freshwater Environmental Problems and Coho Production in Oregon. Information Report 84-11. Oregon Department of Fish and Wildlife, Corvallis.

Jones, W. A. 1887. The Salmon Fisheries of the Columbia River. U.S. Corps of Engineers, Portland, OR.

Jordan, D. S. 1904. The Salmon of the Pacific. *Pacific Fisherman* 2: 1.

Jordan, D. S., and C. H. Gilbert. 1887. The Salmon Fishing and Canning Interests of the Pacific Coast. In *The Fisheries and Fishing Industry of the United States,* section 5, *History and Methods of the Fisheries,* volume I, part 13. Edited by G. B. Goode, pp. 729–751. Washington, DC.

Kline, T. C., Jr., J. J. Goering, O. A. Mathisen, P. H. Poe, and P. L. Parker. 1990. Recycling of Elements Transported Upstream by Runs of Pacific Salmon: I. Evidence in Sashin Creek, Southeastern Alaska. *Canadian Journal of Fisheries and Aquatic Sciences* 47: 136–144.

Kohlstedt, S. G. 1991. *The Origins of Natural Science in America: The Essays of George Brown Goode.* Smithsonian Institution Press, Washington, DC.

Kruckeberg, A. R. 1991. *The Natural History of Puget Sound Country.* University of Washington Press, Seattle.

Kuhler, J. 1940. A History of Agriculture in the Yakima Valley, Washington, from 1880 to 1900. M.S. thesis. University of Washington, Seattle.

Langston, N. 1995. *Forest Dreams, Forest Nightmares.* University of Washington Press, Seattle.

Larkin, G. A., and P. A. Slaney. 1997. Implications of Trends in Marine-Derived Nutrient Influx to South Coastal British Columbia Salmonid Production. *Fisheries* 22(11): 16–24.

Larkin, P. 1974. Play It Again Sam—An Essay on Salmon Enhancement (with appendices by B. Campbell and C. Clay). *Journal of Fisheries Research Board of Canada* 31: 1433–1439.

Lavier, D. 1976. Major Dams on Columbia River and Tributaries. Investigative Reports of Columbia River Fisheries Project. Pacific Northwest Regional Commission, Portland, OR.

Laythe, L. L. 1948. The Fishery Development Program in the Lower Columbia River. *Transactions of the American Fisheries Society,* September 13–15, Atlantic City, NJ.

Leethem, J. 1979. The Western Gold Dredge Company of John Day, Oregon. *Oregon Geology* 41(6): 91–95.

Leffler, R. 1959. The Program of the U.S. Fish and Wildlife Service for the Anadromous Fishes of the Columbia River. *Bulletin of the Oregon State Game Commission* 10: 3–7.

Leibhardt, B. 1990. Law, Environment, and Social Change in the Columbia River Basin: The Yakima Indian Nation as a Case Study, 1840–1933. Ph.D. dissertation. University of California, Berkeley.

Lewis, A. 1994. *Salmon of the Pacific.* Raincoast Books, Vancouver, BC.

Lichatowich, J. 1989. Habitat Alteration and Changes in Abundance of Coho (*Oncorhynchus kisutch*) and Chinook (*O. tshawytscha*) Salmon in Oregon's Coastal Streams. In *Proceedings of the National Workshop on Effects of Habitat Alteration on Salmonid Stocks.* Edited by C. D. Levings, L. B. Holtby, and M. A. Henderson, pp. 92–99. Canadian Special Publication of Fisheries and Aquatic Sciences 105, Ottawa.

Lichatowich, J. A. 1996. Evaluating the Performance of Salmon Management Institutions: The Importance of Performance Measures, Temporal Scales and Production Cycles. In *Pacific Salmon and Their Ecosystems.* Edited by D. J. Stouder, P. A. Bisson, and R. J. Naiman, pp. 69–87. Chapman and Hall, New York.

Lichatowich, J. A., and L. E. Mobrand. 1995. Analysis of Chinook Salmon in the Columbia River from an Ecosystem Perspective. Bonneville Power Administration, DOE/BP-25105-2, Portland, OR.

Lichatowich, J. A., L. E. Mobrand, R. J. Costello, and T. S. Vogel. 1996. A History of Frameworks Used in the Management of Columbia River Chinook Salmon. Report Submitted to Bonneville Power Administration, Portland, OR.

Lichatowich, J. A., G. R. Rahr III, S. M. Whidden, and C. R. Steward. In press. Sanctuaries for Pacific Salmon. In *Sustainable Fisheries Management: Balancing the Conservation and Use of Pacific Salmon.* Edited by E. E. Knutson and four others. Ann Arbor Press, Ann Arbor, MI.

Lichatowich, J., H. Wagner, and T. Nickelson. 1978. Summary of Salmon Task Force Meetings. Oregon Department of Fish and Wildlife, Portland.

Lindsey, C. C., and J. D. McPhail. 1986. Zoogeography of Fishes of the Yukon and Mackenzie Basins. In *The Zoogeography of North American Freshwater Fishes.* Edited by C. H. Hocutt and E. O. Wiley, pp. 639–674. John Wiley and Sons, New York.

Little, A. C. 1901. Tenth and Eleventh Annual Reports of the State Fish Commissioner. Olympia, WA.

Love, R. M. 1970. *The Chemical Biology of Fishes: With a Key to the Chemical Literature.* Academic Press, London.

Lufkin, A., ed. 1991. *California's Salmon and Steelhead: The Struggle to Restore an Imperiled Resource.* University of California Press, Berkeley.

Lyman, W. D. 1919. *History of the Yakima Valley, Washington.* S. J. Clarke Publishing Company.

Lyons, C. 1969. *Salmon: Our Heritage.* Mitchell Press, Vancouver, BC.

MacCrimmon, H. R. 1965. The Beginning of Salmon Culture in Canada. *Canadian Geographical Journal* 71(3): 96–103.

Mace, B. 1960. Chronological History of Fish and Wildlife Administration in Oregon. Oregon Department of Fish and Wildlife, Portland.

Marsh, G. P. 1857. *Artificial Propagation of Fish: Report Made under the Legislative Authority of Vermont.* Free Press Printing, Burlington, VT.

Marsh, M. 1994. Opinion, civil no. 92–973 (lead). United States District Court for the District of Oregon, Portland.

Martin, C. 1978. *Keepers of the Game: Indian-Animal Relationships and the Fur Trade.* University of California Press, Berkeley.

Martin, I. 1994. *Legacy and Testament: The Story of the Columbia River Gillnetters.* Washington State University Press, Pullman.

Martin, J., J. Weber, and G. Edwards. 1992. Hatchery and Wild Stocks: Are They Compatible? *Fisheries* 17(1): 4–8.

Martin, P. S., and H. E. Wright, Jr., eds. 1967. *Pleistocene Extinctions: The Search for a Cause.* Yale University Press, New Haven, CT.

Mathewson, W. 1986. *William L. Finley: Pioneer Wildlife Photographer.* Oregon State University Press, Corvallis.

Matson, R. G., and G. Coupland. 1995. *The Prehistory of the Northwest Coast.* Academic Press, New York.

Mauss, M. 1990. *The Gift: The Form and Reason for Exchange in Archaic Societies.* W. W. Norton, New York.

May, E. C. 1937. *The Canning Clan: A Pageant of Pioneering Americans.* Macmillan, New York.

Mayr, E. 1988. *Toward a New Philosophy of Biology: Observations of an Evolutionist.* Harvard University Press, Cambridge.

McAllister, H. 1908. Annual Report of the Oregon Department of Fisheries. Portland.

McDonald, M. 1894. Address of the Chairman of the General Committee on the World's Fisheries Congress. *Bulletin of the U.S. Fish Commission,* volume 13 (1893), Washington, DC.

McDonald, M. 1895. The Salmon Fisheries of the Columbia Basin. *Bulletin of the U.S. Fish Commission* 14: 153–167.

McDowall, R. M. 1987. The Occurrence and Distribution of Diadromy among Fishes. In *Common Strategies of Anadromous and Catadromous Fishes.* Edited by M. J. Dadswell and five others, pp. 1–13. American Fisheries Society Symposium 1, Bethesda, MD.

McEvoy, A. F. 1986. *The Fisherman's Problem: Ecology and Law in the California Fisheries, 1850–1980.* Cambridge University Press, New York.

McGie, A. M. 1980. Analysis of Relationships Between Hatchery Coho Salmon Transplants and Adult Escapements in Oregon Coastal Watersheds. Information Report Series, Fisheries No. 80-6. Oregon Department of Fish and Wildlife, Charleston, OR.

McGoodwin, J. R. 1990. *Crisis in the World's Fisheries: People, Problems, and Policies.* Stanford University Press, Stanford, CA.

McGregor, A. 1982. *Counting Sheep: From Open Range to Agribusiness on the Columbia Plateau.* University of Washington Press, Seattle.

McGuire, H. 1894. First and Second Annual Reports of the Fish and Game Protector. Salem, OR.

McIntosh, R. 1985. *The Background of Ecology: Concept and Theory.* Cambridge Studies in Ecology. Cambridge University Press, New York.

McKee, B. 1972. *Cascadia: The Geologic Evolution of the Pacific Northwest.* McGraw-Hill Book Company, New York.

McKeown, M. F. 1949. *The Trail Led North: Mont Hawthorne's Story.* Macmillan, New York.

McKervill, H. W. 1967. *The Salmon People: The Story of Canada's West Coast Salmon Fishing Industry.* Gray's Publishing, Sidney, BC.

McPhail, J. D., and C. C. Lindsey. 1986. Zoogeography of Freshwater Fishes of Cascadia (the Columbia System and Rivers North to Stikine). In *The Zoogeography of North American Freshwater Fishes.* Edited by C. H. Hocutt and E. O. Wiley, pp. 615–637. John Wiley and Sons, New York.

Meffe, G. 1992. Techno-arrogance and Halfway Technologies: Salmon Hatcheries on the Pacific Coast of North America. *Conservation Biology* 6(3): 350–354.

Meggs, G. 1991. *Salmon: The Decline of the British Columbia Fishery.* Douglas and McIntyre, Vancouver, BC.

Meine, C. 1987. *Aldo Leopold: His Life and Work.* University of Wisconsin Press, Madison.

Miller, R. J., and E. L. Brannon. 1982. The Origin and Development of Life History Patterns in Pacific Salmon. In *Salmon and Trout Migratory Behavior Symposium.* Edited by E. L. Brannon and E. O. Salo, pp. 296–309. School of Fisheries, University of Washington, Seattle.

Miller, R. R., and R. G. Miller. 1948. The Contribution of the Columbia River System to the Fish Fauna of Nevada: Five Species Unrecorded from the State. *Copeia* 3: 174–187.

Milne, D. 1957. Recent British Columbia Spring and Coho Salmon Tagging Experiments and a Comparison with Those Conducted from 1925 to 1930. Bulletin No. 113. Fisheries Research Board of Canada, Ottawa.

Minckley, W. L., D. A. Hendrickson, and C. E. Bond. 1986. Geography of Western North American Freshwater Fishes: Description and Relationships to Intracontinental Tectonism. In *The Zoogeography of North American Freshwater Fishes.* Edited by C. H. Hocutt and E. O. Wiley, pp. 519–613. John Wiley and Sons, New York.

Moffet, J. W., and S. H. Smith. 1950. Biological Investigations of the Fishery Resources of California. Special Scientific Report 12. U.S. Fish and Wildlife Service, Washington, DC.

Monroe, B. 1995. State Fisheries Chief Martin Might Take Scientific Approach to Politics. *Oregonian* (Portland), April 9.

Moore, E. 1925. Report of the Vice President, Division of Fish Culture, New York State Conservation Commission. *Transactions of the American Fisheries Society,* Fifty-fifth Annual Meeting, Denver, pp. 17–23.

Morrison, S. E. 1927. New England and the Opening of the Columbia River Salmon Trade. *Oregon Historical Quarterly* 28(2): 111–132.

Moulton, G. E. 1989. *The Journals of the Lewis and Clark Expedition. July 28–November 1, 1805.* Volume 5. University of Nebraska Press, Lincoln.

Moyle, P. B., and J. J. Cech, Jr. 1982. *Fishes: An Introduction to Ichthyology.* Prentice-Hall, Englewood Cliffs, NJ.

Mullen, R. 1981. Oregon's Commercial Harvest of Coho Salmon *Oncorhynchus kisutch* (Walbaum), 1892–1960. Information Report Series, Fisheries No. 81-3. Oregon Department of Fish and Wildlife, Portland.

Mumford, J. K. 1964. *John Ledyard's Journal of Captain Cook's Last Voyage.* Oregon State University Press, Corvallis.

Naiman, R. J., J. M. Melillo, and J. E. Hobbie. 1986. Ecosystem Alteration of Boreal Forest Streams by Beaver (*Castor canadensis*). *Ecology* 67(5): 1254–1269.

Naiman, R. J., C. A. Johnston, and J. C. Kelley. 1988. Alteration of North American Streams by Beaver: The Structure and Dynamics of Streams are Changing as Beaver Recolonize Their Historic Habitat. *Bioscience* 38(11): 753–762.

National Fish Hatchery Review Panel. 1994. Report of the National Fish Hatchery Review Panel. The Conservation Fund, Arlington, VA.

National Marine Fisheries Service. 1998. Progress of Species Status Reviews in NMFS Northwest Region. August 5, www.nwr.noaa.gov.

National Research Council. 1996. *Upstream: Salmon and Society in the Pacific Northwest.* National Academy Press, Washington, DC.

Neave, F. 1958. Stream Ecology and Production of Anadromous Fish. In *The Investigation of Fish-power Problems.* Edited by P. Larkin, pp. 43–48. H. R. MacMillan Lectures in Fisheries. University of British Columbia, Vancouver.

Needham, P. N. 1939. Natural Propagation Versus Artificial Propagation in Relation to Angling. *Transactions of the Fourth North American Wildlife Conference,* February 13–15. American Wildlife Institute, Washington, DC.

Nehlsen, W. 1993. Historical Salmon Runs of the Upper Deschutes and Their Environments. Report to Portland General Electric Company. Portland, OR.

Nehlsen, W., J. E. Williams, and J. A. Lichatowich. 1991. Pacific Salmon at the Crossroads: Stocks at Risk from California, Oregon, Idaho, and Washington. *Fisheries* 16(2): 4–21.

Nehlsen, W., J. Lichatowich, and J. Williams. 1992. Pacific Salmon and the Search for Sustainability. *Renewable Resources Journal* 10(2): 20–26.

Nelson, R. K. 1982. A Conservation Ethic and Environment: The Koyukon of Alaska. In *Resource Managers: North American and Australian Hunter-Gatherers.* Edited by N. M. Williams and E. S. Hunn, pp. 211–228. Selected Symposium 67. American Association for the Advancement of Science, Boulder, CO.

Netboy, A. 1974. *The Salmon: Their Fight for Survival.* Houghton Mifflin, Boston.

Neuberger, R. 1938. *Our Promised Land.* Macmillan, New York.

Neuberger, R. 1945. The Great Salmon Experiment. *Harper's Magazine* 190(1137): 229–236.

Newell, D. 1989. *The Development of the Pacific Salmon-Canning Industry: A Grown Man's Game.* McGill–Queen's University Press, Montreal.

Nickelson, T. 1986. Influences of Upwelling, Ocean Temperature and Smolt Abundance on Marine Survival of Coho Salmon (*Oncorhynchus kisutch*) in the Oregon Production Area. *Canadian Journal of Fisheries and Aquatic Sciences* 43(3): 527–535.

Nickelson, T. E., M. F. Solazzi, and S. L. Johnson. 1986. Use of Hatchery Coho Salmon (*Oncorhynchus kisutch*) Presmolts to Rebuild Wild Populations in Oregon Coastal Streams. *Canadian Journal of Fisheries and Aquatic Sciences* 43: 2443–2449.

Nickelson, T. E., J. D. Rodgers, S. L. Johnson, and M. F. Solazzi. 1992. Seasonal Changes in Habitat Use by Juvenile Coho Salmon (*Oncorhynchus kisutch*) in Oregon Coastal Streams. *Canadian Journal of Fisheries and Aquatic Sciences* 49: 783–789.

Nicols, A. 1882. *The Acclimatisation of the Salmonidae at the Antipodes: Its History and Results.* Sampson, Low, Marston, Searle, and Rivington, London.

Nisbet, J. 1994. *Sources of the River: Tracking David Thompson across Western North America.* Sasquatch Books, Seattle.

Northwest Power Planning Council. 1986. Compilation of Information on Salmon and Steelhead Losses in the Columbia River Basin. Portland, OR.

Northwest Power Planning Council. 1994. Columbia River Basin Fish and Wildlife Program. Portland, OR.

Oliphant, J. 1932. The Cattle Trade from the Far Northwest to Montana. *Agricultural History* 6: 1.

Oliver, H. 1967. *Gold and Cattle Country.* Binford and Mort, Portland, OR.

O'Malley, H. 1928. Symposium on Fisheries and Fishery Investigations: Our Opportunities, Our Responsibilities. Part II: Proceedings of the Divisional Conference, January 4–7, 1927. In Progress in Biological Inquiries 1926. Edited by E. Higgins. Appendix 7 to the Report of the U.S. Commissioner of Fisheries for 1927. Bureau of Fisheries Document No. 1029. Washington, DC.

Ontko, G. 1993. *Thunder over the Ochoco: The Gathering Storm.* Volume 1. Maverick Publications, Bend, OR.

Oregon Business Council. 1996. A New Vision for Pacific Salmon. Portland.

Oregon Department of Fish and Wildlife. 1982. Comprehensive Plan for Production and Management of Oregon's Anadromous Salmon and Trout. Part II: Coho Salmon Plan. Oregon Department of Fish and Wildlife, Portland.

Oregon Department of Fish and Wildlife, and Washington Department of Fish and Wildlife. 1995. Status Report: Columbia River Fish Runs and Fisheries, 1938–1994. Portland.

Oregon Department of Fisheries. 1903. Annual Report to the Governor. Salem.

Oregon Fish Commission. 1929. Oregon Fish Commission Minutes, October 8, Portland. Oregon State Archives, Salem.

Oregon Fish Commission. 1933. Biennial Report of the Fish Commission of the State of Oregon for 1931 and 1932. Salem.

Oregon Fish Commission. 1935. Biennial Report for 1933 and 1934. Portland.

Oregon Fish Commission. 1937. Biennial Report of the Fish Commission of the State of Oregon to the Governor and the Thirty-ninth Legislative Assembly, 1937. State of Oregon, Salem.

Oregon Fish Commission. 1939. Biennial Report of the Fish Commission of the State of Oregon to the Governor and the Fortieth Legislative Assembly, 1939. State of Oregon, Salem.

Oregon Fish Commission. 1955. Biennial Report. Portland.

Oregon Fish Commission. 1962. Biennial Report July 1, 1960–June 30, 1962. Portland.

Oregon Fish Commission. 1964. Biennial Report July 1, 1962–June 30, 1964. Portland.

Oregon Fish and Game Commission. 1919. Biennial Report of the Fish and Game Commission of the State of Oregon. Salem.

Oregon Game Commission. 1932. First Progress Report of the Ten-Year Plan. Portland.

Oregon Game Commission. 1942. Biennial Report of the Oregon Game Commission. Salem.

Oregon Journal (Portland). 1908. Master Fish Warden Van Dusen Dismissed. March 26. Oregon Historical Society Archives, Seufert Canning Company, Record 1102.

Oregon Journal. 1919. Aliens Threaten the Salmon Industry. August 8.
Oregon Journal. 1919. School Children Getting Up Big Finley Petition. December 12.
Oregon Journal. 1919. Make Finley Biologist Is [Governor] Ulcott View. December 23.
Oregon Journal. 1919. The Reason for Finley. December 30.
Oregon Journal. 1920. Investigation of Ousting of Finley Proposed. January 12.
Oregon State Board of Fish Commissioners. 1888. Second Report to the Governor. Salem.
Oregon State Board of Fish Commissioners. 1890. Fourth Annual Report. Salem.
Oregon State Fish and Game Protector. 1896. Third and Fourth Annual Reports of the State Fish and Game Protector of the State of Oregon, 1895–1896. Salem.
Oregon State Planning Board. 1938. A Study of Commercial Fishing Operations on the Columbia River. Report Submitted to the Governor of Oregon. Salem.
Oregon Voter (Portland). 1928. Save Rivers for Fish. July 14, pp. 21–24.
Oregonian. (Portland). 1875. Salmon Fisheries in Oregon. March 3.
Oregonian. 1877. Remonstrance. January 1.
Oregonian. 1906. Vanity of Salmon Theories (editorial). November 13.
Oregonian. 1997. Kitzhaber Convenes a Summit on Salmon. June 4.
Oregonian. 1998. Overhauling Fish Factories (editorial). December 7.
Orr, E. L., W. N. Orr, and E. M. Baldwin. 1992. *Geology of Oregon.* Fourth Edition. Kendall-Hunt Publishing Company, Dubuque, IA.
Pacific Fisherman. 1904. Salmon Marking Experiments on the Pacific Coast 2(6): 25.
Pacific Fisherman. 1904. Western Editorial on Important Fishery Questions 2(6): 19.
Pacific Fisherman. 1906. Increasing Use of Gasoline Engines in Fishing Industry 5(6).
Pacific Fisherman. 1916. Fourth Annual Salmon Day Celebrated 15(3): 1.
Pacific Fisherman. 1920. Loss of Salmon Fry in Irrigation 18(2).
Pacific Fisherman. 1931. Federal Reclamation Bureau Arraigned for Ignoring State Fishway Laws 29(4).
Pacific Fisherman. 1936. All Sockeye Hatcheries Closed in British Columbia 34(5): 17.
Pacific Fisherman. 1958. Sixty Years Ago: When a Fish Boat Engine Was a Dream of Pioneers 56(13): 20.
Pacific Fishery Management Council. 1979. Freshwater Habitat, Salmon Produced, and Escapements for Natural Spawning along the Pacific Coast of the United States. Portland, OR.
Pacific Fishery Management Council. 1992. Review of the 1991 Ocean Salmon Fisheries. Portland, OR.
Pacific Fishery Management Council. 1997. Amendment 13 to the Pacific Coast Salmon Plan: Fishery Management Regime to Ensure Protection and Rebuilding of Oregon Coastal Natural Coho. Portland, OR.

Pacific Fishery Management Council. 1999. Review of the 1998 Ocean Salmon Fisheries. Portland, OR.

Pacific Northwest Field Committee. 1950. Reevaluation of the Columbia River Fishery Program—Interim Report, Portland, OR. National Archives, Washington, DC, Record Group 22.

Parman. D. 1983. Inconsistent Advocacy: The Erosion of Indian Fishing Rights in the Pacific Northwest, 1933–1956. *Pacific Historical Review* 53: 163–189.

Parsons, L. 1993. Management of Marine Fisheries in Canada. National Research Council of Canada and the Department of Fisheries and Oceans, Ottawa.

Pearse, P. 1994. An Assessment of the Salmon Stock Development Program of Canada's Pacific Coast. Final Report to Department of Fisheries and Oceans. Vancouver, BC.

Piddocke, S. 1968. The Potlatch System of the Southern Kwakiutl: A New Perspective. In *Economic Anthropology: Readings in Theory and Analysis.* Edited by E. E. LeClair, Jr., and H. K. Schneider. Holt, Rinehart and Winston, New York.

Pielou, E. C. 1991. *After the Ice Age: The Return of Life to Glaciated North America.* University of Chicago Press, Chicago.

Platts, W. 1991. Livestock Grazing. *American Fisheries Society Special Publication* 19: 389–423.

Plummer, F. 1902. Forest Conditions in the Cascade Range, Washington, between the Washington and Mount Rainier Forest Reserves. U.S. Geological Survey, Washington, DC.

Pollock, C. R. 1932. Fortieth and Forty-first Annual Reports of the Washington State Department of Fisheries and Game. Olympia.

Rees, H. 1982. *Shaniko: From Wool Capital to Ghost Town.* Binford and Mort, Portland, OR.

Reiger, J. F. 1986. *American Sportsmen and the Origins of Conservation.* University of Oklahoma Press, Norman.

Reimers, P. E. 1973. The Length of Residence of Juvenile Fall Chinook Salmon in Sixes River, Oregon. *Research Reports of the Fish Commission of Oregon* 4(2): 3–43.

Reisenbichler, R. R. 1988. Relation Between Distance Transferred from Natal Stream and Recovery Rate for Hatchery Coho Salmon. *North American Journal of Fisheries Management* 8: 172–174.

Resource Development Branch of Department of Fisheries, Vancouver. 1971. Hatcheries in British Columbia: Part 1: *Western Fisheries,* April: 12–41.

Resource Development Branch of Department of Fisheries, Vancouver. 1971. Hatcheries in British Columbia, Part 2. *Western Fisheries,* May: 72–78.

Rich, W. H. 1920. Early History and Seaward Migration of Chinook Salmon in the Columbia and Sacramento Rivers. *Bulletin of the Bureau of Fisheries* No. 37, Washington, DC.

Rich, W. H. 1922. A Statistical Analysis of the Results of Artificial Propagation of Chinook Salmon. Manuscript Library, Northwest and Alaska Fisheries Science Center, National Marine Fisheries Service, Seattle.

Rich, W. H. 1925. Progress in Biological Inquiries, July 1 to December 31, 1924. Bureau of Fisheries Document No. 990. Washington, DC.

Rich, W. H. 1935. The Biology of Columbia River Salmon. *Northwest Science* 40(1): 3–14.

Rich, W. H. 1937. "Homing" of Pacific Salmon. *Science* 85(2211): 477–478.

Rich, W. H. 1939. Local Populations and Migration in Relation to the Conservation of Pacific Salmon in the Western States and Alaska. Contribution No. 1. Department of Research, Fish Commission of the State of Oregon, Salem.

Rich, W. H. 1940. Fishery Problems Raised by the Development of Water Resources. Contribution No. 2. Oregon Fish Commission, Salem.

Rich, W. H. 1940. Future of the Columbia River Salmon Fisheries. *Stanford Ichthyological Bulletin* 2(2): 37–47.

Rich, W. H. 1941. The Present State of the Columbia River Salmon Resources. Contribution No. 3. Oregon Fish Commission, Salem.

Rich, W. H. 1942. The Salmon Runs in the Columbia River in 1938. Fishery Bulletin No. 37. U.S. Fish and Wildlife Service, Washington, DC.

Rich, W. H., and H. B. Holmes. 1929. Experiments in Marking Young Chinook Salmon on the Columbia River, 1916 to 1927. *Bulletin of the Bureau of Fisheries,* Document No. 1047, Washington, DC.

Rich, W., F. Fish, M. Hanavan, and H. Holmes. 1944. Memorandum to Elmer Higgins, February 9. Program and Policy for the Fish and Wildlife Service in Respect of the Salmon Fisheries of the Columbia River. National Archives, Pacific Northwest Region, Seattle, Record Group 77, Box 34621.

Richey, J. E., M. A. Perkins, and C. R. Goldman. 1975. Effects of Kokanee Salmon (*Oncorhynchus nerka*) Decomposition on the Ecology of a Subalpine Stream. *Journal of Fisheries Research Board of Canada* 32: 817–820.

Ricker, W. 1947. Hells Gate and the Sockeye. *Journal of Wildlife Management* 11(1): 10–20.

Ricker, W. 1972. Hereditary and Environmental Factors Affecting Certain Salmonid Populations. In *The Stock Concept of Pacific Salmon.* Edited by R. Simon and P. Larkin. H. R. MacMillan Lectures in Fisheries. U.S. Bureau of Fisheries, Seattle.

Ricker, W. 1975. The Fisheries Research Board of Canada—Seventy-five Years of Achievements. *Journal of Fisheries Research Board of Canada* 32(8): 1465–1490.

Ricker, W. 1987. Effects of the Fishery and of Obstacles to Migration on the Abundance of Fraser River Sockeye Salmon (*Oncorhynchus nerka*). Canadian Technical Report of Fisheries and Aquatic Sciences No. 1522. Nanaimo, BC.

Riddell, B. 1993. Spatial Organization of Pacific Salmon: What to Conserve? In *Genetic Conservation of Salmonid Fishes.* Edited by J. Cloud and G. Thorgaard, Plenum Press, New York.

Riseland, J. L. 1907. Sixteenth and Seventeenth Annual Reports of the State Fish Commissioner and Game Warden to the Governor of the State of Washington. Washington Department of Fisheries and Game (1905–1906). Olympia.

Riseland, J. L. 1912. Twenty-second and Twenty-third Annual Reports of the State Fish Commissioner to the Governor of the State of Washington. Washington Department of Fisheries and Game, Olympia.

Rivers, C. *History of the Rogue River Fisheries.* Oregon Department of Fish and Wildlife, Portland.

Robbins, W. G. 1993. Narrative Form and Great River Myths: The Power of Columbia River Stories. *Environmental History Review* 17(2): 1–21.

Robbins, W. G. 1994. *Colony and Empire: The Capitalist Transformation of the American West.* University Press of Kansas, Lawrence.

Robbins, W. G. 1997. *Landscapes of Promise: The Oregon Story, 1800–1940.* University of Washington Press, Seattle.

Robertson, A. 1921. Further Proof of the Parent Stream Theory. *Transactions of the American Fisheries Society* 51: 87–90.

Roos, J. 1991. *Restoring Fraser River Salmon: A History of the International Pacific Salmon Fisheries Commission, 1937–1985.* Pacific Salmon Fisheries Commission, Vancouver, BC.

Roosevelt, T. 1908. Eighth Annual Message. In *The State of the Union Messages of the Presidents, 1790–1966.* Volume III. Edited by F. Israel. Chelsea House Publishers, New York, 1967.

Roppel, P. 1986. Salmon from Kodiak. Alaska Historical Commission Studies in History No. 216, Anchorage.

Rose, A. 1992. High-tech Hatcheries a Failure. *Vancouver (BC) Globe and Mail,* August 1.

Rowley, W. 1985. *U.S. Forest Service Grazing and Rangelands: A History.* Texas A&M University Press, College Station.

Rowley, W. D. 1996. *Reclaiming the Arid West: The Career of Francis G. Newlands.* Indiana University Press, Bloomington.

Royal, L. 1953. The Effects of Regulatory Selectivity on the Productivity of Fraser River Sockeye Salmon. *The Canadian Fish Culturist* 14: 1–12.

Salmon and Steelhead Advisory Commission. 1984. A New Management Structure for Anadromous Salmon and Steelhead Resources and Fisheries of the Washington and Columbia River Conservation Areas. National Marine Fisheries Service, Seattle.

Schalk, R. F. 1977. The Structure of an Anadromous Fish Resource. In *For Theory Building in Archaeology: Essays on Faunal Remains, Aquatic Resources and Spatial Analysis and Systemic Modeling.* Edited by L. R. Binford, pp. 207–249. Academic Press, New York.

Schalk, R. F. 1981. Land Use and Organizational Complexity Among Foragers of Northeastern North America. *In Affluent Foragers.* Edited by S. Koyama and D. H. Thomas, pp. 53–75. Senri, Osaka, Japan.

Scheufele, R. W. No date (perhaps 1970). History of the Columbia Basin Inter-Agency Committee. Pacific Northwest River Basins Commission, Vancouver, WA.

Schluchter, M., and J. A. Lichatowich. 1977. Juvenile Life Histories of Rogue River

Spring Chinook Salmon *Oncorhynchus tshawytscha* (Walbaum), as Determined from Scale Analysis. Information Report Series, Fisheries No. 77-5. Oregon Department of Fish and Wildlife, Corvallis.

Schoettler, R. 1953. Sixty-second Annual Report of the Washington Department of Fisheries. Olympia.

Scofield, C. 1912. The Settlement of Irrigated Lands. In *U.S. Department of Agriculture Yearbook, 1912.* Government Printing Office, Washington, DC, pp. 483–494.

Scofield, N. B. 1925. Minutes of Meeting of Fisheries Executives of the Pacific Coast, p. 124. General stacks, Fisheries Library, University of Washington, Seattle.

Scott, W. B., and E. J. Crossman. 1973. Freshwater Fishes of Canada. Fisheries Research Board of Canada. Bulletin 184. Ottawa.

Sedell, J. R., and K. J. Luchessa. 1981. Using the Historical Record as an Aid to Salmonid Habitat Enhancement. In *Symposium on Acquisition and Utilization of Aquatic Habitat Inventory Information,* October 23–28, 1981. Edited by N. B. Armantrout, pp. 210–223. American Fisheries Society, Western Division, Portland, OR.

Shapovalov, L. 1953. An Evaluation of Steelhead and Salmon Management in California. Inland Fisheries Branch, California Department of Fish and Game, Sacramento.

Simenstad, C. A., J. A. Estes, and K. W. Kenyon. 1978. Aleuts, Sea Otters, and Alternate Stable-State Communities. *Science* 200(28): 403–411.

Smith, C. L. 1979. *Salmon Fishers of the Columbia.* Oregon State University Press, Corvallis.

Smith, E. 1920. The Taking of Immature Salmon in Waters of the State of Washington. Washington Department of Fisheries, Olympia.

Smith, H. M. 1893. Fish Acclimatization of the Pacific Coast. *Science* 22(550): 88–89.

Smith, H. M. 1896. A Review of the History and Results of the Attempts to Acclimatize Fish and Other Water Animals in the Pacific States. *Bulletin of the U.S. Fish Commission for 1895,* volume 15. Government Printing Office, Washington, DC.

Smith, J. 1993. Retrospective Analysis of Changes in Stream and Riparian Habitat Characteristics Between 1935 and 1990 in Two Eastern Cascade Streams. M.S. thesis, University of Washington, Seattle.

Smith, T. 1994. *Scaling Fisheries.* Cambridge University Press, New York.

Smith, V. 1919. Fish Culture Methods in the Hatcheries of the State of Washington. Washington State Fish Commission, Olympia.

Stacey, D. A. 1982. Sockeye and Tinplate: Technological Change in the Fraser River Canning Industry, 1871–1912. British Columbia Provincial Museum Heritage Record No. 15. British Columbia.

Stearley, R. F. 1992. Historical Ecology of Salmonidae, with Special Reference to

Oncorhynchus. In *Systematics, Historical Ecology, and North American Freshwater Fishes.* Edited by R. L. Mayden, pp. 622–658. Stanford University Press, Stanford, CA.

Steinberg, T. 1991. *Nature Incorporated: Industrialization and the Waters of New England.* Cambridge University Press, New York.

Stewart, H. 1977. *Indian Fishing: Early Methods on the Northwest Coast.* University of Washington Press, Seattle.

Stone, L. 1872. Letter to Professor Spencer Baird, August 31. National Archives, Beltsville, MD, Record Group 22, Volume 7, Entry 1.

Stone, L. 1873. Letter to Spencer Baird, February 24. National Archives, Beltsville, MD, Record Group 22.

Stone, L. 1879. Report of Operations at the Salmon-Hatching Station on the Clackamas River, Oregon, in 1877. Report of the U.S. Fish Commissioner for 1877. Washington, DC.

Stone, L. 1884. The Artificial Propagation of Salmon in the Columbia River Basin. *Transactions of the American Fish-Culture Association,* Thirteenth Annual Meeting, May 13–14, Washington, DC, New York.

Stone, L. 1892. A National Salmon Park. *Transactions of the American Fisheries Society* 21: 149–162.

Stone, L. 1896. Artificial Propagation of Salmon on the Pacific Coast of the United States: With Notes on the Natural History of the Quinnat Salmon. U.S. Commission of Fish and Fisheries, Washington, DC.

Stone, L. 1897. Some Brief Reminiscences of the Early Days of Fish-Culture in the United States. *Bulletin of the U.S. Fish Commission* No. 22, Washington, DC.

Stoutemyer, B. E. 1931. Letter to Henry O'Malley, Commissioner of Fisheries, January 28. National Archives, Washington, DC, Record Group 22.

Suttles, W. 1987. *Coast Salish Essays.* University of Washington Press, Seattle.

Swan, J. C. 1857. *The Northwest Coast or, Three Years' Residence in Washington Territory.* University of Washington Press, Seattle.

Taylor, J. 1996. Making Salmon: Economy, Culture, and Science in the Oregon Fisheries, Precontact to 1960. Ph.D. dissertation. University of Washington, Seattle.

Tetlow, R. T., and G. J. Barbey. 1990. *Barbey: The Story of a Pioneer Columbia River Salmon Packer.* Binford and Mort, Portland, OR.

Thomas, W. K., R. E. Withler, and A. T. Beckenbach. 1986. Mitochondrial DNA Analysis of Pacific Salmonid Evolution. *Canadian Journal of Zoology* 64: 1058–1064.

Thomison, P. 1987. When Celilo Was Celilo: An Analysis of Salmon Use During the Past 11,000 Years in the Columbia Plateau. M.A. thesis. Oregon State University, Corvallis.

Thompson, P., T. Wailey, and T. Lummis, 1983. *Living the Fishing.* Routledge and Kegan Paul, London.

Thompson, W. F. 1922. The Marine Fisheries, the State and the Biologist. *The Scientific Monthly* 15: 542–550.

Thompson, W. F. 1951. An Outline for Salmon Research in Alaska. Fisheries Research Institute Circular 18. University of Washington, Seattle.

Thompson, W. F. 1959. An Approach to Population Dynamics of the Pacific Red Salmon. *Transactions of the American Fisheries Society* 88(3): 206–209.

Thompson, W. F. 1965. Fishing Treaties and the Salmon of the North Pacific. *Science* 150: 1786–1789.

Thorpe, J. E. 1982. Migration in Salmonids, with Special Reference to Juvenile Movements in Freshwater. In *Proceedings of the Salmon and Trout Migratory Behavior Symposium.* Edited by E. L. Brannon and E. O. Salo, pp. 86–97. University of Washington, Seattle.

Tims, D. 1990. OreAqua Turning Salmon into Fertilizer. *Oregonian* (Portland), October 5.

Titcomb, J. 1904. Report on the Propagation and Distribution of Food-fishes. Report of the U.S. Commissioners of Fish and Fisheries for 1902. Washington, DC.

U.S. Army Corps of Engineers. 1975. Lower Snake River Fish and Wildlife Compensation Plan. Special Report, Walla Walla [WA] District.

U.S. Commission of Fish and Fisheries. 1874. Report of the Commissioner for 1872 and 1873, part II. Government Printing Office, Washington, DC, pp. 757–772.

U.S. Congressional Record. 1931. Seventy-first Congress, Third Session, volume 74, pp. 392–393.

U.S. Department of Interior and the Lower Elwha Klallam Tribe. 1994. The Elwha Report: Restoration of the Elwha River Ecosystem and Native Anadromous Fisheries. National Park Service, Port Angeles, WA.

U.S. Fish and Wildlife Service. 1951. A Program for the Preservation of the Fisheries of the Columbia River Basin. National Archives, Washington, DC, Record Group 48.

U.S. Fish and Wildlife Service. 1951. Transcript of a conference with state, provincial, and federal agencies to develop and engineering-biological research program that will attempt to solve the problems of migratory fish passage over high dams and related problems. National Archives, Washington, DC, Record Group 48.

U.S. Fish and Wildlife Service. 1998. Proceedings of the Lower Snake River Compensation Plan Status Review Symposium. Boise, ID.

U.S. General Accounting Office. 1992. Endangered Species: Past Actions Taken to Assist Columbia River Salmon. GAO/RCED-92-173BR. Washington, DC.

Van Cleve, R. 1952. The School of Fisheries. *The Progressive Fish-Culturist* 14(4): 159–164.

Van Cleve, R., and R. Ting. 1960. The Condition of Salmon Stocks in the John Day, Umatilla, Walla Walla, Grande Ronde, and Imnaha Rivers as Reported by Various Fisheries Agencies. University of Washington, Department of Oceanography, Seattle.

Van Dusen, H. 1903. Annual Report of the Oregon Department of Fisheries. Salem.

Van Dusen, H. 1908. Annual Report of the Department of Fisheries of the State of Oregon to the Legislative Assembly, 1909. Salem.

Van Dusen, H. G. 1909. Letter to Senator I. H. Bingham, April 19. Oregon Historical Society Archives, Portland, Warren Packing Company, Record 1141, Box 1.

Wahl, R. J., and R. R. Vreeland. 1978. Bioeconomics Contribution of Columbia River Hatchery Fall Chinook Salmon, 1961 through 1964 Broods, to the Pacific Salmon Fisheries. *Fisheries Bulletin* 76(1): 179–208.

Wales, J. 1948. California's Fish Screen Program. *California Fish and Game* 34(2): 45–51.

Wallis, J. 1963. An Evaluation of the Alsea River Salmon Hatchery, Oregon Fish Commission Research Laboratory, 1961. Clackamas, OR.

Wallis, J. 1964. An Evaluation of the Bonneville Salmon Hatchery, Oregon Fish Commission Research Laboratory, 1961 Clackamas, OR.

Walters, C. 1995. Fish on the Line: The Future of Pacific Fisheries. Report to the David Suzuki Foundation. Fisheries Project Phase I, Vancouver, BC.

Waples, R. S. 1991. Definition of "Species" Under the Endangered Species Act: Application to Pacific Salmon. NOAA Technical Memorandum, NMFS F/NWC-194. U.S. Department of Commerce, National Marine Fisheries Service, Seattle.

Ward, F., and P. Larkin. 1964. Cyclic Dominance in Adams River Sockeye Salmon. Progress Report No. 11. International Pacific Salmon Fisheries Commission, New Westminster, BC.

Ward, H. 1912. The Preservation of the American Fish Fauna. *Transactions of the American Fisheries Society* 42: 157–170.

Ware, D., and R. Thompson. 1991. Link Between Long-term Variability in Upwelling and Fish Production in the Northeast Pacific Ocean. *Canadian Journal of Fisheries and Aquatic Sciences* 48(12): 2296–2306.

Washington State Department of Fisheries. 1960. Fisheries. Volume III. Olympia.

Washington State Department of Fisheries. 1992. Salmon 2000 Technical Report. Phase 2: Puget Sound, Washington Coast and Integrated Planning. Olympia.

Washington State Department of Fisheries and Game. 1921. Thirtieth and Thirty-first Annual Reports of the State Fish Commissioner to the Governor of the State of Washington. Olympia.

Washington State Department of Fisheries and Game. 1925. Thirty-fourth and Thirty-fifth Annual Reports. Olympia.

Washington State Fish Commissioner. 1904. Fourteenth and Fifteenth Annual Reports of the Washington Department of Fisheries. Seattle.

Washington State Fish Commissioner. 1920. Twenty-eighth and Twenty-ninth Annual Reports. Olympia.

Washington State Fisheries Board. 1925. Meeting of Fisheries Executives of the Pacific Coast, March 16–17, Seattle. General stacks, Fisheries and Oceanography Library, University of Washington, Seattle.

Washington State Senate. 1943. Minutes of the Washington State Columbia River Salmon Interim Committee, October 20.

Washington State Senate. 1943. Report on the Problems Affecting the Fisheries of the Columbia River. Columbia River Fisheries Interim Investigating Committee. Olympia.

Watanabe, H. 1972. *The Ainu Ecosystem: Environment and Group Structure.* University of Washington Press, Seattle.

Waterman, T. T., and A. L. Kroeber. 1965. The Kepel Fish Dam. *University of California Publications in American Archaeology and Ethnology (1934–1943)* 35: 49–79. Kraus Reprint Corporation, New York (reprint).

Wendler, H., and G. Deschamps. 1955. Logging Dams on Coastal Washington Streams. Washington Department of Fisheries, *Fisheries Research Papers* 1(3): 27–38.

Wendler, H. O. 1966. Regulation of Commercial Fishing Gear and Seasons on the Columbia River from 1859 to 1963. Washington Department of Fisheries, *Fisheries Research Papers* 2(4): 19–31.

Whitaker, H. 1893. Address of the President. *Transactions of the American Fisheries Society,* Twenty-second Annual Meeting, July 15, Chicago. Printed in New York.

Whitaker, H. 1896. Some Observations on the Moral Phases of Modern Fish Culture. *Transactions of the American Fisheries Society,* June 12–13, 1895, New York.

White, L., Jr. 1967. The Historical Roots of Our Ecologic Crisis. *Science* 155(3767): 1203–1206.

White, R. 1995. *The Organic Machine.* Hill and Wang, New York.

Wilderness Society. 1993. The Living Landscape. Volume 2: Pacific Salmon and Federal Lands. Washington, DC.

Wilkinson, C., and D. Conner. 1983. The Law of the Pacific Salmon Fishery: Conservation and Allocation of a Transboundary Common Property Resource. *Kansas Law Review* 32(1): 109.

Wilkinson, C. F. 1992. *Crossing the Next Meridian: Land, Water, and the Future of the West.* Island Press, Washington, DC.

Willson, M. F., and K. C. Halupka. 1995. Anadromous Fish as Keystone Species in Vertebrate Communities. *Conservation Biology* 9(3): 489–497.

Winton, J. 1991. Supplementation of Wild Salmonids: Management Practices in British Columbia. M.S. thesis. University of Washington, Seattle.

Wissmar, R. and five others. 1994. A History of Resource Use and Disturbance in the Riverine Basin of Eastern Oregon and Washington (Early 1880s–1990s). *Northwest Science* 69, Special Issue.

Wood, E. M. 1953. A Century of American Fish Culture, 1853–1953. *The Progressive Fish-Culturist* 15(4): 147–162.

Worster, D. 1977. *Nature's Economy: A History of Ecological Ideas.* Cambridge University Press, New York.

Worster, D. 1985. *Rivers of Empire.* Pantheon Books, New York.

Wright, S. 1993. Fishery Management of Wild Salmon Stocks to Prevent Extinction. *Fisheries* 18(5): 3–4.

Index